AF867898

AUGUSTUS DE MORGAN

FORMAL LOGIC
OR
The Calculus of Inference
NECESSARY AND PROBABLE

Elibron Classics
www.elibron.com

Elibron Classics series.

ISBN 1-4021-0072-8 (paperback)
ISBN 1-4021-5366-X (hardcover)

This Elibron Classics Replica Edition is an unabridged facsimile of the edition published in 1847 by Taylor and Walton, London.

FORMAL LOGIC:

OR,

The Calculus of Inference,

Neceſſary and Probable.

BY

AUGUSTUS DE MORGAN

Of Trinity College Cambridge,

Fellow of the Cambridge Philoſophical Society, Secretary of the Royal Aſtronomical Society, Profeſſor of Mathematics in Univerſity College London.

Καλὸς ὁ νόμος ἐάν τις αὐτῷ νομίμως χρῆται.

LONDON:

TAYLOR AND WALTON,

Bookſellers and Publiſhers to Univerſity College,

28, *Upper Gower Street.*

M DCCC XLVII

PREFACE.

THE ſyſtem given in this work extends beyond that commonly received, in ſeveral directions. A brief ſtatement of what is now ſubmitted for adoption into the theory of inference will be the matter of this preface.

In the form of the propoſition, the copula is made as abſtract as the terms: or is conſidered as obeying only thoſe conditions which are neceſſary to inference.

Every name is treated in connection with its *contrary* or *contradictory* name; the diſtinction between theſe words not being made, and others ſupplied in conſequence. Eight really ſeparable forms of predication are thus obtained, between any two names: the eight of the common ſyſtem amounting only to ſix, when, as throughout my work, the two forms of a convertible propoſition are conſidered as identical.

The complex propoſition is introduced, conſiſting in the coexiſtence of two ſimple ones. The theory of the ſyllogiſm of complex propoſitions is made to precede that of the ſimple or ordinary ſyllogiſm; which laſt is deduced from it. I have only uſed the word *complex*, becauſe *ſimple* was already appropriated (ſee page 85).

By the introduction of contraries, the number of valid syllogistic forms is increased to thirty-two, connected together by many rules of relation, but all shewn to contain, each with reference to its own disposition of names and contraries, only one form of inference.

The distinction of figure is avoided from the beginning by introducing into every proposition an order of reference to its terms.

A simple notation, which includes the common one, gives the means of representing every syllogism by three letters, each accented above or below. By inspection of one of these symbols it is seen immediately, 1. What syllogism is represented, 2. Whether it be valid or invalid, 3. How it is at once to be written down, 4. What axiom the inference contains, or what is the act of the mind when it makes that inference (chapter XIV).

A subordinate notation is used (page 60) in abbreviation of the proposition at length.

Compound names are considered, both when the composition is conjunctive, and when it is disjunctive. Distinct notation and rules of transformation are given, and the compound syllogisms are treated as reducible to ordinary ones, by invention of compound names.

The theory of the numerical syllogism is investigated, in which, upon the hypothesis of numerical quantity in both terms of every proposition, a numerical inference is made.

But, when the numerical relations of the several terms are fully known, all that is unusual in the quantity of the predicate is shown to be either superfluous, or else, as I have called it, spurious.

The old doctrine of modals is made to give place to the numerical theory of probability. Many will object to this theory as extralogical. But I cannot see on what definition, founded on real distinction, the exclusion of it can be maintained. When I am told that logic considers the validity of the inference, independently of the truth or falsehood of the matter, or supplies the conditions under which the hypothetical truth of the matter of the premises gives hypothetical truth to the matter of the conclusion, I see a real definition, which propounds for consideration the forms and laws of inferential thought. But when it is further added that the only hypothetical truth shall be absolute truth, certain knowledge, I begin to see arbitrary distinction, wanting the reality of that which preceded. Without pretending that logic can take cognizance of the probability of any given matter, I cannot understand why the study of the effect which partial belief of the premises produces with respect to the conclusion, should be separated from that of the consequences of supposing the former to be absolutely true. Not however to dispute upon names, I mean that I should maintain, against those who would exclude the theory of probability from logic, that, call it by what name they like, it should accompany logic as a study.

I have, of course, been obliged to express, in my own manner, my own convictions on points of mental philosophy. But any one will see that, in all which I have proposed for adoption, it matters nothing whether my views of the phenomena of thought, or others, be made the basis of the explanation. So far therefore, as I am confidered as propofing forms of fyllogifm, &c. to the logician, and not giving inftruction to the ftudent of the fcience, the reader has nothing to do with my choice of the terms in which mental operations are fpoken of.

In the appendix will be found fome remarks on the perfonal controverfy between Sir W. Hamilton of Edinburgh and myfelf, of which I fuppofe the celebrity of my opponent, and the appearance of part of it in a journal fo widely circulated as the *Athenæum*, has caufed many ftudents of logic to hear or read fomething.

At the end of the contents of fome chapters in the following table, are a few additions and corrections, to which I requeft the reader's attention.

A. De Morgan.

Univerfity College, London,
October 14, 1847.

TABLE OF CONTENTS.

*** The articles entered in *Italic*, are thoſe, the contents of which belong to the peculiar ſyſtem preſented in this work.

Additions and corrections.—*Page* 79, in the firſt diagram, for D_1D_1D, read $D_1D_1D_1$; *page* 88, *line* 23, *inſtead of* has the other two for its opponents, *read* has its opponents in the ſet; *page* 90, *line* 4, *from the bottom, for* premiſs *read* premiſe: the firſt ſpelling has been common enough, but it ſeems ſtrange that the cognate words promiſe, ſurmiſe, demiſe, &c. ſhould not have dictated the ſecond. *Page* 96;

CHAPTER XIII.—*On Fallacies* (pages 237—286).

Additions and corrections. Page 250, *lines* 3 *and* 5*; for* millenium *read* millennium, *and for* Newtonion *read* Newtonian.

CHAPTER XIV.—*On the verbal Deſcription of the Syllogiſm* (pages 286—296).

ELEMENTS OF LOGIC.

CHAPTER I.

Firſt Notions.

THE firſt notion which a reader can form of Logic is by viewing it as the examination of that part of reaſoning which depends upon the manner in which inferences are formed, and the inveſtigation of general maxims and rules for conſtructing arguments, ſo that the concluſion may contain no inaccuracy which was not previouſly aſſerted in the premiſes. It has ſo far nothing to do with the truth of the facts, opinions, or preſumptions, from which an inference is derived; but ſimply takes care that the inference ſhall certainly be true, if the premiſes be true. Thus, when we ſay that all men will die, and that all men are rational beings, and thence infer that ſome rational beings will die, the *logical* truth of this ſentence is the ſame whether it be true or falſe that men are mortal and rational. This logical truth depends upon the *ſtructure of the ſentence*, and not upon the particular matters ſpoken of. Thus,

Inſtead of	Write,
All men will die.	Every Y is X.
All men are rational beings.	Every Y is Z.
Therefore ſome rational beings will die.	Therefore ſome Zs are Xs.

The ſecond of theſe is the ſame propoſition, logically conſidered, as the firſt; the conſequence in both is virtually contained in, and rightly inferred from, the premiſes. Whether the premiſes be true or falſe, is not a queſtion of logic, but of morals, philoſophy, hiſtory, or any other knowledge to which their ſubject-matter belongs: the queſtion of logic is, does the concluſion certainly follow if the premiſes be true?

Every act of reaſoning muſt mainly conſiſt in comparing together different things, and either finding out, or recalling from previous knowledge, the points in which they reſemble or differ from each other. That particular part of reaſoning which is called *inference*, conſiſts in the compariſon of ſeveral and different things with one and the ſame other thing; and aſcertaining the reſemblances, or differences, of the ſeveral things, by means of the points in which they reſemble, or differ from, the thing with which all are compared.

There muſt then be ſome propoſitions already obtained before any inference can be drawn. All propoſitions are either aſſertions or denials, and are thus divided into *affirmative* and *negative*. Thus, X is Y, and X is not Y, are the two forms to which all propoſitions may be reduced. Theſe are, for our preſent purpoſe, the moſt ſimple forms; though it will frequently happen that much circumlocution is needed to reduce propoſitions to them. Thus, ſuppoſe the following aſſertion, 'If he ſhould come to-morrow, he will probably ſtay till Monday;' how is this to be reduced to the form X is Y? There is evidently ſomething ſpoken of, ſomething ſaid of it, and an affirmative connection between them. Something, if it happen, that is, the happening of ſomething, makes the happening of another ſomething probable; or *is* one of the things which render the happening of the ſecond thing probable.

X	is	Y
The happening of his arrival to-morrow	is	an event from which it may be inferred as probable that he will ſtay till Monday.

The forms of language will allow the manner of aſſerting to be varied in a great number of ways; but the reduction to the preceding form is always poſſible. Thus, 'ſo he ſaid' is an affirmation, reducible as follows:

What you have juſt ſaid (or whatever elſe 'ſo' refers to)	is	the thing which he ſaid.

By changing 'is' into 'is not,' we make a negative propoſition; but care muſt always be taken to aſcertain whether a propoſition which appears negative be really ſo. The principal danger is that of confounding a propoſition which is negative with another which is affirmative of ſomething requiring a negative to deſcribe it. Thus, 'he reſembles the man who was not in the room,' is affirmative, and muſt not be confounded with 'he does not reſemble the man who was in the room.' Again, 'if he ſhould come to-morrow, it is probable he will not ſtay till Monday,' does not mean the ſimple denial of the preceding propoſition, but the affirmation of a directly oppoſite propoſition. It is,

X	is	Y
The happening of his arrival to-morrow,	is	an event from which it may be inferred to be *im*probable that he will ſtay till Monday:

whereas the following,

The happening of his arrival to-morrow,	is *not*	an event from which it may be inferred as probable that he will ſtay till Monday,

would be expreſſed thus: 'If he ſhould come to-morrow, that is no reaſon why he ſhould ſtay till Monday.'

Moreover, the negative words not, no, &c., have two kinds of meaning which muſt be carefully diſtinguiſhed. Sometimes they deny, and nothing more: ſometimes they are uſed to affirm the direct contrary. In caſes which offer but two alternatives, one of which is neceſſary, theſe amount to the ſame thing, ſince the denial of one, and the affirmation of the other, are obviouſly equivalent propoſitions. In many idioms of converſation, the negative implies affirmation of the contrary in caſes which offer not only alternatives, but degrees of alternatives. Thus, to the queſtion, 'Is he tall?' the ſimple anſwer, 'No,' moſt frequently means that he is the contrary of tall, or conſiderably under the average. But it muſt be remembered, that, in all logical reaſoning, the negation is ſimply negation, and nothing more, never implying affirmation of the contrary.

The common propoſition that two negatives make an affirmative, is true only upon the ſuppoſition that there are but two poſſible things, one of which is denied. Grant that a man muſt be either able or unable to do a particular thing, and then *not unable* and able are the ſame things. But if we ſuppoſe various degrees of performance, and therefore degrees of ability, it is falſe, in the common ſenſe of the words, that two negatives make an affirmative. Thus, it would be erroneous to ſay, ‘John is able to tranſlate Virgil, and Thomas is not unable; therefore, what John can do Thomas can do,’ for it is evident that the premiſes mean that John is ſo near to the beſt ſort of tranſlation that an affirmation of his ability may be made, while Thomas is conſiderably lower than John, but not ſo near to abſolute deficiency that his ability may be altogether denied. It will generally be found that two negatives imply an affirmative of a weaker degree than the poſitive affirmation.

Each of the propoſitions, ‘X is Y,’ and ‘X is not Y,’ may be ſubdivided into two ſpecies: the *univerſal*, in which every poſſible caſe is included; and the *particular*, in which it is not meant to be aſſerted that the affirmation or negation is univerſal. The four ſpecies of propoſition are then as follows, each being marked with the letter by which writers on logic have always diſtinguiſhed it.

A	*Univerſal Affirmative*	Every X	is	Y
E	*Univerſal Negative*	No X	is	Y
I	*Particular Affirmative*	Some Xs	are	Ys
O	*Particular Negative*	Some Xs	are not	Ys

In common converſation the affirmation of a part is meant to imply the denial of the remainder. Thus, by ‘ſome of the apples are ripe,’ it is always intended to ſignify that ſome are not ripe. This is not the caſe in logical language, but every propoſition is intended to make its amount of affirmation or denial, and no more. When we ſay, ‘Some X is Y,’ or, more grammatically, ‘Some Xs are Ys,’ we do not mean to imply that ſome are not: this may or may not be. Again, the word ſome means, ‘one or more, poſſibly all.’ The following table will ſhew the bearing of each propoſition on the reſt.

Every X is Y	affirms *Some Xs are Ys*	and denies	*No X is Y* *Some Xs are not Ys*

No X is Y affirms *Some Xs are not Ys* and denies { *Every X is Y* / *Some Xs are Ys* }

Some Xs are Ys does not contradict { *Every X is Y* / *Some Xs are not Ys* } but denies *No X is Y*

Some Xs are not Ys does not contradict { *No X is Y* / *Some Xs are Ys* } but denies *Every X is Y*

Contradictory propoſitions are thoſe in which one denies *any thing* that the other affirms; *contrary* propoſitions are thoſe in which one denies *every thing* which the other affirms, or affirms every thing which the other denies. The following pair are contraries,

Every X is Y and No X is Y

and the following are contradictories,

Every X is Y to Some Xs are not Ys
No X is Y to Some Xs are Ys

A contrary, therefore, is a complete and total contradictory; and a little conſideration will make it appear that the deciſive diſtinction between contraries and contradictories lies in this, that contraries may both be falſe, but of contradictories, one muſt be true and the other falſe. We may ſay, 'Either P is true, or *ſomething* in contradiction of it is true;' but we cannot ſay, 'Either P is true, or *every thing* in contradiction of it is true.' It is a very common miſtake to imagine that the *denial* of a propoſition gives a right to *affirm* the contrary; whereas it ſhould be, that the *affirmation* of a propoſition gives a right to *deny* the contrary. Thus, if we deny that Every X is Y, we do not affirm that No X is Y, but only that Some Xs are not Ys; while, if we affirm that Every X is Y, we deny No X is Y, and alſo Some Xs are not Ys.

But, as to contradictories, affirmation of one is denial of the other, and denial of one is affirmation of the other. Thus, either Every X is Y, or Some Xs are not Ys: affirmation of either is denial of the other, and *vice verſâ*.

Let the ſtudent now endeavour to ſatisfy himſelf of the following. Taking the four preceding propoſitions, A, E, I, O, let the ſimple letter ſignify the affirmation, the ſame letter in parentheſes the denial, and the abſence of the letter, that there is neither affirmation nor denial.

From A follow	(E), I, (O)	From (A) follow	O
From E	(A), (I), O	From (E)	I
From I	(E)	From (I)	(A), E, O
From O	(A)	From (O) . . .	A, (E), I

Theſe may be thus ſummed up : The affirmation of a univerſal propoſition, and the denial of a particular one, enable us to affirm or deny all the other three ; but the denial of a univerſal propoſition, and the affirmation of a particular one, leave us unable to affirm or deny two of the others.

In ſuch propoſitions as ‘ Every X is Y,’ ‘ Some Xs are not Ys,’ &c., X is called the *ſubject*, and Y the *predicate*, while the verb ‘ is ’ or ‘ is not,’ is called the *copula*. It is obvious that the words of the propoſition point out whether the ſubject is ſpoken of univerſally or partially, but not ſo of the predicate, which it is therefore important to examine. Logical writers generally give the name of *diſtributed* ſubjects or predicates to thoſe which are ſpoken of univerſally ; but as this word is rather technical, I ſhall ſay that a ſubject or predicate enters wholly or partially, according as it is univerſally or particularly ſpoken of.

1. In A, or ‘ Every X is Y,’ the ſubject enters wholly, but the predicate only partially. For it obviouſly ſays, ‘ Among the Ys are all the Xs,’ ‘ Every X is part of the collection of Ys, ſo that all the Xs make a part of the Ys, the whole it *may* be.’ Thus, ‘ Every horſe is an animal,’ does not ſpeak of all animals, but ſtates that all the horſes make up a portion of the animals.

2. In E, or ‘ No X is Y,’ both ſubject and predicate enter wholly. ‘ No X whatſoever is any one out of all the Ys ;’ ‘ ſearch the whole collection of Ys, and *every* Y ſhall be found to be ſomething which is not X.’

3. In I, or ‘ Some Xs are Ys,’ both ſubject and predicate enter partially. ‘ Some of the Xs are found among the Ys, or make up a part (the whole poſſibly, but not known from the preceding) of the Ys.’

4. In O, or ‘ Some Xs are not Ys,’ the ſubject enters partially, and the predicate wholly. ‘ Some Xs are none of them any whatſoever of the Ys ; every Y will be found to be no one out of a certain portion of the Xs.’

It appears then that,

In affirmatives, the predicate enters partially.

In negatives, the predicate enters wholly.

In contradictory propositions, both subject and predicate enter differently in the two.

The *converse* of a proposition is that which is made by interchanging the subject and predicate, as follows:

	The proposition.	Its converse.
A	Every X is Y	Every Y is X
E	No X is Y	No Y is X
I	Some Xs are Ys	Some Ys are Xs
O	Some Xs are not Ys	Some Ys are not Xs

Now, it is a fundamental and self-evident proposition, that no consequence must be allowed to assert more widely than its premises; so that, for instance, an assertion which is only of some Ys can never lead to a result which is true of all Ys. But if a proposition assert agreement or disagreement, any other proposition which asserts the same, to the same extent and no further, must be a legitimate consequence; or, if you please, must amount to the whole, or part, of the original assertion in another form. Thus, the converse of A is not true: for, in 'Every X is Y,' the predicate enters partially; while in 'Every Y is X,' the subject enters wholly. 'All the Xs make up a part of the Ys, then a part of the Ys are among the Xs, or some Ys are Xs.' Hence, the only *legitimate* converse of 'Every X is Y' is, 'Some Ys are Xs.' But in 'No X is Y,' both subject and predicate enter wholly, and 'No Y is X' is, in fact, the same proposition as 'No X is Y.' And 'Some Xs are Ys' is also the same as its converse 'Some Ys are Xs:' here both terms enter partially. But 'Some Xs are not Ys' admits of no converse whatever; it is perfectly consistent with all assertions upon Y and X in which Y is the subject. Thus neither of the four following lines is inconsistent with itself.

Some Xs are not Ys	and	Every Y is X
Some Xs are not Ys	and	No Y is X
Some Xs are not Ys	and	Some Ys are Xs
Some Xs are not Ys	and	Some Ys are not Xs.

Having thus discussed the principal points connected with the simple assertion, I pass to the manner of making two assertions give a third. Every inftance of this is called a *fyllogifm*, the two affertions which form the bafis of the third are called *premifes*, and the third itfelf the *conclufion*.

If two things both agree with a third in any particular, they agree with each other in the fame; as, if X be of the fame colour as Y, and Z of the fame colour as Y, then X is of the fame colour as Z. Again, if X differ from Y in any particular in which Z agrees with Y, then X and Z differ in that particular. If X be not of the fame colour as Y, and Z be of the fame colour as Y, then X is not of the colour of Z. But if X and Z both differ from Y in any particular, nothing can be inferred; they may either differ in the fame way and to the fame extent, or not. Thus, if X and Z be both of different colours from Y, it neither follows that they agree, nor differ, in their own colours.

The paragraph preceding contains the effential parts of all inference, which confifts in comparing two things with a third, and finding from their agreement or difference with that third, their agreement or difference with one another. Thus, Every X is Y, every Z is Y, allows us to infer that X and Z have all thofe qualities in common which are neceffary to Y. Again, from every X is Y, and 'No Z is Y,' we infer that X and Z differ from one another in all particulars which are effential to Y. The preceding forms, however, though they reprefent common reafoning better than the ordinary fyllogifm, to which we are now coming, do not conftitute the ultimate forms of inference. Simple *identity* or *non-identity* is the ultimate ftate to which every affertion may be reduced; and we fhall, therefore, firft afk, from what identities, &c., can other identities, &c., be produced? Again, fince we name objects in fpecies, each fpecies confifting of a number of individuals, and fince our affertion may include all or only part of a fpecies, it is further neceffary to afk, in every inftance, to what extent the conclufion drawn is true, whether of all, or only of part?

Let us take the fimple affertion, 'Every living man refpires;' or every living man is one of the things (however varied they may be) which refpire. If we were to enclofe all living men in a large triangle, and all refpiring objects in a large circle, the preceding affertion, if true, would require that the whole of the triangle fhould be contained in the circle. And in the fame way we may reduce any aſſertion to the expreſſion of a coincidence, total or partial, between two figures. Thus, a point in a circle may repreſent an individual of one ſpecies, and a point in a triangle an individual of another ſpecies: and we may expreſs that the whole of one ſpecies is aſſerted to be contained or not contained in the other by ſuch forms as, 'All the △ is in the ○'; 'None of the △ is in the ○'.

Any two aſſertions about X and Z, each expreſſing agreement or diſagreement, total or partial, with or from Y, and leading to a concluſion with reſpect to X or Z, is called a ſyllogiſm, of which Y is called the *middle term*. The plaineſt ſyllogiſm is the following:—

Every X is Y	All the △ is in the ○
Every Y is Z	All the ○ is in the □
Therefore Every X is Z	Therefore All the △ is in the □

In order to find all the poſſible forms of ſyllogiſm, we muſt make a table of all the elements of which they can conſiſt; namely—

X and Y		Z and Y
Every X is Y	A	Every Z is Y
No X is Y	E	No Z is Y
Some Xs are Ys	I	Some Zs are Ys
Some Xs are not Ys	O	Some Zs are not Ys
Every Y is X	A	Every Y is Z
Some Ys are not Xs	O	Some Ys are not Zs

Or their ſynonymes,

△ and ○		□ and ○
All the △ is in the ○	A	All the □ is in the ○
None of the △ is in the ○	E	None of the □ is in the ○
Some of the △ is in the ○	I	Some of the □ is in the ○
Some of the △ is not in the ○	O	Some of the □ is not in the ○
All the ○ is in the △	A	All the ○ is in the □
Some of the ○ is not in the △	O	Some of the ○ is not in the □

Now, taking any one of the ſix relations between X and Y, and combining it with either of thoſe between Z and Y, we have ſix pairs of premiſes, and the ſame number repeated for every different relation of X to Y. We have then thirty-ſix forms to consider: but, thirty of these (namely, all but (A, A) (E, E), &c.,) are half of them repetitions of the other half. Thus, 'Every X is Y, no Z is Y,' and 'Every Z is Y, no X is Y,' are of the same form, and only differ by changing X into Z and Z into X. There are then only 15+6, or 21 distinct forms, some of which give a necessary conclusion, while others do not. We shall select the former of these, classifying them by their conclusions; that is, according as the inference is of the form A, E, I, or O.

I. In what manner can a universal affirmative conclusion be drawn; namely, that one figure is entirely contained in the other? This we can only assert when we know that one figure is entirely contained in the circle, which itself is entirely contained in the other figure. Thus,

Every X is Y	All the △ is in the ○	A
Every Y is Z	All the ○ is in the □	A
Every X is Z	All the △ is in the □	A

is the only way in which a universal affirmative conclusion can be drawn.

II. In what manner can a universal negative conclusion be drawn; namely, that one figure is entirely exterior to the other? Only when we are able to assert that one figure is entirely within, and the other entirely without, the circle. Thus,

Every X is Y	All the △ is in the ○	A
No Z is Y	None of the □ is in the ○	E
No X is Z	None of the △ is in the □	E

is the only way in which a universal negative conclusion can be drawn.

III. In what manner can a particular affirmative conclusion be drawn; namely, that part or all of one figure is contained in the other? Only when we are able to assert that the whole circle is part of one of the figures, and that the whole, or part of the circle, is part of the other figure. We have then two forms.

Every Y is X	All the ○ is in the △	A
Every Y is Z	All the ○ is in the □	A
Some Xs are Zs	Some of the △ is in the □	I

Every Y is X	All the ○ is in the △	A
Some Ys are Zs	Some of the ○ is in the □	I
Some Xs are Zs	Some of the △ is in the □	I

The ſecond of theſe contains all that is ſtrictly neceſſary to the concluſion, and the firſt may be omitted. That which follows when an aſſertion can be made as to ſome, muſt follow when the ſame aſſertion can be made of all.

IV. How can a particular negative propoſition be inferred; namely, that part, or all of one figure, is not contained in the other? It would ſeem at firſt ſight, whenever we are able to aſſert that part or all of one figure is in the circle, and that part or all of the other figure is not. The weakeſt ſyllogiſm from which ſuch an inference can be drawn would then ſeem to be as follows.

Some Xs are Ys	Some of the △ is in the ○
Some Zs are not Ys	Some of the □ is not in the ○
∴ Some Zs are not Xs	∴ Some of the △ is not in the □

But here it will appear, on a little conſideration, that the concluſion is only thus far true; that thoſe Xs which are Ys cannot be *thoſe* Zs which are not Ys; but they may be *other* Zs, about which nothing is aſſerted when we ſay that *ſome* Zs are not Ys. And further conſideration will make it evident, that a concluſion of this form can only be arrived at when one of the figures is entirely within the circle, and the whole, or part of the other without; or elſe when the whole of one of the figures is without the circle, and the whole or part of the other within; or laſtly, when the circle lies entirely within one of the figures, and not entirely within the other. That is, the following are the diſtinct forms which allow of a particular negative concluſion, in which it ſhould be remembered that a particular propoſition in the premiſes may always be changed into a univerſal one, without affecting the concluſion. For that which neceſſarily follows from "ſome," follows from "all."

Every X is Y	All the △ is in the ○	A
Some Zs are not Ys	Some of the □ is not in the ○	O
∴ Some Zs are not Xs	Some of the □ is not in the △	O

No X is Y	None of the △ is in the ○	E
Some Zs are Ys	Some of the □ is in the ○	I
∴ Some Zs are not Xs	Some of the □ is not in the △	O

Every Y is X	All the ○ is in the △	A
Some Ys are not Zs	Some of the ○ is not in the □	O
∴ Some Xs are not Zs	Some of the △ is not in the □	O

It appears, then, that there are but ſix diſtinct ſyllogiſms. All others are made from them by ſtrengthening one of the premiſes, or converting one or both of the premiſes, where ſuch converſion is allowable; or elſe by firſt making the converſion, and then ſtrengthening one of the premiſes. And the following arrangement will ſhow that two of them are univerſal, three of the others being derived from them by weakening one of the premiſes in a manner which does not deſtroy, but only weakens, the concluſion.

1. Every X is Y
Every Y is Z
———
Every X is Z

3. Every X is Y
No Z is Y
———
No X is Z

. . . .

2. Some Xs are Ys
Every Y is Z
———
Some Xs are Zs

4. Some Xs are Ys
No Z is Y
———
Some Xs are not Zs

5. Every X is Y
Some Zs are not Ys
———
Some Zs are not Xs

6. Every Y is X
Some Ys are not Zs
———
Some Xs are not Zs

We may ſee how it ariſes that one of the partial ſyllogiſms is not immediately derived, like the others, from a univerſal one. In the preceding, A E E may be conſidered as derived from A A A, by changing the term in which Y enters univerſally into a univerſal negative. If this be done with the other term inſtead, we have

No X is Y
Every Y is Z } from which univerſal premiſes we cannot deduce a univerſal concluſion, but only ſome Zs are not Xs.

If we weaken one and the other of theſe premiſes, as they ſtand, we obtain

Some Xs are not Ys
Every Y is Z
———
No concluſion

and

No X is Y
Some Ys are Zs
———
Some Zs are not Xs

equivalent to the fourth of the preceding: but if we convert the firſt premiſe, and proceed in the ſame manner,

From	No Y is X	we obtain	Some Ys are not Xs
	Every Y is Z		Every Y is Z
	Some Zs are not Xs		Some Zs are not Xs

which is legitimate, and is the ſame as the laſt of the preceding liſt, with X and Z interchanged.

Before proceeding to ſhow that all the uſual forms are contained in the preceding, let the reader remark the following rules, which may be proved either by collecting them from the preceding caſes, or by independent reaſoning.

1. The middle term muſt enter univerſally into one or the other premiſe. If it were not ſo, then one premiſe might ſpeak of one part of the middle term, and the other of another; ſo that there would, in fact, be no middle term. Thus, 'Every X is Y, Every Z is Y,' gives no concluſion: it may be thus ſtated;

All the Xs make up *a part* of the Ys
All the Zs make up *a part* of the Ys

And, before we can know that there is any common term of compariſon at all, we muſt have ſome means of ſhowing that the two parts are to ſome extent the ſame; or the preceding premiſes by themſelves are inconcluſive.

2. No term muſt enter the concluſion more generally than it is found in the premiſes; thus, if X be ſpoken of partially in the premiſes, it muſt enter partially into the concluſion. This is obvious, ſince the concluſion muſt aſſert no more than the premiſes imply.

3. From premiſes both negative no concluſion can be drawn. For it is obvious, that the mere aſſertion of diſagreement between each of two things and a third, can be no reaſon for inferring either agreement or diſagreement between theſe two things. It will not be difficult to reduce any caſe which falls under this rule to a breach of the firſt rule: thus, No X is Y, No Z is Y, gives

Every X is (ſomething which is not Y)
Every Z is (ſomething which is not Y)

in which the middle term is not ſpoken of univerſally in either. Again, 'No Y is X, ſome Ys are not Zs,' may be converted into

Every X is (a thing which is not Y)
Some (things which are not Zs) are Ys

in which there is no middle term.

4. From premiſes both particular no concluſion can be drawn. This is ſufficiently obvious when the firſt or ſecond rule is broken, as in 'Some Xs are Ys, Some Zs are Ys.' But it is not immediately obvious when the middle term enters one of the premiſes univerſally. The following reaſoning will ſerve for exerciſe in the preceding reſults. Since both premiſes are particular in form, the middle term can only enter one of them univerſally by being the predicate of a negative propoſition; conſequently (Rule 3) the other premiſe muſt be affirmative, and, being particular, neither of its terms is univerſal. Conſequently both the terms as to which the concluſion is to be drawn enter partially, and the concluſion (Rule 2) can only be a particular *affirmative* propoſition. But if one of the premiſes be negative, the concluſion muſt be *negative* (as we ſhall immediately ſee). This contradiction ſhows that the ſuppoſition of particular premiſes producing a legitimate reſult is inadmiſſible.

5. If one premiſe be negative, the concluſion, if any, muſt be negative. If one term agree with a ſecond and diſagree with a third, no agreement can be inferred between the ſecond and third.

6. If one premiſe be particular, the concluſion muſt be particular. This may be ſhown as follows. If two propoſitions P and Q, together prove a third, R, it is plain that P and the denial of R, prove the denial of Q. For P and Q cannot be true together without R. Now if poſſible, let P (a particular) and Q (a univerſal) prove R (a univerſal). Then P (particular) and the denial of R (particular) prove the denial of Q. But two particulars can prove nothing.

In the preceding ſet of ſyllogiſms we obſerve one form only which produces A, or E, or I, but three which produce O.

Let an aſſertion be ſaid to be weakened when it is reduced from univerſal to particular, and ſtrengthened in the contrary caſe. Thus, 'Every X is Z' is called ſtronger than 'Some Xs are Zs.'

Every usual form of syllogism which can give a legitimate result is either one of the preceding six, or another formed from one of the six, either by changing one of the assertions into its converse, if that be allowable, or by strengthening one of the premises, without altering the conclusion, or both. Thus,

Some Xs are Ys Every Y is Z	may be written	Some Ys are Xs Every Y is Z

What follows will still follow from { *Every* Y is X
Every Y is Z }

for all which is true when 'Some Ys are Xs,' is not less true when 'Every Y is X.'

It would be possible also to form a legitimate syllogism by weakening the conclusion, when it is universal, since that which is true of all is true of some. Thus, 'Every X is Y, Every Y is Z,' which yields 'Every X is Z,' also yields 'Some Xs are Zs.' But writers on logic have always considered these syllogisms as useless, conceiving it better to draw from any premises their strongest conclusion. In this they were undoubtedly right; and the only question is, whether it would not have been advisable to make the premises as weak as possible, and not to admit any syllogisms in which more appeared than was absolutely necessary to the conclusion. If such had been the practice, then

Every Y is X, Every Y is Z, therefore Some Xs are Zs

would have been considered as formed by a spurious and unnecessary excess of assertion. The minimum of assertion would be contained in either of the following,

Every Y is X, Some Ys are Zs, therefore Some Xs are Zs
Some Ys are Xs, Every Y is Z, therefore Some Xs are Zs

In this chapter, syllogisms have been divided into two classes: first, those which prove a universal conclusion; secondly, those which prove a partial conclusion, and which are (all but one) derived from the first by weakening one of the premises, in such manner as to produce a legitimate but weakened conclusion. Those of the first class are placed in the first column, and of the other in the second.

Universal.		Particular.	
A Every X is Y		Some Xs are Ys	I
A Every Y is Z	——	Every Y is Z	A
A Every X is Z		Some Xs are Zs	I
		Some Xs are Ys	I
		No Y is Z	E
A Every X is Y		Some Xs are not Zs	O
E No Y is Z	——	Every X is Y	A
E No X is Z		Some Zs are not Ys	O
		Some Zs are not Xs	O
		Every Y is X	A
....		Some Ys are not Zs	O
		Some Xs are not Zs	O

In all works on logic, it is customary to write that premise first which contains the predicate of the conclusion. Thus,

Every Y is Z		Every X is Y
Every X is Y	would be written, and not	Every Y is Z
Every X is Z		Every X is Z

The premises thus arranged are called major and minor; the predicate of the conclusion being called the major term, and its subject the minor. Again, in the preceding case we see the various subjects coming in the order Y, Z; X, Y; X, Z: and the number of different orders which can appear is four, namely—

Y Z	Z Y	Y Z	Z Y
X Y	X Y	Y X	Y X
X Z	X Z	X Z	X Z

which are called the four *figures*, and every kind of syllogism in each figure is called a *mood*. I now put down the various moods of each figure, the letters of which will be a guide to find out those of the preceding list from which they are derived. Co means that a premise of the preceding list has been converted; + that it has been strengthened; Co +, that both changes have taken place. Thus,

A	Every Y is Z		A	Every Y is Z
I	Some Xs are Ys	becomes	A	Every Y is X : (Co +)
I	Some Xs are Zs		I	Some Xs are Zs

And Co + points out the following: If ſome Xs be Ys, then ſome Ys are Xs (Co); and all that is true when Some Ys are Xs, is true when Every Y is X (+); therefore the ſecond ſyllogiſm is legitimate, if the firſt be ſo.

Firſt Figure.

A	Every Y is Z	A	Every Y is Z
A	Every X is Y	I	Some Xs are Ys
A	Every X is Z	I	Some Xs are Zs
E	No Y is Z	E	No Y is Z
A	Every X is Y	I	Some Xs are Ys
E	No X is Z	O	Some Xs are not Zs

Second Figure.

E	No Z is Y (Co)	E	No Z is Y (Co)
A	Every X is Y	I	Some Xs are Ys
E	No X is Z	O	Some Xs are not Zs
A	Every Z is Y	A	Every Z is Y
E	No X is Y (Co)	O	Some Xs are not Ys
E	No X is Z	O	Some Xs are not Zs

Third Figure.

A	Every Y is Z	E	No Y is Z
A	Every Y is X (Co +)	A	Every Y is X (Co +)
I	Some Xs are Zs	O	Some Xs are not Zs
I	Some Ys are Zs (Co)	O	Some Ys are not Zs
A	Every Y is X	A	Every Y is X
I	Some Xs are Zs	O	Some Xs are not Zs
A	Every Y is Z	E	No Y is Z.
I	Some Ys are Xs (Co)	I	Some Ys are Xs (Co)
I	Some Xs are Zs	O	Some Xs are not Zs

Fourth Figure.

A Every Z is Y (+)	I Some Zs are Ys
A Every Y is X	A Every Y is X
I Some Xs are Zs	I Some Zs are Xs

A Every Z is Y	E No Z is Y (Co)
E No Y is X	A Every Y is X (Co +)
E No X is Z	O Some Xs are not Zs

E No Z is Y (Co)
I Some Ys are Xs (Co)

O Some Xs are not Zs

The above is the ancient method of dividing ſyllogiſms; but, for the preſent purpoſe, it will be ſufficient to conſider the ſix from which the reſt can be obtained. And ſince ſome of the ſix have X in the predicate of the concluſion, and not Z, I ſhall join to them the ſix other ſyllogiſms which are found by tranſpoſing Z and X. The complete liſt, therefore, of ſyllogiſms with the weakeſt premiſes and the ſtrongeſt concluſions, in which a compariſon of X and Z is obtained by compariſon of both with Y, is as follows:

Every X is Y	Every Z is Y	Some Xs are Ys	Some Zs are Ys
Every Y is Z	Every Y is X	No Y is Z	No Y is X
Every X is Z	Every Z is X	Some Xs are not Zs	Some Zs are not Xs
Every X is Y	Every Z is Y	Every X is Y	Every Z is Y
No Y is Z	No Y is X	Some Zs are not Ys	Some Xs are not Ys
No X is Z	No Z is X	Some Zs are not Xs	Some Xs are not Zs
Some Xs are Ys	Some Zs are Ys	Every Y is X	Every Y is Z
Every Y is Z	Every Y is X	Some Ys are not Zs	Some Ys are not Xs
Some Xs are Zs	Some Zs are Xs	Some Xs are not Zs	Some Zs are not Xs

In the liſt of page 12, there was nothing but recapitulation of forms, each form admitting a variation by interchanging X and Z. This interchange having been made, and the reſults collected as above, if we take every caſe in which Z is the predicate, or can be made the predicate by allowable converſion, we have a collection of all poſſible *weakeſt* forms in which the reſult is one of the four 'Every X is Z,' 'No X is Z,' 'Some Xs are Zs,' 'Some Xs are not Zs;' as follows. The premiſes are written in what appeared the moſt natural order, without diſtinction of major or minor.

Every X is Y
Every Y is Z
———
Every X is Z

Some Xs are Ys
Every Y is Z
———
Some Xs are Zs

Some Zs are Ys
Every Y is X
———
Some Xs are Zs

Every X is Y
No Z is Y
———
No X is Z

Every Z is Y
No X is Y
———
No X is Z

Some Xs are Ys
No Z is Y
———
Some Xs are not Zs

Every Z is Y
Some Xs are not Ys
———
Some Xs are not Zs

Every Y is X
Some Ys are not Zs
———
Some Xs are not Zs

Every aſſertion which can be made upon two things by compariſon with any third, that is, every ſimple inference, can be reduced to one of the preceding forms. Generally ſpeaking, one of the premiſes is omitted, as obvious from the concluſion; that is, one premiſe being named and the concluſion, that premiſe is implied which is neceſſary to make the concluſion good. Thus, if I ſay, "That race muſt have poſſeſſed ſome of the arts of life, for they came from Aſia," it is obviouſly meant to be aſſerted, that all races coming from Aſia muſt have poſſeſſed ſome of the arts of life. The preceding is then a ſyllogiſm, as follows:

That race is 'a race of Aſiatic origin:'
Every 'race of Aſiatic origin' is 'a race which muſt have poſſeſſed ſome of the arts of life:'
Therefore, That race *is* a race which muſt have poſſeſſed ſome of the arts of life.

A perſon who makes the preceding aſſertion either means to imply, antecedently to the concluſion, that all Aſiatic races muſt have poſſeſſed arts, or he talks nonſenſe if he aſſert the concluſion poſitively. 'X muſt be Z, for it is Y,' can only be an inference when 'Every Y is Z.' This latter propoſition may be called the ſuppreſſed premiſe; and it is in ſuch ſuppreſſed propoſitions that the greateſt danger of error lies. It is alſo in ſuch propoſitions that men convey opinions which they would not willingly expreſs. Thus, the honeſt witneſs who ſaid, 'I always thought him a reſpectable man—he kept his gig,' would probably not have admitted in direct terms, 'Every man who keeps a gig muſt be reſpectable.'

I ſhall now give a few detached illuſtrations of what precedes.

"His imbecility of character might have been inferred from his proneneſs to favourites; for all weak princes have this failing." The preceding would ſtand very well in a hiſtory, and many would paſs it over as containing very good inference. Written, however, in the form of a ſyllogiſm, it is,

	All weak princes	are prone to favourites
	He	was prone to favourites
Therefore	He	was a weak prince

which is palpably wrong. (Rule 1.) The writer of ſuch a ſentence as the preceding might have meant to ſay, 'for all who have this failing are weak princes;' in which caſe he would have inferred rightly. Every one ſhould be aware that there is much falſe form of inference ariſing out of badneſs of ſtyle, which is juſt as injurious to the habits of the untrained reader as if the errors were miſtakes of logic in the mind of the writer.

'X is leſs than Y; Y is leſs than Z: therefore X is leſs than Z.' This, at firſt ſight, appears to be a ſyllogiſm; but, on reducing it to the uſual form, we find it to be,

	X is (a magnitude leſs than Y)
	Y is (a magnitude leſs than Z)
Therefore	X is (a magnitude leſs than Z)

which is not a ſyllogiſm, ſince there is no middle term. Evident as the preceding is, the following additional propoſition muſt be formed before it can be made explicitly logical. 'If Y be a magnitude leſs than Z, then every magnitude leſs than Y is alſo leſs than Z.' There is, then, before the preceding can be reduced to a ſyllogiſtic form, the neceſſity of a deduction from the ſecond premife, and the fubftitution of the refult inftead of that premife. Thus,

X is lefs than Y
Lefs than Y is lefs than Z: following from Y is lefs than Z.

Therefore X is lefs than Z

But, if the additional argument be examined—namely, if Y be lefs than Z, then that which is lefs than Y is lefs than Z—it will be found to require precifely the fame confiderations repeated; for the original inference was nothing more. In fact, it may eafily be feen as follows, that the propofition before us involves more than any fimple fyllogifm can exprefs. When we fay that X is lefs than Y, we fay that if X were applied to Y, every part of X would match a part of Y, and there would be parts of Y remaining over. But when we fay, 'Every X is Y,' meaning the premife of a common fyllogifm, we fay that every inftance of X is an inftance of Y, without faying any thing as to whether there are or are not inftances of Y ftill left, after thofe which are alfo X are taken away. If, then, we wifh to write an ordinary fyllogifm in a manner which fhall correfpond with 'X is lefs than Y, Y is lefs than Z, therefore X is lefs than Z,' we muft introduce a more definite amount of affertion than was made in the preceding forms. Thus,

Every X is Y, and there are Ys which are not Xs
Every Y is Z, and there are Zs which are not Ys

Therefore Every X is Z, and there are Zs which are not Xs

Or thus:

The Ys contain all the Xs, and more
The Zs contain all the Ys, and more

The Zs contain all the Xs, and more

The moft technical form, however, is,

From Every X is Y; [Some Ys are not Xs]
Every Y is Z; [Some Zs are not Ys]
Follows Every X is Z; [Some Zs are not Xs]

This fort of argument is called *à fortiori* argument, becaufe the premifes are more than fufficient to prove the conclufion, and the extent of the conclufion is thereby greater than its mere form would indicate. Thus, 'X is lefs than Y, Y is lefs than Z, therefore, *à fortiori*, X is less than Z,' means that the extent to which X is less than Z must be greater than that to which X is less than Y, or Y than Z. In the syllogism last written, either of the bracketted premises might be struck out without destroying the conclusion; which last would, however, be weakened. As it stands, then, the part of the conclusion, 'Some Zs are not Xs,' follows *à fortiori*.

The argument *à fortiori* may then be defined as a universally affirmative syllogism, in which both of the premises are shewn to be less than the whole truth, or greater. Thus, in 'Every X is Y, Every Y is Z, therefore Every X is Z,' we do not certainly imply that there are more Ys than Xs, or more Zs than Ys, so that we do not know that there are more Zs than Xs. But if we be at liberty to state the syllogism as follows,

All the Xs make up part (and part only) of the Ys
Every Y is Z;

then we are certain that

All the Xs make up part (and part only) of the Zs.

But if we be at liberty further to say that

All the Xs make up part (and part only) of the Ys
All the Ys make up part (and part only) of the Zs

then we conclude that

All the Xs make up *part of part* (only) of the Zs

and the words in Italics mark that quality of the conclusion from which the argument is called *à fortiori*.

Most syllogisms which give an affirmative conclusion are generally meant to imply *à fortiori* arguments, except only in mathematics. It is seldom, except in the exact sciences, that we meet with a proposition, 'Every X is Z,' which we cannot immediately couple with 'some Zs are not Xs.'

When an argument is completely established, with the exception of one assertion only, so that the inference may be drawn as soon as that one assertion is established, the result is stated in a form which bears the name of an *hypothetical* syllogism. The word hypothesis means nothing but supposition; and the species of syllogism just mentioned first lays down the assertion that a consequence will be true if a certain condition be fulfilled, and then either asserts the fulfilment of the condition, and thence the consequence, or else denies the consequence, and thence denies the fulfilment of the condition. Thus, if we know that

When X is Z, it follows that P is Q;

then, as soon as we can ascertain that X is Z, we can conclude that P is Q; or, if we can shew that P is not Q, we know that X is not Z. But if we find that X is not Z, we can infer nothing; for the preceding does not assert that P is Q *only* when X is Z. And if we find out that P is Q we can infer nothing. This conditional syllogism may be converted into an ordinary syllogism, as follows. Let K be any 'case in which X is Z,' and V, a 'case in which P is Q;' then the preceding assertion amounts to 'Every K is V.' Let L be a particular instance, the X of which may or may not be Z. If X be Z in the instance under discussion, or if X be not Z, we have, in the one case and the other,

	Every K is V	Every K is V
	L is a K	L is not a K
Therefore	L is a V	No conclusion

Similarly, according as a particular case (M) is or is not V, we have

Every K is V	Every K is V
M is a V	M is not a V
No conclusion	M is not a K

That is to say: the assertion of an hypothesis is the assertion of its necessary consequence, and the denial of the necessary consequence is the denial of the hypothesis: but the assertion of the necessary consequence gives no right to assert the hypothesis, nor does the denial of the hypothesis give any right to deny the truth of that which would (were the hypothesis true) be its necessary consequence.

Demonstration is of two kinds: which arises from this, that every proposition has a contradictory; and of these two, one must be true and the other must be false. We may then either prove a proposition to be true, or its contradictory to be false. 'It is true that every X is Z,' and 'it is false that there are some Xs which are not Zs,' are the same proposition; and the proof of either is called the indirect proof of the other.

But how is any propoſition to be proved falſe, except by proving a contradiction to be true? By proving a neceſſary conſequence of the propoſition to be falſe. But this is not a complete anſwer, ſince it involves the neceſſity of doing the ſame thing; or, ſo far as this anſwer goes, one propoſition cannot be proved falſe unleſs by proving another to be falſe. But it may happen, that a neceſſary conſequence can be obtained which is obviouſly and ſelf-evidently falſe, in which caſe no further proof of the falſehood of the hypotheſis is neceſſary. Thus the proof which Euclid gives that all equiangular triangles are equilateral is of the following ſtructure, logically conſidered.

(1.) If there be an equiangular triangle not equilateral, it follows that a whole can be found which is not greater than its part.*

(2.) It is falſe that there can be any whole which is not greater than its part (ſelf evident).

(3.) Therefore it is falſe that there is any equiangular triangle which is not equilateral; or all equiangular triangles are equilateral.

When a propoſition is eſtabliſhed by proving the truth of the matters it contains, the demonſtration is called *direct;* when by proving the falſehood of every contradictory propoſition, it is called *indirect.* The latter ſpecies of demonſtration is as logical as the former, but not of ſo ſimple a kind; whence it is deſirable to uſe the former whenever it can be obtained.

The uſe of indirect demonſtration in the Elements of Euclid is almoſt entirely confined to thoſe propoſitions in which the converſes of ſimple propoſitions are proved. It frequently happens that an eſtabliſhed aſſertion of the form

Every X is Z . . . . (1)

may be eaſily made the means of deducing,

Every (thing not X) is not Z . . (2)

which laſt gives

Every Z is X . . . . (3)

* This is the propoſition in proof of which nearly the whole of the demonſtration of Euclid is ſpent.

The converfion of the fecond propofition into the third is ufually made by an indirect demonftration, in the following manner: If poffible, let there be one Z, which is not X, (2) being true. Then there is one thing which is not X and is Z; but every thing not X is not Z; therefore there is one thing which is Z and is not Z: which is abfurd. It is then abfurd that there fhould be one fingle Z which is not X; or, Every Z is X.

The following propofition contains a method which is of frequent ufe.

Hypothesis.—Let there be any number of propofitions or affertions,—three for inftance, X, Y, and Z,—of which it is the property that one or the other muft be true, *and one only*. Let there be three other propofitions, P, Q, and R, of which it is alfo the property that one, and one only, muft be true. Let it alfo be a connexion of thofe affertions, that

When X is true, P is true
When Y is true, Q is true
When Z is true, R is true

Consequence: then it follows that

When P is true, X is true
When Q is true, Y is true
When R is true, Z is true

For, when P is true, then Q and R muft be falfe; confequently, neither Y nor Z can be true, for then Q or R would be true. But either X, Y, or Z muft be true, therefore X muft be true; or, when P is true, X is true. In a fimilar way the remaining affertions may be proved.

Cafe 1. If	When P is Q,	X is Z
	When P is not Q,	X is not Z
It follows that	When X is Z,	P is Q
	When X is not Z,	P is not Q

Cafe 2. If	When X is greater than Z,	P is greater than Q
	When X is equal to Z,	P is equal to Q
	When X is lefs than Z,	P is lefs than Q
It follows that	When P is greater than Q,	X is greater than Z
	When P is equal to Q,	X is equal to Z
	When P is lefs than Q,	X is lefs than Z

CHAPTER II.

On Objects, Ideas, and Names.

LOGIC is derived from a Greek word (λόγος) which signifies communication of thought, usually by speech. It is the name which is generally given to the branch of inquiry (be it called science or art), in which the act of the mind in reasoning is considered, particularly with reference to the connection of thought and language. But no definition yet given in few words has been found satisfactory to any considerable number of thinking persons.

All existing things upon this earth, which have knowledge of their own existence, possess, some in one degree and some in another, the power of *thought*, accompanied by *perception*, which is the awakening of thought by the effect of external objects upon the senses. By thought I here mean, all mental action, not only that comparatively high state of it which is peculiar to man, but also that lower degree of the same thing which appears to be possessed by brutes.

With respect to the mind, considered as a complicated apparatus which is to be studied, we are not even so well off as those would be who had to examine and decide upon the mechanism of a watch, merely by observation of the functions of the hands, without being allowed to see the inside. A mechanician, to whom a watch was presented for the first time, would be able to give a good guess as to its structure, from his knowledge of other pieces of contrivance. As soon as he had examined the law of the motion of the hands, he might conceivably invent an instrument with similar properties, in fifty different ways. But in the case of the mind, we have manifestations only, without the smallest power of reference to other similar things, or the least knowledge of structure or process, other than what may be derived from those manifestations. It is the problem of the watch to those who have never seen any mechanism at all.

We have nothing more to do with the ſcience of mind, uſually called *metaphyſics*,* than to draw a very few neceſſary diſtinctions, which, whatever names we uſe to denote them, are matters of fact connected with our ſubject. Some modes of expreſſing them favor one ſyſtem of metaphyſics, and ſome another; but ſtill they are matters of obſerved fact. Our words muſt be very imperfect ſymbols, drawn from compariſon of the manifeſtations of thought with thoſe of things in corporeal exiſtence. For inſtance, I juſt now ſpoke of the mind as an apparatus, or piece of mechaniſm. It is a ſtructure of ſome ſort, which has the means of fulfilling various purpoſes; and ſo far it reſembles the hand, which by the diſpoſition of bone and muſcle, can be made to perform an immenſe variety of different motions and graſps. Where the reſemblance begins to be imperfect, and why, is what we cannot know. In all probability we ſhould need new modes of perception, other ſenſes beſides ſight, hearing, and touch, in order to know thought as we know colour, ſize, or motion. But the purpoſe of the preſent treatiſe is only the examination of ſome of the manifeſtations of thinking power in their relation to the language in which they are expreſſed. Knowledge of thought and knowledge of the reſults of thought,

* All ſyſtems make an aſſumption of the uniformity of proceſs in all minds, carried to an extent the propriety of which ought to be a matter of ſpecial diſcuſſion. There are no writers who give us ſo much *muſt* with ſo little *why*, as the metaphyſicians. If perſons who had only ſeen the outſide of the timepiece, were to invent machines to anſwer its purpoſe, they might arrive at their object in very different ways. One might uſe the pendulum and weight, another the ſprings and the balance: one might diſcover the combination of toothed wheels, another a more complicated action of lever upon lever. Are we *ſure* that there are not differences in our minds, ſuch as the preceding inſtance may ſuggeſt by analogy; if ſo, *how* are we ſure? Again, if our minds be as tables with many legs, do we know that a weight put upon different tables will be ſupported in the ſame manner in all. May not the ſame leg ſupport much or all of a certain weight in one mind, and little or nothing in another? I have ſeen ſtriking inſtances of ſomething like this, among thoſe who have examined for themſelves the grounds of the mathematical ſciences.

I would not diſſuade a ſtudent from metaphyſical inquiry; on the contrary, I would rather endeavour to promote the deſire of entering upon ſuch ſubjects: but I would warn him, when he tries to look down his own throat with a candle in his hand, to take care that he does not ſet his head on fire.

are very different things. The watch abovementioned might have the functions of its hands discovered, might be used in finding longitude (and even latitude) all over the world, without the parties using it having the smallest idea of its interior structure.

That our minds, souls, or thinking powers (use what name we may) exist, is the thing of all others of which we are most certain, each for himself. Next to this, nothing can be more certain to us, each for himself, than that other things also exist; other minds, our own bodies, the whole world of matter. But between the character of these two certainties there is a vast difference. Any one who should deny his own existence would, if serious, be held beneath argument: he does not know the meaning of his words, or he is false or mad. But if the same man should deny that any thing exists except himself, that is, if he should affirm the whole creation to be a dream of his own mind, he would be absolutely unanswerable. If I (who *know* he is wrong, for *I* am certain of *my own* existence) argue with him, and reduce him to silence, it is no more than might* happen in his dream. A celebrated metaphysician, Berkeley, maintained that with regard to *matter*, the above is the state of the case: that our impressions of matter are only impressions, communicated by the Creator without any intervening cause of communication.

Our most convincing communicable proof of the existence of other things, is, not the appearance of objects, but the necessity of admitting that there are *other minds* besides our own. The external inanimate objects might be creations of our own thought, or thinking and perceptive function: they are so sometimes, as in the case of insanity, in which the mind has frequently the appearance of making the whole or part of its own external world. But when we see other beings, performing similar functions to those which we ourselves perform, we come so irresistibly to the conclusion that there must be other sentients like ourselves, that we should rather compare a person who doubted it to one who denied his own existence, than to one who simply denied the real external existence of the *material* world.

* It is not impossible that in a real dream of sleep, some one may have created an antagonist who beat him in an argument to prove that he was awake.

When once we have admitted different and independent minds, the reality of external objects (external to all those minds) follows as of course. For different minds receive impressions at the same time, which their power of communication enables them to know are similar, so far as any impressions, one in each of two different minds, can be known to be similar. There must be a *somewhat* independent of those minds, which thus acts upon them all at once, and without any choice of their own. This *somewhat* is what we call an external object: and whether it arise in Berkeley's mode, or in any other, matters nothing to us here.

We shall then, take it for granted that external *objects* actually exist, independently of the mind which perceives them. And this brings us to an important distinction, which we must carry with us throughout the whole of this work. Besides the actual external object, there is also the mind which perceives it, and what (for want of better words or rather for want of knowing whether they be good words or not) we must call the *image of that object in the mind*, or the *idea* which it communicates. The term *subject* is applied by metaphysicians to the perceiving mind: and thus it is said that a thing may be considered *subjectively* (with reference to what it is in the mind) or *objectively* (with reference to what it is independently of any particular mind). But logicians use the word subject in another sense. In a proposition such as 'bread is wholesome', the thing spoken of, 'bread', is called the subject of the proposition: and in fact the word *subject* is in common language so frequently confounded with *object*, that it is almost hopeless to speak clearly to beginners about themselves as *subjects*. I shall therefore adopt the words *ideal* and *objective*, *idea* and *object*, as being, under explanations, as good as any others: and better than *subject* and *object* for a work on logic.

The word *idea*, as here used, does not enter in that vague sense in which it is generally used, as if it were an opinion that might be right or wrong. It is that which the object gives to the mind, or the state of the mind produced by the object. Thus the idea of a horse is *the horse in the mind:* and we know no other horse. We admit that there is an external *object*, a horse, which may give a *horse in the mind* to twenty different persons: but no one of these twenty knows the object; each one only knows his *idea*.

There is an object, because each of the twenty persons receives an idea without communicating with the others : so that there is something external to give it them. But when they talk about it, under the name of a horse, they talk about their ideas. They all refer to the object, as being the thing they are talking about, until the moment they begin to differ: and then they begin to speak, not of external horses, but of impressions on their minds ; at least this is the case with those who know what knowledge is ; the positive and the unthinking part of them still talk of the *horse*. And the latter have a great advantage* over the former with those who are like themselves.

Why then do we introduce the term *object* at all, since all our knowledge lies in ideas ? For the same reason as we introduce the term *matter* into natural philosophy, when all we know is form, size, colour, weight, &c., no one of which is matter, nor even all together. It is convenient to have a word for that external source from which *sensible* ideas are produced : and it is just as convenient to have a word for the external source, material or not, from which *any* idea is produced. Again, why do we speak of our power of considering things either ideally or objectively, when as we can know nothing but ideas, we can have no right to speak of any thing else ? The answer is that, just as in other things, when we speak of an object, we speak of the *idea of an object*. We learn to speak of the external world, because there are others like ourselves who evidently draw ideas from the same sources as ourselves : hence we come to have the idea of those sources, the idea of external objects, as we call them. But we do not know those sources ; we know only our ideas of them.

We can even use the terms ideal and objective in what may appear a metaphorical sense. When we speak of ourselves in the manner of this chapter, we put ourselves, as it were, in the position of spectators of our own minds : we speak and think of our

* One man asserts *a fact* on his own knowledge, another asserts his full *conviction* of the contrary fact. Both use the evidence of their senses : but the second knows that *full conviction* is all that man can have. The first will carry it hollow in a court of justice, in which persons are constantly compelled to swear, not only that they have an impression, but that the impression is correct ; that is to say, is the impression which mankind in general would have, and must have, and ought to have.

own minds objectively. And it must be remembered that by the word object, we do not mean *material* object only. The mind of another, any one of its thoughts or feelings, any relation of minds to one another, a treaty of peace, a battle, a discussion upon a controverted question, the right of conveying a freehold, —are all objects, independently of the persons or things engaged in them. They are things external to our minds, of which we have ideas.

An object communicates an idea: but it does not follow that every idea is communicated by an object. The mind can create ideas in various ways; or at least can derive, by combinations which are not found in external existence, new collections of ideas. We have a perfectly distinct idea of a unicorn, or a flying dragon: when we say there are no such things, we speak objectively only: ideally, they have as much existence as a horse or a sheep; to a herald, more. Add to this, that the mind can separate ideas into parts, in such manner that the parts alone are not ideas of any existing separate material objects, any more than the letters of a word are constituent parts of the meaning of the whole. Hence we get what are called *qualities* and *relations*. A ball may be hard and round, or may have hardness and roundness: but we can not say that hardness and roundness are separate external material objects, though they are objects the ideas of which necessarily accompany our perception of certain objects. These ideas are called *abstract* as being removed or abstracted from the complex idea which gives them: the abstraction is made by comparison or observation of resemblances. If a person had never seen any thing round except an apple, he would perhaps never think of roundness as a distinct object of thought. When he saw another round body, which was evidently not an apple, he would immediately, by perception of the resemblance, acquire a separate idea of the thing in which they resemble one another.

Abstraction is not performed upon the ideas of material objects only. For instance, from conduct of one kind, running through a number of actions, performed by a number of persons, we get the ideas of goodness, wickedness, talent, courage. But we must not imagine that we can make ideally external representation of these words. They are *objects*, that is to say, the mind considers them as external to itself: but they are not material objects.

Some people deny their exiſtence, and look upon them as only abſtract words, or words under which we ſpeak of minds or bodies without ſpecifying any more than one of the ideas produced by theſe minds or bodies. For inſtance, they aſſert that when we ſay 'knowledge gives power' it is really that perſons with knowledge are therefore able, or have power, to produce, or to do, what perſons without it cannot. This is a queſtion which it does not concern me here to diſcuſs.

Seeing that the mind poſſeſſes a power of originating new combinations of ideas, and alſo of abſtracting from complex ideas the more ſimple ones of which, it ſeems natural to ſay, they are compoſed, it has long been a queſtion among metaphyſicians whether the mind has any ideas of its own which it poſſeſſes independently of all ſuggeſtion from external objects. It is not neceſſary that I ſhould attempt to lead the ſtudent to any concluſion* on this ſubject: for our purpoſe, the diſtinction between ideas and objects, though it were falſe, is of more importance than that between innate and acquired ideas, though it be true. But one of theſe two things muſt be true: either we have ideas which we do *not* acquire from or by means of communication with the external world (experience, trial of our ſenſes) or there is a power in the mind of acquiring a certainty and a generality which experience alone could not properly give. For inſtance, we are ſatisfied as of our own exiſtence that ſeven and three collected are the ſame as five and five, *whatever the objects may be*

* It has always appeared to me much ſuch a queſtion as the following. There are hooks which certainly catch fiſh if put into the water; and moſt certainly they have been put into the water. There are then fiſh upon them. But theſe fiſh might have been on ſome of them when they were put into the water. It is to no purpoſe to inquire whether it was ſo or not, unleſs there be ſome diſtinction between the fiſhes which may make it a queſtion whether ſome of them could have been bred in the river into which the hooks were put. The mind has certainly a power of acquiring and retaining ideas, which power, when put into communication with the external world, it muſt exerciſe. There is no mind to experiment upon, except thoſe which have had ſuch communication. Are there found any ideas which we have reaſon to think *could* not have been acquired by this communication? any fiſhes which *could* not have come out of the river? Metaphyſicians ſeem to admit that if any ideas be innate, they are thoſe of ſpace, time, and of cauſe and effect: they ſeem alſo to admit, that if there be any ideas, which, not being innate, are ſure to be acquired, they are theſe very ones.

which are counted: the thing is true of fingers, pebbles, counters, ſheep, trees, &c. &c. &c. We cannot have aſſured ourſelves of this by experience: for example, we *know* it to be true of pebbles at the North Pole, though we have never been there; we are as ſure of it as of our own exiſtence. I do not mean that we have a rational conviction only, fit to act upon, that it is ſo at the North Pole, becauſe it is ſo in every place in which it has been tried: if we had nothing elſe, we ſhould have this; but we feel that this leſſer conviction is ſwallowed up by a greater. We have the leſſer conviction that the pebbles at the pole fall to the ground when they are let go: we are very ſure of this, without aſſerting that it cannot be otherwiſe: we ſee no *impoſſibility* in thoſe pebbles being ſuch as always to remain in air wherever they are placed.* But that ſeven and three are no other than five and five is a matter which we are prepared to affirm as poſitively of the pebbles at the North Pole as of our own fingers, both that it is ſo, and that it *muſt* be ſo. Whence ariſes this actual difference in point of fact, between our mode of viewing and knowing

* Metaphyſicians, in their ſyſtems, have often taken this diſtinction to be one of ſyſtem only, treating it as a thing to be accepted or rejected with the ſyſtem, inſtead of an actual and indiſputable phenomenon which requires explanation under any ſyſtem. Dr. Whewell, of all Engliſh writers on natural ſcience I know, is the one who has made the fact, as a fact, pervade his writings, ſometimes attached to a ſyſtem, ſometimes not. The following remarks on the general ſubject are worth conſideration: "It is indeed, extremely difficult to find, in ſpeaking of this ſubject, expreſſions which are ſatisfactory. The reality of the objects which we perceive is a profound, apparently an inſoluble problem. We cannot but ſuppoſe that exiſtence is ſomething different from our knowledge of exiſtence:—that what exiſts, does not exiſt merely in our knowing that it does:—truth is truth whether we know it or not. Yet how can we conceive truth, otherwiſe than as ſomething known? How can we conceive things as exiſting, without conceiving them as objects of perception? Ideas and Things are conſtantly oppoſed, yet neceſſarily coexiſtent. How they are thus oppoſite and yet identical, is the ultimate problem of all philoſophy. The ſucceſſive phaſes of philoſophy have conſiſted in ſeparating and again uniting theſe two oppoſite elements; in dwelling ſometimes upon the one and ſometimes upon the other, as the principal or original or only element; and then in diſcovering that ſuch an account of the ſtate of the caſe was inſufficient. Knowledge requires ideas. Reality requires things. Ideas and things coexiſt. Truth *is*, and is known. But the complete explanation of theſe points appears to be beyond our reach."

different ſpecies of aſſertions? the truth of the laſt named aſſertion is not born with us, for children are without it, and learn it by experience, as we know. The *muſt be ſo* cannot be acquired from experience in the common way, for that ſame experience on which we rely tells us that however often a thing may have been found true, whatever rule may have been eſtabliſhed by repeated inſtances, an exception may at laſt occur. There ſeems then to be in the mind a power of developing, from the ideas which experience gives, a real and true diſtinction of neceſſary and not neceſſary, poſſible and impoſſible. The things which are without us always confirm our *neceſſary* propoſitions: but how we derive that complete aſſurance that they *will* do ſo as faithfully as hitherto they *have* done ſo, is not within our power to ſay.

Connected with ideas are the *names* we give them; the ſpoken or written ſounds by which we think of them, and communicate with others about them. To have an idea, and to make it the ſubject of thought as an idea, are two perfectly diſtinct things: the *idea of an idea* is not the idea itſelf. I doubt whether we could have made thought itſelf the ſubject of thought without language. As it is, we give names to our ideas, meaning by a name not merely a ſingle word, but any collection of words which conveys to one mind the idea in another. Thus a-man-in-a-black-coat-riding-along-the high-road-on-a-bay-horſe is as much the name of an idea as man, black, or horſe. We can coin words at pleaſure; and, were it worth while, might invent a ſingle word to ſtand for the preceding phraſe.

Names are uſed indifferently, both for the objects which produce ideas, and for the ideas produced by them. This is a diſadvantage, and it will frequently be neceſſary to ſpecify whether we ſpeak ideally or objectively. In common converſation we ſpeak ideally and think we ſpeak objectively: we take for granted that our own ideas are fit to paſs to others, and will convey to them the ſame ideas as the objects themſelves would have done. That this may be the caſe, it is neceſſary firſt, that the object ſhould really give us the ſame ideas as to others; ſecondly, that our words ſhould carry from us to our correſpondents the ſame ideas as thoſe which we intended to expreſs by them. How, and in what caſes, the firſt or the ſecond condition is not fulfilled, it is impoſſible to know or to enumerate. But we have nothing to do here except to obſerve that we are only incidentally concerned with this queſtion in a work of logic. We preſume fixed and, if objective, objectively true ideas, with certain names attached: ſo that it is never in doubt whether a name be or be not properly attached to any idea. This method muſt be followed in all works of ſcience: a conceivably attainable end is firſt preſumed to be attained, and the conſequences of its attainment are ſtudied. Then, afterwards, comes the queſtion whether this end is always attained, and if not, why. The way to mend bad roads muſt come at the end, not at the beginning, of a treatiſe on the art of making good ones.

Every name has a reference to every idea, either affirmative or negative. The term *horſe* applies to every thing, either poſitively or negatively. This (no matter what I am ſpeaking of) either *is* or *is not* a horſe. If there be any doubt about it, either the idea is not preciſe, or the term *horſe* is ill underſtood. A name ought to be like a boundary, which clearly and undeniably either ſhuts in, or ſhuts out, every idea that can be ſuggeſted. It is the imperfection of our minds, our language, and our knowledge of external things, that this clear and undeniable incluſion or excluſion is ſeldom attainable, except as to ideas which are *well within the boundary*: at and near the boundary itſelf all is vague. There are decided greens and decided blues: but between the two colours there are ſhades of which it muſt be unſettled by univerſal agreement to which of the two colours they belong. To the eye, green paſſes into blue by imperceptible gradations: our ſenſes will ſuggeſt no place on which all agree, at which one is to end and the other to begin.

But the advance of knowledge has a tendency to ſupply means of preciſe definition. Thus, in the inſtance above cited, Wollaſton and Fraunhofer have diſcovered the black lines which always exiſt in the ſpectrum of ſolar colours given by a glaſs priſm, in the ſame relative places. There are definite places in the ſpectrum, by the help of which the place of any ſhade of colour therein exiſting may* be aſcertained, and means of definition given.

When a name is complex, it frequently admits of definition,

* It is quite within the poſſibilities of the application of ſcience to the

nominal or real. A name may be said to be *defined nominally* when we can of right substitute for it other terms. In such a case, a person may be made to know the meaning of the word without access to the object of which it is to give the idea. Thus, an *island* is completely defined in 'land surrounded by water.' In definition, we do not mean that we are necessarily to have very precise terms in which to explain the name defined: but, as the terms of the definition so is the name which is defined; according as the first are precise or vague, clear or obscure, so is the second. Thus there may be a question as to the meaning of *land*: is a marsh sticking up out of the water an island? Some will say that, as opposed to water, a marsh is land, others may consider marsh as intermediate between what is commonly called [dry] land and water. If there be any vagueness, the term island must partake of it: for island is but short for 'land surrounded by water,' whether this phrase be vague or precise. This sort of definition is *nominal*, being the substitution of names for names. It is complete, for it gives all that the name is to mean. An island, as such, can have nothing necessarily belonging to it except what necessarily belongs to 'land surrounded by water.' By *real* definition, I mean such an explanation of the word, be it the whole of the meaning or only part, as will be sufficient to separate the things contained under that word from all others. Thus the following, I believe, is a complete definition of an *elephant*; 'an animal which naturally drinks by drawing the water into its nose, and then spirting it into its mouth.' As it happens, the animal which does this is the elephant only, of all which are known upon the earth: so long as this is the case, so long the above definition answers every purpose; but it is far from involving all the ideas which arise from the word. Neither sagacity, nor utility, nor the production of ivory, are necessarily connected with drinking by help of the nose. And this definition is purely objective; we do not mean that every idea we could form of an animal so drinking is to be called an elephant. If a new animal were to be discovered, having the same mode of drinking, it would be a matter of pure choice whether it should be called elephant or not. It

arts that the time should come when the spectrum, and the lines in it, will be used for matching colours in every linen-draper's shop.

must then be settled whether it shall be called an elephant, and that race of animals shall be divided into two species, with distinctive definitions; or whether it shall have another name, and the definition above given shall be incomplete, as not serving to draw an entire distinction between the elephant and all other things.

It will be observed that the nominal definition includes the real, as soon as the terms of substitution are really defined: while the real definition may fall short of the nominal.

When a name is clearly understood, by which we mean when of every object of thought we can distinctly say, this name does or does not, contain that object—we have said that the name applies to everything, in one way or the other. The word man has an application both to Alexander and Bucephalus: the first *was* a man, the second *was not*. In the formation of language, a great many names are, as to their original signification, of a purely negative character: thus, parallels are only lines which do *not* meet, aliens are men who are *not* Britons (that is, in our country). If language were as perfect and as copious as we could imagine it to be, we should have, for every name which has a positive signification, another which merely implies all other things: thus, as we have a name for a tree, we should have another to signify every thing that is not a tree. As it is, we have sometimes a name for the positive, and none for the negative, as in *tree:* sometimes for the negative and none for the positive, as in *parallels:* sometimes for both, as in a frequent use of *person* and *thing*. In logic, it is desirable to consider names of inclusion with the corresponding names of exclusion: and this I intend to do to a much greater extent than is usual: inventing names of exclusion by the prefix not, as in tree and not-tree, man and not-man. Let these be called *contrary*,* or *contradictory*, names.

Let us take a pair of contrary names, as man and not-man. It is plain that between them they represent everything imaginable or real, in the universe. But the contraries of common language usually embrace, not the whole universe, but some one general idea. Thus, of men, Briton and alien are contraries: every man must be one of the two, no man can be both. Not-Briton and alien are identical names, and so are not-alien and Briton.

* I intend to draw no distinction between these words.

The same may be said of integer and fraction among numbers, peer and commoner among subjects of the realm, male and female among animals, and so on. In order to express this, let us say that the whole idea under consideration is *the universe* (meaning merely the whole of which we are considering parts) and let names which have nothing in common, but which between them contain the whole idea under consideration, be called contraries *in, or with respect to, that universe.* Thus, the universe being mankind, Briton and alien are contraries, as are soldier and civilian, male and female, &c.: the universe being animal, man and brute are contraries, &c.

Names may be represented by the letters of the alphabet: thus A, B, &c., may stand for any names we are considering, simple or complex. The contraries may be represented by not-A, not-B, &c., but I shall usually prefer to denote them by the small letters *a*, *b*, &c. Thus, everything in the universe (whatever that universe may embrace) is either A or not-A, either A or *a*, either B or *b*, &c. Nothing can be both B and *b*; every not-B is *b*, and every not-*b* is B: and so on.

No language, as may well be supposed, has been constructed beforehand with any intention of providing for the wants of any metaphysical system. In most, it is seen that the necessity of providing for the formation of contrary terms has been obeyed. Our own language has borrowed from the Latin as well as from its parent: thus we have *imperfect*, *disagreeable*, as well as *unformed* and *witless*. There is a choice of contraries without very well settled modes of appropriation: standing for different degrees of contrariety. Thus we have *not perfect* which is not so strong a term as *imperfect*; and *not imperfect*, the contrary of a contrary, which is not so strong as *perfect*. The wants of common conversation have sometimes retained a term and allowed the contrary to sink into disuse; sometimes retained the contrary and neglected the original term; sometimes have even introduced the contrary without introducing any term for the original notion, and allowed no means of expressing the original notion except as the contrary of a contrary. If we could imagine a perfect language, we should suppose it would contain a mode of signifying the contrary of every name: this indeed our own language may be said to have, though sometimes in an awkward and unidiomatic manner. One inflexion, or one additional word, may ſerve to ſignify a contrary of any kind: thus *not man* is effeƈtive to denote all that is other than man. But there is a wider want, which can only be partially ſupplied, for its complete ſatisfaƈtion would require words almoſt beyond the power of arithmetic to count: and all that has been done to make it leſs conſiſts, in our language and in every other, moſtly in the formation of compound terms, be they ſubſtantive and adjeƈtive, double ſubſtantives, or any others. A claſs of objeƈts has a ſub-claſs contained within it, the individuals of which are diſtinguiſhed from all others of the claſs by ſomething common to them and them only. If the diſtinguiſhing charaƈteriſtic have been ſeparated, and a word formed to ſignify the abſtraƈt idea, that word, or an adjeƈtive formed from it (if it be not an adjeƈtive) is joined with the general name of the claſs. Thus we have ſtrong men, white horſes, &c. Or it may happen that the individuals of the ſub-claſs take, in right of the diſtinguiſhing charaƈteriſtic, a perfeƈtly new name, and by the moſt varied rules. A corn-grinding man is called from the implement he uſes, a *miller*; a meat-killing man from the organ which he ſupplies, a *butcher*, (if the firſt idea of the etymology of this word be correƈt). Other men uſe mills and other trades feed the mouth: ſtill cuſtom has ſettled theſe terms, though the firſt is only conneƈted with its origin by the ſpelling, and the ſecond by a derivation which muſt be ſought in another language. But again, it will more often happen that a diſtinƈtive charaƈteriſtic, belonging to ſome only, gives no diſtinƈtive name to thoſe *ſome*, which ſtill remain an unnamed *ſome* out of the whole, to be ſeparated by the deſcription of their charaƈteriſtic when wanted; inſtead of being the *all* of a name invented to expreſs them, and them alone of their claſs. In ſuch a predicament, for inſtance, are men who have never ſeen the ſea, as diſtinguiſhed from thoſe who have ſeen it. Hence it appears that particular propoſitions are not ſo diſtinƈt from univerſal ones in real charaƈter as they are generally made to be. If I ſay 'ſome As are Bs' the reader may well ſuppoſe that it is not often neceſſary to advert to this faƈt: had it been ſo, a name would have been invented ſpecially to ſignify 'As which are Bs.' If this name had been C, the propoſition would have been '*every* C is B.'

The ſame convenience which dictates the formation of a name for one ſub-claſs and not for another, rules in the formation of contrary terms, as already noted. And theſe caprices of language —for logically conſidered they are nothing elſe, though their formation is far from lawleſs—make it deſirable to include in a formal treatiſe the moſt complete conſideration of all propoſitions, with reference not only to their terms, but alſo to the contraries of thoſe terms. Every negative propoſition is affirmative, and every affirmative is negative. Whatever completely does one of the two, include or exclude, alſo does the other. If I ſay that 'no A is B,' then, *b* being the name of every thing not B in the univerſe of the propoſition, I ſay that 'every A is *b*:' and if I ſay that 'every A is B,' I ſay that 'no A is *b*.' Whether a language will happen to poſſeſs the name B, or *b*, or both, depends on circumſtances of which logical preference is never one, except in treatiſes of ſcience. The Engliſh may poſſeſs a term for B, the French only for *b*: ſo that the ſame idea muſt be preſented in an affirmative form to an Engliſhman, as in 'every A is B,' and in a negative one to a Frenchman, as 'no A is *b*.' From all this it follows that it is an accident of language whether a propoſition is univerſal or particular, poſitive or negative. We, having the names A and B, may be able to ſay 'every A is B:' another language, which only names the contrary of B, muſt ſay 'no A is *b*.' A third language, in which As have not a ſeparate name, but are only individuals of the claſs C, muſt ſay 'ſome Cs are Bs;' while a fourth, which is in the further predicament of naming only *b*, muſt have it 'ſome Cs are not *b*s.' When we come to conſider the ſyllogiſm, we ſhall have full confirmation of the correctneſs and completeneſs of this view.

It may be objected that the introduction of terms which are merely negations of the poſitive ideas contained in other terms is a ſpecies of fiction. I anſwer, that, firſt, the fiction, if it be a fiction, exiſts in language, and produces its effects: nor will it eaſily be proved more fictitious than the invention of ſounds to ſtand for things. But, ſecondly, there is a much more effective anſwer, which will require a little development.

When writers on logic, up to the preſent time, uſe ſuch contraries as man and not-man, they mean by the alternative, man and everything elſe. There can be little effective meaning, and no uſe, in a claſſification which, becauſe they are not men, includes in one word, *not-man*, a planet and a pin, a rock and a featherbed, bodies and ideas, wiſhes and things wiſhed for. But if we remember that in many, perhaps moſt, propoſitions, the range of thought is much leſs extenſive than the whole univerſe, commonly ſo called, we begin to find that the whole extent of a ſubject of diſcuſſion is, for the purpoſe of diſcuſſion, what I have called a *univerſe*, that is to ſay, a range of ideas which is either expreſſed or underſtood as containing the whole matter under conſideration. In ſuch univerſes, contraries are very common: that is, terms each of which excludes every caſe of the other, while both together contain the whole. And, it muſt be obſerved that the contraries of a limited univerſe, though it be a ſufficient real definition of either that it is not the other, are frequently both of them the objects from which poſitive ideas are obtained. Thus, in the univerſe of property, perſonal and real are contraries, and a definition of either is a definition of the other. But though each be a negative term as compared with the other, no one will ſay that the idea conveyed by either is that of a mere negation. Money is *not* land, but it *is* ſomething. And even when the contrary term is originally invented merely as a negation, it may and does acquire poſitive properties. Thus alien is ſtrictly not-Briton: but ſuppoſe a man taken in arms againſt the crown on ſome ſpot within its dominions, and claiming to be a priſoner of war. The anſwer that he *is* a Britiſh ſubject is a negation: to eſtabliſh his poſitive claim he firſt muſt prove himſelf an alien, and moreover that he is in another poſitive predicament, namely, that he is the ſubject of a power at war with Great Britain. Accordingly, of two contraries, neither muſt be conſidered as *only* the negation of the other: except when the univerſe in queſtion is ſo wide, and the poſitive term ſo limited, that the things contained under the contrary name have nothing but the negative quality in common.

Perception of agreements and diſagreements is the foundation of all *aſſertion:* the acquirement of ſuch perception with reſpect to any two ideas by the compariſon of both with a third, is the proceſs of all *inference.* To infer, by compariſon of abſtract ideas, is the peculiar privilege of man; to need inference is his imperfection. To what point man would carry inference if he wanted language, how much further the lower animals could carry what they have of it if they had language, are questions on which it is vain to speculate. The words *is* and *is not*, which imply the agreement or disagreement of two ideas, must exist, explicitly or implicitly, in every assertion. And what we call agreement or disagreement, may be reduced to identity or non-identity. When we say John is a man, we have the first and most objective form of assertion. Looked at in the most objective point of view it is only this, *John* is one of the individual objects who are called *man*. Looked at ideally, the proposition is more general. The idea of man, gathered from instances, presents itself as a collective mass of ideas, of which we can figure to ourselves an instance without necessarily calling up the idea of any man that ever existed. In the ideal conception of man, Achilles is a man as much as the Duke of Wellington, whether the former ever existed objectively or no: of all the ideas of man which the mind can imagine, the former is one as well as the latter.

The separation of ideas, or formation of abstract ideas, and assertion by means of them, presents nothing, for our purpose, which differs from the former case. If we say 'this picture is beautiful,' the mere phrase is incomplete, for 'beautiful' is only an attribute, a purely ideal reference to a classification which the mind makes, dictated by its own judgment. The picture being a material object, cannot *be* anything but an object, cannot belong to any class of notions, unless that class contain objects. What the proposition may mean is to a certain extent dependent upon the implied substantive to which beautiful belongs: that is, to the class of objects which the proposition implies the mind to have separated into beautiful and not beautiful. It may be that the picture is a beautiful picture: or a beautiful work of art, taking its place in that division by which not only pictures, but statues, buildings, reliefs, &c. are separated into beautiful and otherwise: or a beautiful creation of human thought, placed among works of art, imagination, or science, &c. in the subdivision beautiful: or finally, it may be a beautiful *thing*, placed with all objects of perception in a similar subdivision.

In all assertions, however, it is to be noted, once for all, that *formal logic*, the object of this treatise, deals with *names* and not with either the *ideas* or *things* to which these names belong. We are concerned with the properties of 'A is B' and 'A is not B' so far as they present an idea independently of any specification of what A and B mean: with such ideas upon propositions as are presented by their *forms*, and are common to all forms of the same kind. The reality of logic is the examination of the use of *is* and *is not:* the tracing of the consequences of the application of these words. The argument 'when the sun shines it is day: but it is not day, therefore the sun does not shine,' contains a theory and two facts, the latter of which is made to follow from the former by the theory. That inference is made is seen in the word *therefore:* and the sentence is capable of being put upon its trial for truth or falsehood by logical examination. But this examination rejects the meaning of sun and day, the truth of the theory and of the facts; and only inquires into the right which the sentence, of its own structure, gives us to introduce the word *therefore.* It merely enters upon 'when A *is*, B *is;* but B *is not;* therefore A *is not:*' and decides that this is a correct junction of precedents and consequent, an exhibition of necessary connexion between what goes before and after *therefore*, and a development, in the latter, of what is virtually, though not actually, expressed in the former. What A and B may mean is of no consequence to the *inference*, or right to *bring in* 'A is *not.*'

Thus A and B, divested of all specific meaning, are really names as names, independently of things: or at least may be so considered. For the truth of the proposition, under all meanings, gives us a right to suppose, if we like, that *names are the meanings*—that is to say, that we may put it thus, 'When the name A is, the name B is: but the name B is not; therefore the name A is not.

It is not therefore the object of logic to determine whether conclusions be true or false; but whether what are asserted to be conclusions are *conclusions.* By a *conclusion* is meant that which is and must be *shut in with* certain other preceding things put in first: it is that which must have been put into a sentence because certain other things were put in. To *infer a conclusion* is to *bring in*, as it were, the direct statement of that which has been virtually stated already—has been *shut in.* When we say 'A is B, B is C; we *conclude* A is C'; it would be more correct to say 'A is B, B is C; we *have concluded* A is C'. We should never think of saying 'we have put into a box a man's upper dress of the colour of the trees; therefore we *must put in* a green coat'; —we should say '*we have put in*.' To infer the conclusion then is to bring in a statement that we have *concluded*.

Inference does not give us more than there was before: but it may make us *see* more than we *saw* before: ideally speaking, then, it does give us (in the mind) more than there was before. But the homely truth that no more can come out than was in, though accepted as to all material objects even by metaphysicians —who are generally well pleased to find the key of a box which contains what they want, though sure that it will put in no more than was there already—has been applied to logic, and even to mathematics, in depreciation of their rank as branches of knowledge. Those who have made this strangest of human errors must have assumed an ideal omniscience, and looked at human imperfection objectively. Omniscience need neither compare ideas, nor draw inferences: the conclusion which we deduce *from* premises, is always present *with* them; truths are *concomitants*, not *consequences*. When we say that one assertion *follows* from another, we speak purely ideally, and describe an imperfection of our own minds: it is not that the consequence follows from the premises, but that our *perception* of the consequence follows our *perception* of the premises: the consequence, objectively speaking, is in, and with, and of, the premises. We speak wrongly if we speak ideally, when we say that 'A is C,' *is* in 'A is B and B is C': in fact, it is only by giving an objective view to the argument, that we can even assert that it *will be* seen. To uncultivated minds, this simple conclusion is never concomitant with the premises, and only with some difficulty a consequence.

From the certainty that a consequence may be made to come out, which is an allegorical use of the word *out*, we assume a right to declare, by the same sort of allegory,* that it was *in*. The premises therefore contain the conclusion: and hence some have spoken as if in studying how to draw the conclusion, we were studying to know what we knew before. All the propositions of pure geometry, which multiply so fast that it is only a small

* I am of opinion that it is more consistent with analogy to say that the hypothesis *is contained in* its necessary consequence, than to say that the former contains the latter. My reason will appear in the course of the work.

and isolated class even among mathematicians who know all that has been done in that science, are certainly contained in, that is necessarily deducible from, a very few simple notions. But *to be known from* these premises is very different from being *known with* them.

Another form of the assertion is that consequences are *virtually* contained in the premises, or (I suppose) *as good as* contained in the premises. Persons not spoiled by sophistry will smile when they are told that knowing two straight lines cannot enclose a space, the whole is greater than its part, &c.—they as good as knew that the three intersections of opposite sides of a hexagon inscribed in a circle must be in the same straight line. Many of my readers will learn this now for the first time: it will comfort them much to be assured, on many high authorities, that they virtually knew it ever since their childhood. They can now ponder upon the distinction, as to the state of their own minds, between virtual knowledge and absolute ignorance.

There must always be some contention as to the relative value of their knowledge between the students of the things which we can see must have been, and of the things which, for what we can see, might have been otherwise. How much of the distinction is due to our ignorance, no one can tell. In the mean time, it is of more use to point out the advantage, as things are, of studying both kinds of knowledge, than to attempt to institute a rivalry between them. Those who have undervalued the study of necessary consequences, have allowed themselves, in illustrating their argument, phrases * which taken literally, mean more perhaps than they intended.

* We might sometimes take them to mean that the study of necessary connexion in logic, mathematics, &c., is at least useless, if not pernicious. Now we should suppose, if this be what they mean, that close connexion, short of absolute necessity, must partake somewhat of the same character. If the absolute mathematical necessity that three angles of a triangle are equal to two right angles is therefore to be avoided, the study of physics, in which there are the necessities which we express by the term *laws of nature*, must do some harm. History, in which we may so often count upon the actions which motives will produce, cannot be quite faultless: and there are laws of formation in language which might as well be kept out of sight, for they act almost with the uniformity of laws of nature. True knowledge must consist in the study of the actions of madmen: that a certain man imagined

The study of logic, then, considered relatively to human knowledge, stands in as low a place as that of the humble rules of arithmetic, with reference to the vast extent of mathematics and their physical applications. Neither is the less important for its lowliness: but it is not every one who can see that. Writers on the subject frequently take a scope which entitles them to claim for logic one of the highest places: they do not confine themselves to the connexion of premises and conclusion, but enter upon the *periculum et commodum* of the formation of the premises themselves. In the hands of Mr. Mill, for example (and to some extent in those of Dr. Whateley) logic is the science of distinguishing truth from falsehood, so as both to judge the premises and draw the conclusion, to compare name with name, not only as to identity or difference, but in all the varied associations of thought which arise out of this comparison.

CHAPTER III.

On the Abstract Form of the Proposition.

IN the preceding chapter, I have endeavoured to put together such notions on the actual sources of our knowledge as may give the reader the means of thinking upon points which any system of logic, however restricted, must necessarily suggest. We cannot attempt to connect our use of words with our notions of things, without the occurrence of a great many difficulties, a great many sources of adverse theories, and of never-ending disputes. We cannot even represent phænomena, as phænomena, except in the language of some system, and it may be of a wrong one. The confidence which the favourers of these several theories place in their correctness is a sufficient reason for keeping the account of the process of the understanding, so far as it can be made an exact science, as distinct as possible from all of them: for they differ widely, and if they agree in anything which can

himself to be Cæsar, when he might just as well have been Newton or Nebuchadnezzar, must be a real bit of knowledge, not virtually contained in anything else, wholly or partially.

be diſtinctly apprehended, it is only in having names of great authority enrolled among the partiſans of every one.

In order to examine the laws of inference, of the way of diſtinctly perceiving the right to ſay 'therefore,' 'ſo that,' 'whence it muſt be,' &c., &c., in a manner which may be admitted, ſo far as we go, by all, we muſt make this ſeparation very complete. All admit propoſitions, as 'man is animal,' 'no man is faultleſs;' all are, after a little thought, agreed upon the modes of inference: but upon the import of a ſimple propoſition, there is every kind of difference. How much we mean, when we ſay 'man is animal,' and how we arrive at our meaning, is matter for volumes on different ſides of unſettled queſtions.

In order properly to examine the laws of inference, or of any thing elſe, we muſt firſt endeavour to arrive at a diſtinct abſtraction of ſo much of the idea we are concerned with, as is itſelf the precedent reaſon, if it be right ſo to ſpeak, of the law in queſtion. This is an eaſy proceſs upon familiar things. We do not give the carriers of goods much credit for profundity, in ſeeing that, on a given road, there is only the difference of weight by which they are concerned to know how one parcel differs from another; and further that, as long as they have to carry a pound, it matters nothing whether it be of ſugar or iron. It is this proceſs which we want to perform to the utmoſt, upon the ſimple propoſition. Writers on logic, from Ariſtotle downwards, have made a large and important ſtep in ſubſtituting for ſpecific names, with all their ſuggeſtions about them, the mere letters of the alphabet, A, B, C, &c. Theſe letters are *ſymbols*, and *general* ſymbols: each of them ſtands for any one we pleaſe of its claſs. But what are they ſymbols of, names, ideas,* or the objects which give thoſe ideas? The anſwer is, that this is preciſely one of thoſe conſiderations which we may leave behind, in abſtracting what is neceſſary to an examination of the laws of inference. The only condition is, that we are to confine ourſelves to one or the other. When we ſay man is animal, it may be that the name man is contained in the name animal, that the idea of man is contained in that of animal, or that the object man is in the object animal. Or if there were twenty more different appropriations of the

* Meaning of courſe (page 30) ideas of ideas, and ideas of objects.

ſymbols, the ſame thing might be ſaid of each. This is, I believe, the firſt uſe of the general ſymbol in order of time; the algebraical uſe of letter or other ſymbol, to deſignate number, being both ſubſequent and derived.

When therefore we ſay 'Every X is Y', we underſtand that X is a ſymbol which repreſents an inſtance of a name, idea, object, &c., as the caſe may be. There may be more or fewer of ſuch inſtances; they may be numerable or innumerable. And the ſame of Y. The language of logicians has generally been unfavourable to the diſtinct perception of their terms being diſtributively applicable to claſſes of inſtances. They have rather been *quantitative* than *quantuplicitative:* expreſſing themſelves as if, in ſaying that animal is a larger or wider term than man, they would rather draw their language from the idea of two areas, one of which is larger than the other, than from two collections of indiviſible units, one of which is in number more than the other. They have even carried this ſo far as to make it doubtful, except from context, whether their diſtinction between univerſal and particular is that of *all* and *ſome*, or of *the whole* and *part*. If their inſtances had been *white ſquares*, their 'all A is B' and 'ſome A is B' might have applied as well to 'All the ſquare is white' and 'Some of the ſquare is white' as to 'All the ſquares are white' and 'Some of the ſquares are white.' I ſhall take particular care to uſe numerical language, as diſtinguiſhed from magnitudinal, throughout this work, introducing of courſe, the plurals Xs, Ys, Zs, &c.

I may mention here another mode of ſpeaking, which will, I think, appear objectionable to all who are much uſed to conſideration of quantity. When a compound idea contains two or more ſimpler ones, ſome logicians have ſpoken as if the combination were legitimately repreſented by *arithmetical addition.* Thus the combination of the ideas of *animal* and *rational* muſt give the idea of *man:* for the two notions co-exiſt in nothing elſe that we know of. Accordingly, ſome write *animal* + *rational* = *man*. If this be intended as an abſtraction of the notation of arithmetic, for the purpoſe of fitting to it entirely different meaning, there is of courſe no objection which I need conſider here: but it ſeems to me that more is meant, and that thoſe who have uſed this notation imagine a great reſemblance between *combining* ideas, and *cumulating* them. What the difference is, I cannot pretend to ſay, any more than I can pretend to ſay what the difference is between chemically combining volumes of oxygen and hydrogen, ſo as to produce water, and ſimple cumulation of them in the ſame veſſel, ſo as to produce a mixed gas: every beginner knows that the electric ſpark, or ſome other inexplicable agency, is neceſſary to turn the mixed gas into a new chemical combination. But that the difference exiſts in the former caſe alſo, ſeems to me as clear as any thing I can imagine. Even in chemiſtry the cumulative notation, which was once thought an all-ſufficient mode of expreſſing the reſults of the atomic theory, has failed with the progreſs of knowledge. To a conſiderable extent, the introduction of modes of cumulation as yet anſwers the purpoſe: but there ſtill remain *iſomeric* compounds, differing in properties, but of the ſame compoſition. For example, the tartaric and racemic acids: of which Profeſſor Graham ſays (*Elements of Chemiſtry* p. 158), "A nearer approach to identity could ſcarcely be conceived than is exhibited by theſe bodies, which are, indeed, the ſame both in form and compoſition. . . . . . But by no treatment can the one acid be tranſmuted into the other." If the above mode of confounding cumulation and combination be admiſſible, I ſuppoſe we might eaſily give ourſelves a right to ſay that

$$2 + 2 + \text{addition} = 4$$

an equation at which the mathematician would ſtare.

So much for the characteriſtics of the *terms* of a propoſition, as wanted for the abſtract forms of inference. It remains to conſider thoſe of the connecting copulæ *is* and *is not*.

The complete attempt to deal with the term *is* would go to the form and matter of every thing in *exiſtence*, at leaſt, if not to the poſſible form and matter of all that does not exiſt, but might. As far as it could be done, it would give the grand Cyclopædia, and its yearly ſupplement would be the hiſtory of the human race for the time. That logic exiſts as a treated ſcience, ariſes from the characteriſtics of the word, requiſite to be abſtracted in ſtudying inference, being few and eaſily apprehended. It may be uſed in many ſenſes, all having a common property. Names, ideas, and objects, require it in three different ſenſes. Speak of *names*, and ſay 'man *is* animal': the *is* is here an *is* of applicability; to whatſoever (idea, object, &c.) man is a name to be applied, to that ſame (idea, object, &c.) animal is a name to be applied. As to ideas, the *is* is an *is* of poſſeſſion of all eſſential characteriſtics; *man* is an idea which poſſeſſes, contains, preſents, all that is conſtitutive of the idea *animal*. As to abſolute external objects, the *is* is an *is* of identity, the moſt common and poſitive uſe of the word. Every man *is* one of the animals; touch him, you touch an animal, deſtroy him, you deſtroy an animal.

Theſe ſenſes are not all interchangeable. Take the *is* of identity, and the name *man* is not, as a name, the name *animal:* the idea man is not, as an idea, the idea animal. Now we muſt aſk, what common property is poſſeſſed by each of theſe three notions of *is*, on which the common laws of inference depend. Common laws of inference there certainly are. If the applicability of the name A be always accompanied by that of B, and that of B by that of C, then that of A is always accompanied by that of C. If the idea A contain all that is eſſential to the idea B, and B all that is eſſential to C, then A contains all that is eſſential to C. If the object A be actually the object B, and if B be actually C, then A is actually C.

The following are the characteriſtics of the word *is* which, exiſting in any propoſed meaning of it, make that meaning ſatisfy the requirements of logicians when they lay down the propoſition 'A is B.' To make the ſtatement diſtinct, let the propoſition be doubly ſingular, or refer to one inſtance of each, one A and one B: let it be 'this one A is this one B.'

Firſt, the double ſingular propoſition above mentioned, and every ſuch double-ſingular, muſt be indifferent to converſion: the 'A is B,' and the 'B is A' muſt have the ſame meaning, and be both true or both falſe.

Secondly, the connexion *is*, exiſting between one term and each of two others, muſt therefore exiſt between thoſe two others; ſo 'A is B' and 'A is C' muſt give 'B is C.'

Thirdly, the eſſential diſtinction of the term *is not* is merely that *is* and *is not* are *contradictory alternatives*, one muſt, both cannot, be true.

Every connexion which can be invented and ſignified by the terms *is* and *is not*, ſo as to ſatisfy theſe three conditions, makes all the rules of logic true. No doubt abſolute identity was the ſuggeſting connexion from which all the others aroſe: juſt as arithmetic was the medium in which the forms and laws of algebra were ſuggeſted. But, as now we *invent algebras* by abſtracting the forms and laws of operation, and fitting new meanings to them, ſo we have power to invent new meanings for all the forms of inference, in every way in which we have power to make meanings of *is* and *is not* which ſatisfy the above conditions. For inſtance, let X, Y, Z, each be the ſymbol attached to every inſtance of a claſs of *material* objects, let *is* placed between two, as in 'X is Y' mean that the two are tied together, ſay by a cord, and let X be conſidered as tied to Z when it is tied to Y which is tied to Z, &c. There is no ſyllogiſm but what remains true under theſe meanings. Thus

The ſyllogiſm	Is true in the ſenſe
Every X is Y	Every X is tied to a Y
Some Zs are not Ys	Some Zs are not tied to Ys
∴ Some Zs are not Xs	∴ Some Zs are not tied to Xs

This laſt inſtance might be conſidered as a material repreſentation of attachment together of ideas in the mind.

We muſt diſtinctly obſerve that it is not every caſe of inference which demands all the characteriſtics to be ſatisfied. Thus in the moſt common caſe of all, 'Every A is B, every B is C, therefore every A is C,' of all the three conditions only the ſecond is wanted to ſecure the validity of this caſe. Though it be ſeldom thought worth while to make this obſervation, yet it is univerſal practice to act upon it, and ſo as to introduce into formal logic apparent contradictions of its own rules. For example, the following are allowed to paſs for ſyllogiſms, in the ordinary definition of that word.

'Every A is greater than ſome one B; every B is greater than ſome one C, therefore every A is greater than ſome one C.' And the ſame when inſtead of *greater than* is read *equal to* or *leſs than*. The form which moſt commonly appears is the pair of doubly ſingular propoſitions, 'A (one thing) is greater than B; B is greater than C; therefore A is greater than C.' Here 'greater than greater' is 'greater,' the ſecond rule is ſatisfied, and no other is wanted. But this meaning for *is* (or this ſubſtitute for it, if the reader like it better) will not ſatisfy all the conditions, and therefore will not apply to all the forms of inference.

But *is* in the sense 'is equal to' *does* satisfy all the conditions. This sense of *is*, namely agreement in magnitude, is the copula of the mathematician's syllogism, when he is reasoning on quantity only.

It will probably be affirmed that the generalization thus made, or shown to be possible, in the conception of the word *is* for purposes of inference, amounts only to a very frequent, if not most usual, use of the word, namely, as signifying a certain mode, not of identity, but of agreement in quality. As when we say 'these two things are the same—in colour' or 'the one thing *is* the other—in colour:' that the name man *is* the name animal, in a certain respect, namely, in what the latter can be applied to: that the idea man is animal, in both possessing certain characteristics: that every object man is an object animal, in actual substance: that A is B in magnitude,—when we say A equals B; and so on. But I admit only the converse, namely, that all these uses satisfy the conditions. It would hardly be for any one to say, that every possible use of *is* which satisfies three such simple requirements, has been or can be exhausted. Even the material example which was just now given, cannot be identified with any common use, or easily imaginable one, of the common verb. But if no invented meaning, proper to satisfy the conditions, can be found, other than already exists in more or less of use, still, these conditions are the laws to which the word must submit in its logical acceptation.

There are common uses of the word which are not admitted in logic: and among them, one of the most common, connection of an object with its quality, and of an idea with one of its constituent or associated ideas. As when we say, the rose *is* red, prudence *is* desirable. Here the logical conditions are not satisfied. For example, 'red is the rose,' though a poetical inversion of the first assertion, is not logically true. It is usual to consider such propositions, in logic, as elliptical; thus 'the rose is red' is considered as 'the rose is a red object, or an object of red colour;' in which the *is* now takes one of the senses which allows of conversion. Similarly, in all other cases, the subject and predicate are made to take the same character; both names, both ideas, or both objects. This reduction renders unnecessary both the study of the varieties of meaning of the word is (meaning varieties out of the pale of the conditions above enumerated), and alfo that of the tranfitions of meaning within the circle of which the inference remains good.

The moft common ufes of the verb are;—firft abfolute identity, as in 'the thing he fold you *is* the one I fold him:' fecondly, agreement in a certain particular or particulars underftood, as in 'He is a negro' faid of a European in reference to his colour: thirdly, poffeffion of a quality, as in 'the rofe is red:' fourthly, reference of a fpecies to its genus, as in 'man is an animal.' All thefe ufes are independent of the ufe of the verb alone, denoting exiftence, as in 'man is [i. e. exifts].' In all thefe fenfes, and in all which might be added confiftently with the conditions in page 50, fome propofitions fometimes admit of having the fenfe of *is* fhifted, and fome do not. Thus, in negative propofitions, the *is* of agreement in particulars may be lawfully converted into that of identity: if 'No A is B in colour,' then abfolutely 'No A is B.' But 'Every A is B' in colour, does not prove 'Every A is B. But the firft pair might be connected by a fyllogifm.

The *is* of agreement in particulars may always be reduced to the *is* of identity, by alteration of the predicate; thus 'Every A is B in colour' is 'Every A is a thing having the colour of one of the Bs.' When a fyllogifm has a negative conclufion, and the middle term is, or can be made, the predicate of both premifes, then the whole fyllogifm can be transformed from one in which there is only the *is* of agreement to one in which there is no *is* but that of identity. For example, fuppofe the premifes to be 'No X is Y (in colour); every Z is Y (in colour),' not meaning neceffarily that all the Ys are of one colour, but reading it as 'No X is of the colour of any one of the Ys; every Z is of the colour of one of the Ys.' The conclufion is that 'no Z is X (in colour),' or 'no Z is of the colour of any one of the Xs.' But from this it follows that no Z is X, for if any one Z were abfolutely X, it would have * the colour of that X. This

* The reader muft not paint any of the letters during the procefs. The fenfe in which we fay a door *is* the fame door as before, after it has been painted of a different colour, is not the fenfe of logical identity: it is the fame in all but colour and colouring matter; and the *is* is one of agreement. Except as a joke in fufficient anfwer to a captious objection or a trap, no

laſt concluſion can be brought directly from altered premiſes: thus, *is* being that of identity, we have 'No X is [a thing having the colour of one of the Ys]; every Z is [a thing having the colour of one of the Ys]; therefore no Z is X.' But ſuppoſe we take the following premiſes, 'Some Ys are not Xs (in colour); every Y is Z (in colour).' From this it follows that ſome Zs are not Xs (in colour), and thence that ſome Zs *are not* Xs. But we cannot now alter the premiſes, ſo as to produce the laſt concluſion from X, Z, and a middle term.

CHAPTER IV.

On Propoſitions.

A NAME is a ſymbol which is attached to one or more objects of thought, on account of ſome reſemblance, or community of properties. Or elſe it is a ſymbol attached to ſome one or more objects of thought, to diſtinguiſh them from others having the ſame properties. Objects of the ſame name are, ſo far as that name is concerned, undiſtinguiſhable. And one object may have many names, as being one in each of many claſſes of objects of thought.

Names, as explained in chapter II, are excluſively the objects of formal logic. The identity and difference of things is deſcribed by aſſerting the right to aſſert, or the right to deny, the application of names. And names, whether ſimple or complex, will be repreſented by letters of the alphabet, as X, Y, Z.

A propoſition is the aſſertion of agreement, more or leſs, or diſagreement, more or leſs, between two names. It expreſſes that of the objects of thought called Xs, there are ſome which are, or are not, found among the objects of thought called Ys:

change whatever muſt take place in the terms of concluſion, during inference. The American calculating boy, Zerah Colburn, was aſked how many black beans it would take to make ten white ones; to which he very properly anſwered 'Ten, if you ſkin 'em:' but the ten ſkinned beans would not be the *ſame beans* as before: except, indeed, to thoſe to whom black is white.

that there are objects which have both names, or which have one but not the other, or which have neither.

For the most part, the objects of thought which enter into a proposition are supposed to be taken, not from the whole universe of possible objects, but from some more definite collection of them. Thus when we say "All animals require air," or that the name *requiring air* belongs to every thing to which the name *animal* belongs, we should understand that we are speaking of things on this earth: the planets, &c., of which we know nothing, not being included. By the *universe* of a proposition, I mean the whole range of names in which it is expressed or understood that the names in the proposition are found. If there be no such expression nor understanding, then the universe of the proposition is the whole range of possible names. If, the universe being the name U, we have a right to say 'every X is Y,' then we can only extend the universe so as to make it include all possible names, by saying 'Every X which is U is one of the Ys which are Us,' or something equivalent.

Contrary names, with reference to any one universe, are those which cannot both apply at once, but one or other of which always applies. Thus, the universe being *man*, *Briton* and *alien* are contraries; the universe being *property*, *real* and *personal* are contraries. Names which are contraries in one universe, are not necessarily so in a larger one. Thus in geometry, when the universe is *one plane*, pairs of straight lines are either parallels or intersectors, and never both: parallels and intersectors are then contraries. But when the student comes to solid geometry, in which *all space* is the universe, there are lines which are neither parallels nor intersectors; and these words are then not contraries. But names which are contraries in the larger and containing universe, are necessarily contraries in the smaller and contained, unless the smaller universe absolutely exclude one name, and then the other name is *the* universe.

In future, I always understand some one universe as being that in which all names used are wholly contained: and also (which it is very important to bear in mind) that no one name mentioned in a proposition fills this universe, or applies to everything in it. Nothing is more easy than to treat the supposition of a name being the universe as an extreme case. And I shall denote contraries by large and fmall letters : thus, X being a name, x is the contrary name. And everything (in the univerfe underftood) is either X or x: and nothing is both.

A propofition may be either *fimple* and *incomplete*, or *complex* and *complete*. The fimple propofition only afferts that Xs are Ys, or are not Ys: the complex propofition, which always confifts of two fimple ones, difpofes in one manner or the other of every X and every Y. Thus 'Every X is Y' is a fimple propofition: but it forms a part of two complex propofitions. It may belong either to 'every X is Y and every Y is X,' or to 'Every X is Y and fome Ys are not Xs.'

The propofitions advanced in common life are ufually complex, with one fimple propofition expreffed and one underftood: but books of logic have hitherto confidered only the fimple propofition. And this laft fhould be confidered before the complex form.

The fimple propofition muft be confidered with refpect to *fign*, *relative quantity*, and *order*.

Simple propofitions are of two *figns: affirmative* and *negative*. It is either 'Xs are Ys,' or 'Xs are not Ys.' The phrafes *are* and *are not*, or *is* and *is not*, which mark the diftinction, are called *copulæ*.

The *relative quantity* of a propofition has reference to the numbers of inftances of the different names which enter it. The diftinctions of quantity ufually recognized are *all* and *fome**: leading to the diftinction of *univerfal* and *particular*. Thus 'Every X is Y' and 'Every X is not Y' are the univerfal affirmative and negative propofitions: the latter is ufually ftated as 'No X is Y.' And 'fome Xs are Ys' and 'fome Xs are not Ys' are the particular affirmative and negative propofitions. And when the propofitions are reduced ftrictly to thefe four forms,

* *Some*, in logic, means *one or more*, *it may be all*. He who fays that *fome are*, is not to be held to mean that *the reft are not*. 'Some men breathe,' 'fome horfes are diftinguifhable by fhape from their riders' would be held falfe in common language. The reafon is, as above noted, that common language ufually adopts the complex particular propofition, and implies that fome are not in faying that fome are. The ftudent cannot be too careful to remember this diftinction. A particular propofition is only a 'may be particular.'

the firft named, X, is called the *fubject*, and the fecond named, Y, the *predicate*.

It has been propofed to confider the univerfal propofitions as *definite* with refpect to *quantity*: but this is not quite correct. The phrafe 'all Xs are Ys' does not tell us how many Xs there are, but that, be the unknown number of Xs in *exiftence* what it may, the unknown number mentioned in the propofition is the fame. That which is definite is the *ratio* of the number of Xs of the propofition to the Xs of the univerfe. So underftood, however, the 'definite quantity,' as an abbreviation, may be faid to belong to univerfals. And the indefinitenefs of the particular propofition is only hypothetical. It is in our power to fuppofe the *fome* to be one half of the whole, or two-thirds, or any other fraction.

The quantity of the fubject is expreffed; that of the predicate, though not expreffed, is neceffarily implied by the meaning of language. The predicate of an *affirmative* is *particular*: the predicate of a *negative* is *univerfal*. If I fay 'Xs are Ys,' even though I fpeak of all the Xs, I only really fpeak of fo many Ys as are compared with Xs and found to agree: and thefe need not be all the Ys. 'Every horfe is an animal,' declares that fo many horfes as there are to fpeak of, fo many animals are fpoken of: and leaves it wholly unfettled whether there be or be not more animals left. But if I fhould fay 'Xs are not Ys,' though it fhould be only one X, as in 'this X is not a Y,' yet I fpeak of every Y which exifts. The affertion is 'this X is not any one whatfoever of all the Ys in exiftence.' A perfon who fhould wifh to verify by actual infpection, 'thefe 20 Xs are Ys' might, perchance, be enabled to affirm the refult upon the examination of only 20 Ys, if he came firft upon the right ones. But he could not verify 'this one X is not a Y' until he had examined every Y in exiftence. This is the common doctrine, but though admitting of courfe that the affirmative propofition only enables us to infer of fome inftances of the predicate, yet I think it more correct to fay that the predicate itfelf is fpoken of univerfally, but *indivifibly*, and that in the negative propofition the predicate is fpoken of univerfally *and divifibly*. 'Some Xs are Ys' tells us that each X mentioned is *either* the firft Y, *or* the fecond Y, *or* the third Y, &c., no Y being excluded from comparifon. But

'Some Xs are not Ys' tells us that each X mentioned is abfolutely *not* the firft Y, *nor* the fecond, *nor* the third, &c; is not, in fact, any one of all the Ys. Still, however, the predicate of an affirmative yields no more than it would do if the Ys finally accepted as Xs were fpecially feparated, and confidered as the only Ys fpoken of.

The relation of the univerfal quantity to the whole quantity of inftances in exiftence is *definite*, being that whole quantity itfelf. But the particular quantity is wholly *indefinite*: 'Some Xs are Ys' gives no clue to the fraction of all the Xs fpoken of, nor to the fraction which they make of all the Ys. Common language makes a certain conventional approach to definitenefs, which has been thrown away in works of logic. 'Some,' ufually means a rather fmall fraction of the whole; a larger fraction would be expreffed by 'a good many'; and fomewhat more than half by 'moft'; while a ftill larger proportion would be 'a great majority' or 'nearly all'. A perfectly *definite particular*, as to quantity, would exprefs how many Xs are in exiftence, how many Ys, and how many of the Xs are or are not Ys: as in '70 out of the 100 Xs are among the 200 Ys'. In this chapter I fhall treat only the *indefinite particular*, leaving the *definite particular* for future confideration.

The *order* of a propofition has relation to the choice of fubject and predicate. Thus 'Every X is Y' and 'every Y is X' though both eftablifh a univerfal affirmative relation between X and Y, yet are in fact two different propofitions. They are called *converfe* forms. When the fubject and predicate are of the fame fort of quantity, both univerfal or both particular, the converfe forms give the fame propofition. Thus 'No X is Y' and 'No Y is X' are the fame; neither has any meaning, except perhaps of emphafis, which the other has not. And 'Some Xs are Ys' is the fame as 'Some Ys are Xs'. The univerfal negative, then, in which both terms are univerfal, and the particular affirmative, in which both are particular—are *neceffarily convertible propofitions*. But the univerfal affirmative, in which the fubject is univerfal and the predicate particular, and the particular negative, in which the fubject is particular, and the predicate univerfal—are not neceffarily convertible, and are generally called *inconvertible*. They *may* be convertible, in one cafe, and inconvertible in another. But the term *inconvertible* is not incorrect, for the following reason.

The agreements and disagreements which are treated in logic are of this character; there can only be agreement with one, but there may be disagreement with all. If 'this X be a Y' it is one Y only: it is 'this X is *either* the first Y, *or* the second Y, *or* the third Y, &c. If there be 100 Ys, there is, to those who can know it, 99 times as much negation as affirmation in the proposition: and yet most assuredly it is properly called *affirmative*. But if it be 'this X is not a Y,' we have 'this X is not the first Y, *and* it is not the second Y, *and* it is not the third Y, &c.' The affirmation is what is commonly called disjunctive, the negation conjunctive. A disjunctive negation would be no proposition at all, except that one and the same thing cannot be two different things: any X is either not the first Y *or* not the second Y. And in like manner a conjunctive affirmation would be an impossibility: it would state that one thing is two or more different things.

We must be prepared, then, to consider cases of opposition in which on the one side there is fixed necessity, and on the other side possibility of alternatives: and we must be prepared to denote these by opposite terms, which, looking to etymology only, denote fixed necessities of opposite characters. This happens in the case above: *convertible* means absolutely and necessarily convertible, inconvertible means *convertible or inconvertible as the case may be*. Taking the four forms of one order, we find that each of the universals cannot exist with either proposition of opposite form. Thus 'Every X is Y' cannot be true if either 'No X is Y' or 'Some Xs are not Ys:' while 'No X is Y' cannot be true if either 'Every X is Y' or 'Some Xs are Ys.' But each of the particulars is necessarily inconsistent with nothing but the universal of opposite form. That 'Some Xs are Ys' cannot be true if 'No X is Y' but it may be true if 'Some Xs are not Ys.' And 'Some Xs are not Ys' cannot be true if 'Every X is Y,' but it may be true though 'Some Xs are Ys.'

The pair 'Every X is Y' and 'some Xs are not Ys' are called *contradictory*: and so are the pair 'No X is Y' and 'Some Xs are Ys.' Of each pair of contradictories, one must be true and one muſt be falſe: ſo that the affirmation of either is the denial of the other, and the denial of either is the affirmation of the other. The pair 'Every X is Y' and 'No X is Y' are uſually called *contraries;* contrariety implying the utmoſt extreme of contradiction. Contraries may both be falſe, but cannot both be true. The pair 'Some Xs are Ys,' and 'Some Xs are not Ys,' which may both be true, but cannot both be falſe, are uſually called *ſubcontraries.* But, for reaſons hereafter to be given, I intend to abandon the diſtinction between the words *contrary* and *contradictory*, and to treat them as ſynonymous. And the propoſitions uſually called *contraries*, 'Every X is Y' and 'No X is Y' I ſhall call *ſubcontraries:* while thoſe uſually called *ſubcontraries* 'Some Xs are Ys' and 'Some Xs are not Ys' I ſhall call *ſupercontraries.*

I ſhall now proceed to an enlarged view of the propoſition, and to the ſtructure of a notation proper to repreſent its different caſes.

As uſual, let the univerſal affirmative be denoted by A, the particular affirmative by I, the univerſal negative by E, and the particular negative by O. This is the extent of the common ſymbolic expreſſion of propoſitions: I propoſe to make the following additions for this work. Let one particular choice of order, as to ſubject and predicate, be ſuppoſed eſtabliſhed as a ſtandard of reference. As to the letters X, Y, Z, let the order be always that of the alphabet, XY, YZ, XZ. Let x, y, z, be the contrary names of X, Y, Z; and let the ſame order be adopted in the ſtandard of reference. Let the four forms, when choice is made out of X, Y, Z, be denoted by A_1, E_1, I_1, O_1; but when the choice is made from the contraries, let them be denoted by A', E', I', O'. Thus, with reference to Y and Z, 'Every Y is Z' is the A_1 of that pair and order: while 'Every y is z' is the A'. I ſhould recommend A_1 and A' to be called the *ſub*-A and the *ſuper*-A of the pair and order in queſtion: the helps which this will give the memory will preſently be very apparent. And the ſame of I_1 and I', &c.

Let the following abbreviations be employed;—

X)Y means 'Every X is Y'	X.Y means 'No X is Y'
X:Y — 'Some Xs are not Ys'	XY — 'Some Xs are Ys'

There are eight diſtinct modes, independent of contraries, in which a ſimple propoſition may be made by means of X and Y. Theſe eight modes are X)Y and Y)X, X:Y and Y:X, X.Y and Y.X, and XY and YX. But the eight are equivalent only to ſix: for X.Y and Y.X are the ſame, and ſo are XY and YX. Again, there are ſix ſimple propoſitions between x and y, ſix between X and y, ſix between x and Y. Taking in contraries, there are then twenty-four apparent modes of forming a ſimple propoſition from X and Y: but theſe are not all diſtinct. Eight of them contain all the reſt: theſe eight being the $A_{,}$, $E_{,}$, $I_{,}$, $O_{,}$, A', E', I', O', above deſcribed. This is ſeen in the following table, the ſtudy of which ſhould be carefully made,

$A_{,}$	X)Y	= X.y	= y)x	A'	x)y	= x.Y	= Y)X
$O_{,}$	X:Y	= Xy	= y:x	O'	x:y	= xY	= Y:X
$E_{,}$	X.Y	= X)y	= Y)x	E'	x.y	= x)Y	= y)X
$I_{,}$	XY	= X:y	= Y:x	I'	xy	= x:Y	= y:X

I ſuppoſe moſt readers will readily ſee the truth of the identities here affirmed: if not, the following mode of illuſtration (which will be very uſeful when I come to treat of the ſyllogiſm) may be tried. Let U be the name which is the univerſe of the propoſition: and write down in a line as many Us as there are diſtinct objects to which this name applies. A dozen will do as well for illuſtration as a million. Under every U which is an X write down X: and x, of courſe, under all the reſt. Follow the ſame plan with Y. The occurrence of letters in the ſame column ſhows that they are names of the ſame object. The following are ſpecimens of the eight ſtandard varieties of aſſertion, to which all the reſt may be referred.

$A_{,}$ UUUUUUUUUUUU
XXXXXxxxxxxx
YYYYYYYYyyyy

A' UUUUUUUUUUUU
XXXXXXXXxxxx
YYYYYyyyyyyy

$O_{,}$ UUUUUUUUUUUU
XXXXXXXxxxxx
$I_{,}$ yyyyYYYYYYyy

O' UUUUUUUUUUUU
XXXXXxxxxxxx
I' YYyyyyyyYYYY

$E_{,}$ UUUUUUUUUUUU
XXXXxxxxxxxx
yyyyyyyYYYYY

E' UUUUUUUUUUUU
XXXXXXXXxxxx
yyyyyYYYYYYY

In the firſt ſcheme, $A_{\prime}$, there exiſt twelve Us, the firſt five of which are both Xs and Ys, the next three Ys but not Xs, the laſt four neither Xs nor Ys. This caſe, ſo conſtructed that X)Y is true, ſhows X.y and y)x.

The propoſitions $A_{\prime}$ and $A^{\prime}$, X)Y and x)y, may be called *contranominal*, as having each names contrary of thoſe in the other. It appears, then, that as to inconvertibles, contranominal and converſe are terms of the ſame meaning, for X)Y and y)x are the ſame, and x:y and Y:X. And ſince it is more natural to ſpeak of direct names than of their contraries, it will be beſt to attach to $A^{\prime}$ and $O^{\prime}$ the ideas of Y)X and Y:X; but not ſo as to forget their derivation from x)y and x:y. Obſerve alſo that each univerſal propoſition has converted contranominals for its *affirmative* forms. Thus X)Y = y)x: and though X.Y is not y.x, yet if we make X.Y take the affirmative form X)y, it is equivalent to Y)x. In particular propoſitions, the negative forms have the ſame property. The contranominals of the convertible propoſitions $E_{\prime}$ and $I_{\prime}$ are of totally different meaning. They have never till now been introduced into logic, and a few words of explanation are wanted.

Firſt as to $I^{\prime}$ or xy. We here expreſs that ſome not-Xs are not-Ys, or that there are things in the univerſe which are neither Xs nor Ys. That is, X and Y are not contraries. Next as to $E^{\prime}$ or x.y. We here expreſs that no not-X is not-Y, or that everything in the univerſe is either X or Y, *or both*. Theſe laſt words are important: by omitting them, we ſhould imagine that x.y ſignifies that X and Y are contraries; which is not neceſſarily true.

Accordingly, the eight ſtandard forms of expreſſion, with reference to the order XY, and exhibited in the form in which it will be moſt convenient to think and ſpeak of them, are as follows,

$A_{\prime}$ or X)Y	Every X is Y		$A^{\prime}$ or Y)X	Every *Y* is *X*
$O_{\prime}$ or X:Y	Some Xs are not Ys		$O^{\prime}$ or Y:X	Some *Y*s are not *X*s
$E_{\prime}$ or X.Y	No X is Y		$E^{\prime}$ or x.y	Everything is either *X* or *Y*
$I_{\prime}$ or XY	Some Xs are Ys		$I^{\prime}$ or xy	Some things are neither *X*s nor *Y*s.

Returning to the table, we now ſee the following general laws. 1. Each triad of equivalents contains two inconvertibles and one convertible. 2. Of the four, X, Y, x, y, each of the eight forms ſpeaks univerſally of two, and particularly of two. 3. A propoſition ſpeaks in different ways of each name and its contrary; univerſally of one and particularly of the other. 4. The propoſitions called *contradictory*, from the common meaning of this word, may be ſo called in another ſenſe: for they ſpeak in the ſame manner of contraries. Thus X)Y ſpeaks univerſally of X, and particularly of Y: its denial, X:Y or y:x, ſpeaks univerſally of x, and particularly of y.

Any two of the eight forms being taken, it is clear either that they cannot exiſt together, or that one muſt exiſt when the other exiſts, or that one may exiſt either with or without the other. The alternatives of each caſe are preſented in the following table.

	Denies	Contains	Is indifferent to		Denies	Is contained in	Is indifferent to
$A_,$	$O_,E_,E'$	$I_,I'$	$A'O'$	$O_,$	$A_,$	$E_,E'$	$A'O'I_,I'$
A'	$O'E'E_,$	$I'I_,$	$A_,O_,$	O'	A'	$E'E_,$	$A_,O_,I'I_,$
$E_,$	$I_,A_,A'$	$O_,O'$	$E'I'$	$I_,$	$E_,$	$A_,A'$	$E'I'O_,O'$
E'	$I'A'A_,$	$O'O_,$	$E_,I_,$	I'	E'	$A'A_,$	$E_,I_,O'O_,$

Let the *concomitants* of a propoſition be thoſe to which it is wholly indifferent. Then it appears that each univerſal has for concomitants its contranominal and the contradictory of the laſt: but each particular has all for concomitants except only its own contradictory. Each univerſal denies, beſides its own contradictory, the two univerſals of oppoſite name; and contains the two particulars of the ſame name. The two concomitants of a univerſal may be deſcribed as its univerſal and its particular concomitant.

There is a certain ſort of repetition in our choice of the four forms, combined with the four ſelections XY, Xy, xy, xY. If *any one* of the four forms $A_,$ $E_,$. A' E' be applied to all the above, it will give the four forms derived from XY. Thus the $A_,$ of XY, Xy, xy, xY, are ſeverally the $A_,$, $E_,$, A' and E' of XY; and the E' of XY, Xy, xy, and xY are ſeverally the E', A', $E_,$, and $A_,$ of XY: and ſo on. It will ſerve for exerciſe to verify the above, and ſtill more the caſes contained in the following.

There are four things in a propoſition, each of which may be changed into its contrary: ſubject, predicate, order, and copula. Let S be the direction to change the ſubject into its contrary: P the ſame for the predicate : let T be the direction to transform the order : and F the direction to change the form, from affirmative to negative, or from negative to affirmative. When T enters, let it be done laſt, to avoid confuſion. Thus SPT performed upon X)Y gives x)Y from S, x)y, from P, and y)x from T ; which is X)Y, ſo that in this caſe alteration of ſubject, predicate, and order, is no alteration at all. Let L be the repreſentation of no alteration at all. To inveſtigate equivalent alterations, obſerve, firſt, that F and P, ſingly, are identical : thus F performed on X.Y gives X)Y, and P on X.Y gives X.y. And X)Y = X.y. This perfect identity of F and P in effect, remains in all combinations into which T does not enter. But when T enters, it is S and F which are identical. Thus ST performed on Y)X gives X)y or X.Y : and FT performed on Y)X gives X.Y. The reaſon is, that T interchanges ſubject and predicate ; ſo that F, after T, makes a change which is counterbalanced by a change in what was the ſubject. Accordingly, remembering that each operation performed twice is no operation at all (thus PP is L, and TT is L), we have in all caſes

$$P=F,\ SP=SF,\ PF=L,\ SPF=S$$
$$ST=FT,\ SPT=FPT,\ SFT=T,\ SPFT=PT$$

all which ſhould be tried for exerciſe. Again, in a *convertible* propoſition, transformation is no alteration or T=L : in an *inconvertible* one, transformation changes it into its contranominal ; or T=SP. Now ſet out as follows ;—L, in *convertible* propoſitions is T ; which in *inconvertibles*, is SP ; which, in *convertibles* again, is SPT ; which, in *inconvertibles* again, is TT, or L. Put theſe down as follows, writing under them the operations which are always equivalent to them, as ſhewn above,

$$\begin{array}{c||c|c||c|c} L & T & SP & SPT & L \\ PF & SFT & SF & PFT & PF \end{array}$$

The combinations written under one another are always the ſame in effect : thoſe ſeparated by double lines have the ſame effect on convertibles : thoſe ſeparated by ſingle lines, have the ſame effect on inconvertibles. Again P, for convertibles, is the ſame as PT ; which, for inconvertibles is the ſame as PSP, or S ; which, for convertibles again, is the ſame as ST ; which, for inconvertibles, is SSP or P. Theſe treated as before, give the table

P	PT	S	ST	P
F	SPFT	SPF	FT	F

In theſe two cycles there are L and all the fifteen ſelections which can be made out of S, P, F, T. And every poſſible caſe of equivalent changes is contained in theſe two tables. Thus PT is in all caſes equivalent to SPFT; in convertible caſes, to P and to F; in inconvertible ones, to S and to SPF. And no other combination is in any caſe equivalent to PT. In verification of theſe tables, obſerve that the operation F always occurs in the lower line, and never in the upper; and that this operation changes convertibles into inconvertibles, and *vice verſâ*. We ought then to expect, that the equivalences which, containing F, apply to inconvertibles, will be thoſe which when F is ſtruck out, apply to convertibles; and *vice verſâ*. And ſo we ſhall find it: for inſtance, SPFT and SPF are equivalent when performed on inconvertibles; ſtrike out F and we have SPT and SP, which are equivalent when performed on convertibles.

It appears, then, that any change which can be made on a propoſition, amounts in effect to L, P, S, or PS.. This is another verification of the preceding table: for all our forms may be derived from applying thoſe which relate to XY in the caſes of Xy, xY, and xy.

We have seen that $A_,$ and A' both contain $I_,$ and I'; and that $E_,$ and E' both contain $O_,$ and O'. Hence each of the univerſals may be ſaid to be the *ſtrengthened form* of either of its particulars of the ſame ſign: and each of the particulars the *weakened form* of its univerſals of the ſame ſign. The only diſtinction which appears between the two forms of the convertible particulars, XY and YX, xy and yx, is that the ſtrengthened forms derived from extending the ſubjects are different. Thus xy gives x)y or Y)X; but yx gives y)x or X)Y.

A *complex propoſition* is one which involves within itſelf the aſſertion or denial of each and all of the eight ſimple propoſitions. If theſe eight propoſitions were all concomitants, or if any number of them might be true, and the reſt falſe, there would be 256 poſſible caſes of the complex propoſition. As it is, owing to the connexion eſtabliſhed in the table of page 63, there are but *ſeven*.

Firft, let the names X and Y be fo related that neither of the four univerfals are true. Then all the four particulars are true: and this is the firft cafe. Let it be called a *complex particular*, and denoted by P. Then, denoting coexiftence of fimple propofitions by writing + between their feveral letters, we have

$$P = O' + O_{\iota} + I' + I_{\iota}$$

This cafe is of the leaft frequent mention in the theory of the fyllogifm.

Next, let one of the univerfal propofitions be true. Then five of the other propofitions are fettled, either by affirmation or denial. There remain the two concomitants, which are contradictory; fo that only one is true. Accordingly, with the exception of the complex particular juft defcribed, every complex propofition muft confift of the coexiftence of a univerfal and one of its concomitants. But there are not therefore eight more fuch propofitions: for $A' + A_{\iota}$ and $A_{\iota} + A'$ are the fame, and fo are $E_{\iota} + E'$ and $E' + E_{\iota}$. The remaining number is then reduced to fix, which are

$$\begin{array}{lll} A_{\iota} + O', & A_{\iota} + A', & A' + O_{\iota} \\ E_{\iota} + I', & E_{\iota} + E', & E' + I_{\iota}, \end{array}$$

Thefe muft be feparately examined.

Firft, take $A_{\iota} + A'$ (the order XY always underftood). We have then X)Y and Y)X. That is, there is no object whatfoever which has one of thefe names, but what alfo has the other. The names X and Y are then *identical*, not as names, but as fubjects of application. Where either can be applied, there can the other alfo. Thus, in geometry (the univerfe being plane rectilinear figure) equilateral and equiangular are identical names. Not that they agree in etymology nor in meaning: more than this, a few words would explain the firft to many who could not comprehend the fecond without difficulty. But they agree in that what figure foever has a right to either name, it has the fame right to the other. It will tend to uniformity of language, if we call X, in this cafe, *an identical* of Y, and Y *an identical* of X. Let the fymbol of an identical be D: then we have

$$D = A_{\iota} + A'$$

Next, take $A_1 + O'$. We have then X)Y and Y:X. Every X is Y, and so far there is a character of identity. But some Ys are not Xs; there are more Ys than Xs, and X stops short of a complete claim of identity with Y. Let X be called a *subidentical* of Y (thus *man* is a subidentical of *animal*), and let D_1 denote this case. Then

$$D_1 = A_1 + O'$$

Let $A' + O_1$ exist. We have then Y)X and X:Y. Every Y is X, and so far there is identity. But some Xs are not Ys, there are more Xs than Ys, or X goes beyond a claim of identity with Y. Let X be now called a *superidentical* of Y, and let it be denoted by D'. Then

$$D' = A' + O_1$$

The terms superidentical and subidentical are obviously correlative. If X be either of Y, Y is the other of X. Now let us consider $E_1 + E'$. We have then X.Y and x.y. There is nothing which is both X and Y, there is nothing which is neither. Consequently X and Y are *contraries*, or just fill up the universe. Let C be the mark of this relation. Then

$$C = E_1 + E'$$

Next, take $E_1 + I'$. We have then X.Y and xy. Nothing is both X and Y, but there are things which are neither. X and Y are clear of one another, but do not amount to contraries, for they do not fill up the universe. Let them be called *subcontraries*, (thus in the universe *metal*, *gold* and *silver* are subcontraries, and let C_1 denote the relation. Then

$$C_1 = E_1 + I'$$

Lastly, take $E' + I_1$. We have x.y and XY. The names fill the universe; for there is nothing but what is either X or Y. But they *overfill* it; for some things are both Xs and Ys. There is then all the completeness of a contrary and more. Let X and Y be called *supercontraries*,* and let C' denote the relation. Then we have

$$C' = E' + I_1$$

* The *supercontrary* relation, though essential to a complete system of syllogism, is not frequently met with. The other extreme of the supercon-

To complete our language, let A_{ι} or X)Y, with reference to the order XY, be called *sub-affirmative;* and $A^{\prime}$ or Y)X, *superaffirmative.* Let E_{ι} or X.Y be called *subnegative;* and E or x.y, *supernegative.* Let the particulars I_{ι}, $I^{\prime}$, and O_{ι}, $O^{\prime}$, have also these several names. This extension of our language will require a little explanation.

When I say that X is a subidentical of Y, I mean that the etymological suggestions are actually satisfied. The whole name X, and more, is contained in Y. But when I say that X is a universal sub*affirmative* of Y, or X)Y, I mean no more than that we have the proposition whose form is not superaffirmative, according to the etymology of that word. An algebraist would well understand the distinction at a glance. He has often to distinguish the case in which *a* is less than *b* from that in which *a* is less than or equal to *b*: the case in which the extreme limit of the assertion is not included from that in which it is included.

Again, the word negative had better be viewed as not so much presenting *exclusion* for its first idea, as *inclusion in the contrary.* Thus a subnegative, when universal, is to suggest complete inclusion in the contrary, meaning the extreme case, possibly; namely, that the subnegative names may be contraries. Again, supernegative is to suggest the idea of supercontrary, with the lowest extreme, the relation of contrary, possibly included.

For exercise in this language, and in the ideas which it is meant to present, I now state the following results.

Universal affirmation, though as a general term, it is to include super and sub affirmation, yet looked at as one of the three, and distinguished from the rest, it means identity. The same of negation and contrariety. Subidentity requires universal subaffirmation and particular supernegation. Identity is universal sub and super affirmation, both. Superidentity requires universal superaffirmation and particular subnegation. Subcontrariety requires universal subnegation and particular superaffirmation. Con-

trary, or the subidentical, is so much the easiest of all our complex relations, that the latter rarely allows the former to appear. The first instance that suggested itself to me was *man* and *irrational* (as descriptive of the quality of the individual and not of the species) in the universe *animal.* These more than fill that universe, *idiot* being common to both. But it is more natural to say that *rational* (in this sense) is subidentical of *man.*

trariety is univerſal ſub and ſuper negation, both. Supercontrariety requires univerſal ſupernegation and particular ſubaffirmation. Again, univerſal ſubaffirmation is either ſubidentity or identity: particular ſubaffirmation is a denial of contrariety and ſubcontrariety. Univerſal ſuperaffirmation is either ſuperidentity or identity: particular ſuperaffirmation denies contrariety and ſupercontrariety. Univerſal ſubnegation is either ſubcontrariety or contrariety: particular ſubnegation denies ſubidentity and identity. Univerſal ſupernegation is either ſupercontrariety or contrariety: particular ſupernegation denies ſuperidentity and identity. All this is expreſſed in the following table,

$D_\prime$	affirms	$A_\prime$ *and* O'	$A_\prime$	affirms	$D_\prime$ *or* D
D	—	$A_\prime$ and A'	A	—	$D_\prime$ or D or D'
D'	—	A' and $O_\prime$	A'	—	D' or D
$C_\prime$	—	$E_\prime$ and I'	$E_\prime$	—	$C_\prime$ or C
C	—	$E_\prime$ and E'	E	—	$C_\prime$ or C or C'
C'	—	E' and $I_\prime$	E'	—	C' or C
Denial of $D_\prime$	—	A' *or* $O_\prime$	$O_\prime$	denies	$D_\prime$ *and* D
— D	—	O' or $O_\prime$	O	—	$D_\prime$ and D, or D' and D
— D'	—	$A_\prime$ or O'	O'	—	D' and D
— $C_\prime$	—	E' or $I_\prime$	$I_\prime$	—	$C_\prime$ and C
— C	—	I' or $I_\prime$	I	—	$C_\prime$ and C or C' and C
— C'	—	$E_\prime$ or I'	I'	—	C' and C

Every ſubidentical of a name is the ſubcontrary of its contrary; every ſubcontrary is the ſubidentical of the contrary. Treat the word contrary as negative, the word identical as poſitive; and the two as of different ſigns. Then the algebraical rule 'like ſigns give a poſitive, unlike ſigns a negative,' holds in every caſe: including the variety of it ſo well known as 'two negatives make an affirmative.' When the modifying prepoſition comes firſt it muſt be retained; when it comes ſecond, it muſt be changed. Thus the ſubcontrary of a contrary is a ſubidentical: but the contrary of a ſubcontrary is a ſuperidentical. In putting two relations together, however, we have got into ſyllogiſm, as we ſhall preſently ſee.

The following tables will ſhow a connexion between the expreſſions, for different orders and ſelections, which it may be uſeful to verify.

XY	YX	xY	Yx	Xy	yX	xy	yx
$A_1O'D_1$	$A'O_1D'$	$E'I_1C'$	$E'I_1C'$	$E_1I'C_1$	$E_1I'C_1$	$A'O_1D'$	$A_1O'D_1$
$A'O_1D'$	$A_1O'D_1$	$E_1I'C_1$	$E_1I'C_1$	$E'I_1C'$	$E'I_1C'$	$A_1O'D_1$	$A'O_1D'$
$E_1I'C_1$	$E_1I'C_1$	$A'O_1D'$	$A_1O'D_1$	$A_1O'D_1$	$A'O_1D'$	$E'I_1C'$	$E'I_1C'$
$E'I_1C'$	$E'I_1C'$	$A_1O'D_1$	$A'O_1D'$	$A'O_1D'$	$A_1O'D_1$	$E_1I'C_1$	$E_1I'C_1$

This table only contains ſome of the rules already laid down in pp. 64, 65. It expreſſes that, for inſtance, the A_1, O', and D_1 of XY, are ſeverally the ſame as the E_1, I', and C_1 of yX. This table may be exhibited thus, the identicals counting as inconvertibles, the contraries as convertibles.

Change of	In Convertibles, changes	In Inconvertibles, changes
Subject	Sign and Prepoſition	Sign and Prepoſition
Predicate	Sign	Sign
Subject and Predicate	Prepoſition	Prepoſition
Order	Neither	Prepoſition
Subject and Order	Sign	Sign and Prepoſition
Predicate and Order	Sign and Prepoſition	Sign
Subject, Predicate, and Order	Prepoſition	Neither

In all caſes, change of ſubject is change both of ſign and prepoſition; change of predicate is change of ſign; change of ſubject and predicate is change of prepoſition. Theſe three caſes are of great importance in the ſyllogiſm: and the reader would do well to connect in his mind

Subject	with	*Sign and prepoſition*
Subject and Predicate	—	*Prepoſition*
Predicate	—	*Sign*

It is deſirable to conſider the ſeveral complex relations as to the continuous tranſition from one into another: *the growth of names* concerns not only the etymologiſt, but the logician alſo.

With the analogies and affinities by which the dominion of one name is extended to inſtance after inſtance, and claſs after claſs—and ſometimes, in ſcientific language at leaſt, deprived of a part of what it has held—I have here nothing to do. It is enough that the phenomena exiſt which may be deſcribed as the gradual transformation of one relation into another. The words *butt* and *bottle*, for example, are now ſubcontraries in the univerſe *receptacle*: but the etymology of the ſecond word ſhows that it *was* a subidentical of the first, being a diminutive. And if we were to take the whole class butt, buss, boot, bushel, box, boat, bottle, pottle, &c, which are all of one origin, the number of transitions would be found to be very large.

I assume that all the instances of a name are counted and arranged in its universe: a conceivable, though not attainable, supposition. Also, that the instances of the name are arranged contiguously, as in page 61. Whatever the reason may be which dictates the particular arrangement chosen, it will generally happen that the instances near to the boundary possess the characteristics of the name in a smaller degree than those nearer the middle. Let the contiguous arrangement be made of all the instances of the name Y, the universe being U. Let another name X begin to grow, commencing with one instance, that is, being applied to one of the objects in the universe U, be it a Y or not; then to another contiguous, and so on. We are to enumerate the ways in which such changes, whether of increase or diminution, may cause one name to change its relation to another. According as the change is made by accession or retrenchment, it may be denoted by (+) or (−).

Let the name X begin within the limits of the name Y: its initial relation to Y is then $D_{\prime}$. And the possibility of the following continuous changes is obvious:

$$D_{\prime}\,(+)\,D\,(+)\,D' \qquad D_{\prime}\,(+)\,P\,(+)\,C'$$
$$D_{\prime}\,(+)\,P\,(+)\,D' \qquad D_{\prime}\,(+)\,P\,(-)\,C_{\prime}$$

Hence $D_{\prime}$ may become D' through either D or P, but $C_{\prime}$ or C' only through P. Next, let X begin without the limits of Y: the initial relation is $C_{\prime}$. We may have then

$$C_{\prime}\,(+)\,C\,(+)\,C' \qquad C_{\prime}\,(+)\,P\,(+)\,D'$$
$$C_{\prime}\,(+)\,P\,(+)\,C' \qquad C_{\prime}\,(+)\,P\,(-)\,D_{\prime}$$

Let X begin both within and without Y: its initial relation is then P. And we have

$$P\,(+)\,D', \quad P\,(+)\,C', \quad P\,(-)\,D_{\prime}, \quad P\,(-)\,C_{\prime}$$

But when (−) follows $D_{\prime}$ or D, $C_{\prime}$ or C, we have nothing except

$$D_{\prime}\,(-)\,D_{\prime}, \quad D\,(-)\,D_{\prime}, \quad C_{\prime}\,(-)\,C_{\prime}, \quad C\,(-)\,C_{\prime}$$

If we begin at the other extreme, with the name U, we have

$$U\,(-)\,D' \qquad\qquad U\,(-)\,C'$$

Beginning from D' and C' we have

$$\begin{array}{ll} D'\,(-)\,D\,(-)\,D_, & D'\,(-)\,P\,(-)\,C_, \\ D'\,(-)\,P\,(-)\,D_, & D'\,(-)\,P\,(+)\,C' \\ C'\,(-)\,C\,(-)\,C_, & C'\,(-)\,P\,(-)\,D_, \\ C'\,(-)\,P\,(-)\,C_, & C'\,(-)\,P\,(+)\,D' \end{array}$$

But when (+) follows D' or D, C' or C, we have only

$$D'\,(+)\,D' \quad D\,(+)\,D', \quad C'\,(+)\,C', \quad C\,(+)\,C'$$

From the above liſt it appears that the tranſition which is accompanied by a change of prepoſition only can be made either through the letter without prepoſition or through P: and in all caſes with one continued mode of alteration. But when the tranſition involves change of letter, it can only be made through P: with continuation of the mode of alteration when the prepoſitions are different, and change in the mode when they are the ſame. The following ſucceſſions contain the arrangement of the reſults.

With one alteration (+)	With one alteration (—)	With two alterations (+—)
$D_,\ D\ D'$	$D'\ D\ D_,$	$D_,\ P\ C_,$
$D_,\ P\ D'$	$D'\ P\ D_,$	$C_,\ P\ D_,$
$C_,\ C\ C'$	$C'\ C\ C_,$	———
$C_,\ P\ C'$	$C'\ P\ C_,$	(— +)
———	———	$D'\ P\ C'$
$D_,\ P\ C'$	$D'\ P\ C_,$	$C'\ P\ D'$
$C_,\ P\ D'$	$C'\ P\ D_,$	

The following conſiderations will further ſerve to illuſtrate the want of the extenſion of the doctrine of propoſitions made in this chapter, and alſo the completeneſs of it. Among our moſt fundamental diſtinctions is that of *neceſſity* and *ſufficiency*; of what we *cannot do without*, and what we *can do with*; of that which *muſt precede*, and that which *can follow*. The contraries of theſe are *non-neceſſity* and *non-ſufficiency*. In theſe four words, applied to both Y and y, we have the deſcription of the eight relations of X to Y. For inſtance $A_{\prime}$ or X)Y tells us that to have an X, we muſt take a Y, or to be X, it is *neceſſary* to be Y. Treating all in the ſame way, we have

$A_{\prime}$	X)Y	To take an	X	it is *neceſſary*	to take a	Y
$A^{\prime}$	Y)X	. . .	X	. . *ſufficient*	. . .	Y
$E_{\prime}$	X.Y	. . .	X	. . *neceſſary*	. . .	y
$E^{\prime}$	x.y	. . .	X	. . *ſufficient*	. . .	y
$I_{\prime}$	XY	. . .	X	. *not neceſſary*	. .	y
$I^{\prime}$	xy	. . .	X	. *not ſufficient*	. .	y
$O_{\prime}$	X:Y	. . .	X	. *not neceſſary*	. .	Y
$O^{\prime}$	Y:X	. . .	X	. *not ſufficient*	. .	Y

And the convertibility of the ordinary mode of deſcription with this new one may be eaſily ſhown in any caſe. For example, what can we mean by ſaying that to take a X, it is not ſufficient to take what is not Y? Clearly that by taking not Y, or y, we may at the ſame time take a x, or that there are xs which are ys. And ſo on for the reſt.

Of the four pairs XY, Xy, xy, xY, we know that each propoſition may be expreſſed by three, and refuſes to be expreſſed by one. If we now admit the two words *impoſſible* and *contingent*, meaning by the latter that which, as the caſe may be, is poſſible or impoſſible, we ſhall eaſily ſee the following table for the univerſals:

		XY	Xy	xy	xY
$A_{\prime}$	X)Y	N	I	S	C
$E_{\prime}$	X.Y	I	N	C	S
$A^{\prime}$	Y)X	S	C	N	I
$E^{\prime}$	x.y	C	S	I	N

The letters N, I, S, C, are the initials of neceſſary, &c. And we read in the firſt line, that if X)Y, then to be X it is neceſſary to be Y; to be X, it is impoſſible to be y; to be x it is ſufficient to be y; and to be x, it is contingently poſſible or impoſſible to be Y. Again, if by *n* and *s* we mean *not neceſſary* and *not ſufficient*; by P, actually poſſible; and by C, as before (C being its own contrary), we have the following table for the particulars:

		XY	Xy	xy	xY.
O_1	X : Y	n	P	s	C
I_1	X Y	P	n	C	s
O'	Y : X	s	C	n	P
I'	xy	C	s	P	n

Of the four contrary pairs, n, P, s, C, are related to the particulars precifely as N, I, S, C, are to the univerfals. The interchange of Y and y is always accompanied by the interchange of N and I, S and C, n and P, s and C; the interchange of X and x is that of N and C, S and I, n and C, s and P; of both X and x, Y and y, is that of N and S, C and I, n and s, C and P.

The complex relations may be thus defcribed. According as X is fubidentical, identical, or fuperidentical of Y, to be X it is neceffary and not fufficient, neceffary and fufficient, or not neceffary and fufficient, to be Y: according as X is fubcontrary, contrary, or fupercontrary of Y, to be X it is neceffary and not fufficient, neceffary and fufficient, or not neceffary and fufficient, to be y. Or, as in the following table:

	XY	Xy	xy	xY
D_1	Ns	I	Sn	P
C_1	I	Ns	P	Sn
D'	Sn	P	Ns	I
C'	P	Sn	I	Ns
D	NS	I	NS	I
C	I	NS	I	NS
P	nsP	nsP	nsP	nsP

Inftead of IC and PC, write I and C: for "impoffible, and poffible or impoffible as the cafe may be" is "impoffible" &c.

The names of the complex relations, fubidentity, identity, &c I fuppofe will be held tolerably fatisfactory: thofe of the fimple relations fuggefted in page 68, fubaffirmative &c. have nothing in their favor except analogy with the former, and clofe connexion with the notation. A little practice in their ufe might render thefe laft names available: but it will be advifable to connect them with names more deſcriptive of the meaning, and to adopt theſe laſt, whether we reject or maintain their ſynonymes.

When X) Y, the relation of X to Y is well underſtood as that of the *ſpecies* to the *genus*. We may adopt theſe words, with the underſtanding that the word ſpecies includes the extreme caſe in which the ſpecies is as extenſive as the genus. When X : Y, we may call X a *non-ſpecies* of Y, and Y a *non-genus* of X. When X . Y .we may call X an *excluſive* or *excludent* of Y, or elſe a *non-participant*; and alſo Y of X. When XY, we may ſay that each is *participant*, or *non-excluſive*, of the other. When x . y, which means that X and Y together fill up, or more than fill up, the univere, we may ſay that they are *complemental* names. When x y, which only means that X and Y do not between them contain the univerſe, we may call them *non-complemental*. We have then

Inconvertibles.		Name of X with reſpect to Y.
$A_,$	X)Y	ſpecies, or ſubaffirmative.
$O_,$	X:Y	non-ſpecies, or particular ſubnegative.
A'	Y)X	genus, or ſuperaffirmative.
O'	Y:X	non-genus, or particular ſupernegative.

Convertibles.		Name of X and Y with reſpect to each other.
$E_,$	X.Y	Excluſives, or non-participants, or ſubnegatives.
$I_,$	XY	Non-excluſives, or participants, or particular ſubaffirmatives.
E'	x.y	Complements, or ſupernegatives.
I'	xy	Non-complements, or particular ſuperaffirmatives.

The following exerciſes in theſe terms, really contain the deſcription of all the ſyllogiſms in the next chapter.

Incluſion in the ſpecies is incluſion in the genus; and incluſion of the genus is incluſion of its parts (ſpecies or not).

Excluſion from the genus is excluſion from the ſpecies; and excluſion of the genus is excluſion of its parts (ſpecies or not).

Incluſion or excluſion of the ſpecies is part incluſion or excluſion of the genus.

When the ſpecies is complemental, ſo is the genus: and when the genus is not complemental, neither is the ſpecies.

Excluſion from one complement is incluſion in the other.

Complements of the ſame are participants.

Two fpecies of one genus, are not complements ; neither are two exclufions from the fame.

The complement of a genus is a non-fpecies ; and the complement is a non-fpecies of the non-complement.

CHAPTER V.

On the Syllogifm.

A SYLLOGISM is the inference of the relation between two names from the relation of each of thofe names to a third. Three names therefore are involved, the two which appear in the conclufion, and the third or *middle term*, with which the names, or terms, of the conclufion are feverally compared. The ftatements expreffing the relations of the two *concluding* terms to the *middle* term, are the two *premifes*. In this chapter, no ratio of quantities is confidered except the definite *all* and the indefinite *fome*.

A fyllogifm may be either *fimple* or *complex*. A fyllogifm is *fimple* when in it two fimple propofitions produce the affirmation or denial of a third: or the affirmation of a third, we may fay, fince every denial of one fimple propofition is the affirmation of another. A *complex* fyllogifm is one in which two complex propofitions produce the affirmation or denial of a third complex propofition.

It might be fuppofed that we ought to begin with the fimple fyllogifm, and from thence proceed to the complex. On this point I have fome remarks to offer, in juftification of following precifely the reverfe plan.

Hitherto the complex fyllogifm has never made its appearance in a work on logic, except in one particular cafe, in which it is allowed to be treated as a fimple fyllogifm, though moft obvioufly it is not fo. I allude to the common *à fortiori* argument, as in 'A is greater than B, B is greater than C, therefore A is greater than C.' There is no middle term here: the predicate of the firft propofition is 'a thing greater than B,' the fubject of the fecond propofition is 'B.'

Admitting fully that the quality of the premifes,—that which entitles the concluſion to be made, as it is ſaid, *à fortiori*—marks this argument out as, if anything, ſtronger, clearer, and (could ſuch a thing be) truer, than a ſimple ſyllogiſm; yet it is plain that the very additional circumſtance on which this additional clearneſs depends, takes the argument out of a ſyllogiſm, as defined by all writers. By beginning with the complex ſyllogiſm, and thence deſcending to the ſimple one, it will be ſeen that we begin with caſes which preſent this *à fortiori* and clearer character. I think I ſhall ſhew that the complex ſyllogiſm is eaſier than the ſimple one.

Next, the ſyllogiſm hitherto conſidered has never involved any contrary terms; the conſequence of which has been that various legitimate modes of inference have been neglected. Moreover, ſeveral of the uſual ſyllogiſms are more ſtrong than need be in the premiſes, in order to produce the concluſion. Thus Y)X and Y)Z being admitted as premiſes, the neceſſary concluſion is XZ. But if Y)X be weakened into YX, the ſame concluſion follows. If we call a ſyllogiſm *fundamental*, when neither of its premiſes are ſtronger than is neceſſary to produce the concluſion, it is obvious that every fundamental ſyllogiſm which has a particular premiſe, gives at leaſt as ſtrong a concluſion when that particular is ſtrengthened into a univerſal. But, except when ſtrengthening the premiſe alſo enables us to ſtrengthen the concluſion, in which caſe we have a new and different ſyllogiſm, it ſeems hardly ſyſtematic to mix with fundamental arguments ſyllogiſms which have quality or quantity more than is neceſſary for the concluſion.

The uſe of the complex ſyllogiſm will, as we ſhall ſee, give an independent and ſyſtematic derivation to theſe ſtrengthened ſyllogiſms, as well as to the reſt.

Let X and Z be the terms of the concluſion; and let Y be the middle term. Let the premiſe in which X and Y are compared come firſt of the two. Let the order of reference in each caſe be that of the alphabet

XY YZ XZ

So that by ſtating what X is with reſpect to Y, and what Y is with reſpect to Z, our ſyllogiſm involves the ſtatement of what X therefore muſt be, or therefore cannot be, with reſpect to Z. We can, in every caſe, expreſs the reſult in ſimple words. Thus, one of our ſyllogiſms being what I ſhall repreſent by $D_{,}D_{,}D_{,}$ is as follows. If X be a ſubidentical of Y, and Y a ſubidentical of Z, then X is a ſubidentical of Z. But all this merely amounts to the following 'A ſubidentical of a ſubidentical is a ſubidentical.'

We have then to examine every way in which $D_{,}$ or D' or $C_{,}$ or C' can be combined with $D_{,}$ or D' or $C_{,}$ or C', giving ſixteen caſes in all, and all concluſive in one way or the other. Inſtead of taking an accidental order, and afterwards claſſifying the reſults, it will be better to predict the order which will give claſſification. That order will be to take 1. a D followed by another of the ſame prepoſition 2. a C followed by another of different prepoſition 3. a D followed by another of a different prepoſition. 4. a C followed by another of a like prepoſition. This arrangement gives us

1. $D_{,}D_{,}$	$D'D'$	$D_{,}C_{,}$	$D'C'$	3.	$D_{,}D'$	$D'D_{,}$	$D_{,}C'$	$D'C_{,}$
2. $C_{,}D'$	$C'D_{,}$	$C_{,}C'$	$C'C_{,}$	4.	$C_{,}D_{,}$	$C'D'$	$C_{,}C_{,}$	$C'C'$

Each of theſe caſes will be examined by a method ſimilar to that propoſed in page 61. But a clear perception of the meaning of the words will at once dictate the ſixteen reſults, which are as follows, preceded by the mode in which the ſyllogiſms are to be expreſſed.

$D_{,}D_{,}D_{,}$ Subidentical of ſubidentical is ſubidentical.
$D'D'D'$ Superidentical of ſuperidentical is ſuperidentical.
$D_{,}C_{,}C_{,}$ Subidentical of ſubcontrary is ſubcontrary.
$D'C'C'$ Superidentical of ſupercontrary is ſupercontrary.

$C_{,}D'C_{,}$ Subcontrary of ſuperidentical is ſubcontrary.
$C'D_{,}C'$ Supercontrary of ſubidentical is ſupercontrary.
$C_{,}C'D_{,}$ Subcontrary of ſupercontrary is ſubidentical.
$C'C_{,}D'$ Supercontrary of ſubcontrary is ſuperidentical.

$D_{,}D':C'$ Subidentical of ſuperidentical is *not* ſupercontrary.
$D'D_{,}:C_{,}$ Superidentical of ſubidentical is not ſubcontrary.
$D_{,}C':D'$ Subidentical of ſupercontrary is not ſuperidentical.
$D'C_{,}:D_{,}$ Superidentical of ſubcontrary is not ſubidentical.

$C_{,}D_{,}:D'$ Subcontrary of ſubidentical is *not* ſuperidentical.
$C'D':D_{,}$ Supercontrary of ſuperidentical is not ſubidentical.
$C_{,}C_{,}:C'$ Subcontrary of ſubcontrary is not ſupercontrary.
$C'C':C_{,}$ Supercontrary of ſupercontrary is not ſubcontrary.

In the denials, the extreme limit is included: in the affirmations it is not. Thus 'not ſuperidentical' and 'not ſubidentical' both include 'not identical;' and the ſame of contraries. In the affirmations, extreme limitation of *one* premiſe does not alter the concluſion: but that of *both* reduces the concluſion to its extreme limit. Thus

> Subcontrary of identical is ſubcontrary.
> Contrary of ſuperidentical is ſubcontrary.
> Contrary of identical is contrary.

and ſo on. The rules of this ſpecies of ſyllogiſm are as follows. *For affirmatory concluſions;*—(1.) Like names in the premiſes give D in the concluſion, and unlike names C. (2.) D in the firſt premiſe requires premiſes of the ſame prepoſition; C in the firſt premiſe, of different prepoſitions. (3.) The prepoſition of the concluſion agrees with that of the firſt premiſe. *For negatory concluſions*, the preceding rules are reverſed. Theſe rules will do for the preſent, but they afterwards merge in others.

The ſixteen forms of complex concluſion above given are of the clearneſs of axioms, as ſoon as the terms are diſtinctly apprehended. The following diagrams will aſſiſt, and ſhould be uſed until the propoſitions ſuggeſt their own meaning. Though there be four, yet theſe four are really but one, as will be ſhown.

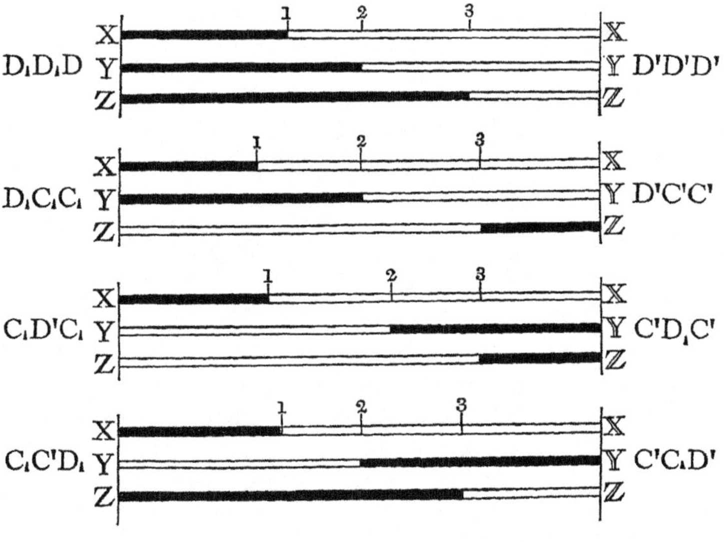

In each diagram are three lines, partly thick and partly open: theſe are meant to be laid over one another, but are kept ſeparate for diſtinctneſs. A point on the firſt line ſignifies a X or a x; and one on the ſecond or third, a Y or a y, and a Z or a z. The univerſe of the propoſitions is ſuppoſed to be the whole breadth. Points which come under one another are ſuppoſed to repreſent the ſame object of thought, variouſly named. Thus in the firſt diagram, when the *thick* lines contain the points named X, Y, and Z, it is ſhown that we mean to ſay there are objects to which all the three names apply: for there are points under one another in the thick part of all the three lines.

When we read by the letters on the left, the thick lines are meant to repreſent the parts in which the Xs, Ys, and Zs muſt be placed: and when by thoſe on the right, the open lines. Accordingly, looking at the third diagram, and at the left, we ſee $C_\prime\ D'\ C_\prime$: while in the diagram, it is clear that X is a ſubcontrary of Y, or that X . Y and x y; and that Y is a ſuperidentical of Z, or that Z) Y and Y : Z. And the concluſion is equally manifeſt, namely, that X is a ſubcontrary of Z. But, looking at the left, and ſeeing $C'\ D_\prime\ C'$, we take the open parts to repreſent the ſpaces in which Xs, Ys, and Zs are found; and the thick parts for thoſe in which xs, ys, and zs are found. Here then we ſee that X is a ſupercontrary of Y, that Y is a ſubidentical of Z, and that, *conſequently*, X is a ſupercontrary of Z.

Some attempts at laying down the premiſes ſo as to evade the concluſions, will be inſtructive to any one who does not immediately ſee the latter. And formal demonſtration is always practicable. Thus if X be a ſubcontrary of Y, that is, if X and Y do not fill the univerſe, and have nothing in common; and if Y be a ſuperidentical of Z, or entirely contain Z, without being filled by it: then it is clear that X muſt be more a ſubcontrary of Z than of Y, by all the inſtances which there are of a Y not being a Z. The diagram, however, is ſo much clearer than this ſort of demonſtration, that the reader, until he has great command of the language, may as well look to the former to ſee that he is right in the latter.

It may be convenient, as a matter of language, to ſpeak of a name as a kind of collective whole, conſiſting of inſtances. And thus we may talk of one name being entirely in another, or partly in and partly out &c, as in fact we have already done.

All the complex ſyllogiſms which conclude by affirmation are obviouſly of the *à fortiori* character: I ſhould rather ſay, thoſe of the firſt three diagrams properly and obviouſly, thoſe of the fourth by an eaſy extenſion of language. The marks 1 2 3 in the middle of the diagrams ſhow how this is. In the firſt, on the left, X is more of a ſubidentical of Z than it is of Y: the inſtances in which its *ſub*-identity appears conſiſt of all thoſe which prove the ſubidentity of X to Y, *together with* all thoſe which prove the ſubidentity of Y to Z. In the third, read from the right, X is more ſupercontrary to Z than it is to Y, by all the inſtances which ſhow the ſubidentity of Y to Z. In the fourth diagram (from the left) we cannot ſay that X is more ſubidentical of Z than of ſomething elſe, ſimply becauſe there is no previous ſubidentity among the relations. But ſtill the diſtinguiſhing characteriſtic of the concluſion takes its quantity from the addition of thoſe of both the premiſes.

If either of the premiſes be brought to the limit which ſeparates it from the relation of an oppoſite prepoſition; that is, if C' or C_1 be changed into C, or elſe D' or D_1 into D: the nature of the concluſion is not altered, except by the loſs of the *à fortiori* character. One of the quantities which have hitherto contributed to the quantity of the concluſion, now diſappears. Thus C_1 D gives C_1 as well as C_1 D'; and C D' gives C_1 as well as C_1 D'; C_1 C gives D_1 as well as C_1 C'.

Let one of the premiſes paſs over the limit, and take the oppoſite prepoſition. Chooſe C_1 D', which gives C_1, and continues to give it, though weakened, when the firſt C_1 becomes C. Then let C_1 become C': ſo that our premiſes are C' D'. The diagram is then as follows

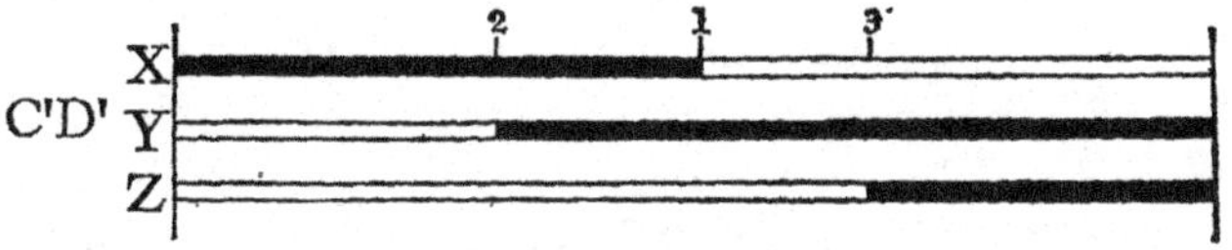

The *quantity* of the concluſion now depends upon the *difference* between the number of inſtances in (12) and (23) and its *quality* upon whether (12) has fewer inſtances than (23), or the ſame number, or more. As I have drawn it, C_1 is the concluſion, ſtill: ſtrengthen the firſt premiſe ſtill more, and the concluſion will pass through C into C' or else into P, and in the second case may pass into D', as in the following diagram

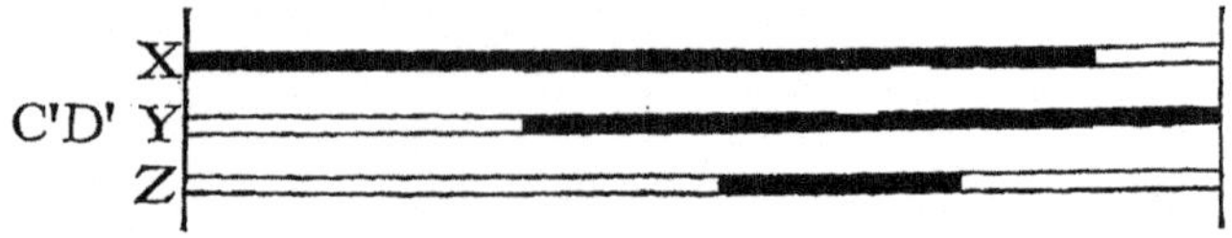

Nothing is impossible except $D_,$ or D. Hence C' D' enables us only to deny $D_,$ and its limit D. Treat the other cases in the same manner, and, remembering that denial is to include denial up to the limit (while affirmation only affirms to any thing short of the limit) we have

$D_, D'$	denies C'	$D' D_,$	denies $C_,$
$D_, C'$	. . D'	$D' C_,$	. . $D_,$
$C_, D_,$	. . D'	$C' D'$	. . $D_,$
$C_, C_,$	. . C'	$C' C'$	. . $C_,$

The rules given above in page 79 may be collected from the instances.

As long as we keep contraries out of view, the ultimate element of inference is of a twofold character. It is either 'X and Z are both Y; therefore X is Z' or else 'X is Y and Z is not Y; therefore X is not Z': X, Y, Z, being single instances of three names; and Y the same instance in both premises. But the use of contraries enables us to give an affirmative form to the latter case. It is 'X is Y, and not-Z is Y'; therefore 'X is not-Z'.

Connected with this change of expression is the following theorem: that all the eight affirmatory complex syllogisms are reducible to any one among them: and the same of the negatory ones. The reader may trace this theorem to the order of the figures 1, 2, 3, being the same in all the four diagrams. Taking $D_,D_,D_,$ as the most simple and natural form, and looking at the diagram of $C_,D'C_,$, we see the last as $D_,D_,D_,$ in 'X is subidentical of y; y is subidentical of z; therefore X is subidentical of z.' If we write the terms of the syllogism after its descriptive letters, as in $D_,D_,D_,$ (XYZ) we have the following results;—

$D_,D_,D_,$ (XYZ)$=D_,D_,D_,$ (XYZ)	$D'D'D'$ (XYZ)$=D_,D_,D_,$ (xyz)
$D_,C_,C_,$ (XYZ)$=D_,D_,D_,$ (XYz)	$D'C'C'$ (XYZ)$=D_,D_,D_,$ (xyZ)
$C_,D'C_,$ (XYZ)$=D_,D_,D_,$ (Xyz)	$C'D_,C'$ (XYZ)$=D_,D_,D_,$ (xYZ)
$C_,C'D_,$ (XYZ)$=D_,D_,D_,$ (XyZ)	$C'C_,D'$ (XYZ)$=D_,D_,D_,$ (xYz)

Thinking of the firſt deſcription only as to relations, and of the ſecond only as to terms, we ſee the following rules of connexion. In the firſt and ſecond premiſes and terms, there are X and Y in the terms, or their contraries, according as there are ſub-accents or ſuperaccents in the relations. But in the concluſion, the term is Z for $D_\prime$ and C', z for D' and $C_\prime$. And we may thus reduce any ſyllogiſm involving any one of the eight varieties of relation combined with any one of the varieties of terms, either to $D_\prime D_\prime D_\prime$ or to XYZ. Thus $C_\prime D' C_\prime$ (XyZ) is $D_\prime D_\prime D_\prime$ (XYz), or $D_\prime C_\prime C_\prime$ (XYZ). Not to load the ſubject with demonſtration of forms, I will give at once the general rules by which changes of accent and letter are governed: remarking that they apply throughout the whole of my ſyſtem.

The varieties in queſtion are eight:

XYZ, xyz; xYZ, Xyz; XyZ, xYz; XYz, xyZ.

in which (thinking of XYZ) all are kept; or all changed; or one only kept; or one only changed. Learn to connect each letter with the propoſitions in which it occurs; marking the propoſitions, premiſes and concluſion, as 1, 2, 3. Connect X with 1,3; Y with 1,2; Z with 2,3. Keeping all, or changing all, makes no alteration of letters: keeping only one, or changing only one, alters the letters in the premiſes in which that one occurs. Thus, be the accents what they may, if in DDD we change only the firſt letter into its contrary, the ſyllogiſm becomes CDC; and the ſame if we keep only the firſt letter unchanged.

As to accents, remember that change of Z produces no effect: look then only at X and Y. When either letter is changed into its contrary, change the accents belonging to the premiſes in which that letter comes firſt; 13 for X, 2 for Y, 123 for XY. For example, what is $C_\prime C' D_\prime$ (Xyz). Here, as to letters, X alone (1,3) is unchanged: then CCD becomes DCC. As to accents, Y is changed, which comes firſt only in 2: change C' into $C_\prime$. Hence $C_\prime C' D_\prime$ (Xyz)=$D_\prime C_\prime C_\prime$ (XYZ). Here we have paſſed from a ſyllogiſm in Xyz to the correſponding equivalent in XYZ: the rules equally hold for the inverſe proceſs, and for all combinations of letters. For the change of XYZ into Xyz, and that of Xyz into XYZ, have only one deſcription: the firſt only left unchanged. Now ſuppoſe it required to know what fyllogifm in xYz anfwers to $D_{\prime}C_{\prime}C_{\prime}$(Xyz). The key words are, *the third only unchanged.* Alter then DCC into DDD by the firft rule, and change all the accents. Thus $D_{\prime}C_{\prime}C_{\prime}$(Xyz)= D'D'D'(xYz). The independent rules are that change of fubject only, changes *both* letter and accent; predicate only, letter; fubject and predicate, accent. Thus to find what D'C'C'(xYz) is, expreffed in XYz, the changes are, in the three premifes S, neither, S, and D'C'C'(xYz)=$C_{\prime}$C'$D_{\prime}$(XYz). The following table may be verified for exercife: it fhows the effect of all changes except that of the middle term.

XYZ	xYZ	XYz	xYz
$D_{\prime}D_{\prime}D_{\prime}$	C'$D_{\prime}$C'	$D_{\prime}C_{\prime}C_{\prime}$	C'$C_{\prime}$D'
C'$D_{\prime}$C'	$D_{\prime}D_{\prime}D_{\prime}$	C'$C_{\prime}$D'	$D_{\prime}C_{\prime}C_{\prime}$
$D_{\prime}C_{\prime}C_{\prime}$	C'$C_{\prime}$D'	$D_{\prime}D_{\prime}D_{\prime}$	C'$D_{\prime}$C'
C'$C_{\prime}$D'	$D_{\prime}C_{\prime}C_{\prime}$	C'$D_{\prime}$C'	$D_{\prime}D_{\prime}D_{\prime}$

Similarly, D'D'D' would have $C_{\prime}$D'$C_{\prime}$ D'C'C' &c. When the middle term only is changed, the table may ftand thus;—

XYZ	$D_{\prime}D_{\prime}D_{\prime}$	C'$D_{\prime}$C'	$D_{\prime}C_{\prime}C_{\prime}$	C'$C_{\prime}$D'
XyZ	$C_{\prime}$C'$D_{\prime}$	D'C'C'	$C_{\prime}$D'$C_{\prime}$	D'D'D'

It will of courfe have been obferved that the eight fyllogifms go in pairs, each one of a pair differing from the other in accentuation, and nothing elfe. When we take fets of four, the ones put together fhould be thofe in which the firft premife, or the fecond, or the conclufion (whichever we take for a ftandard) has $D_{\prime}$ and C', or elfe has D' and $C_{\prime}$.

The fame rules of transformation apply to negatory complex fyllogifms; thus D'$D_{\prime}$:$C_{\prime}$(XYZ) is C'D':$D_{\prime}$(Xyz). In fact thefe rules do not depend upon the character of the inference, nor even upon its validity, but merely on the effects produced in the fingle propofitions by changes of term. Thus the ftatement D'$D_{\prime}C_{\prime}$ (XYZ), an invalid inference, is the fame ftatement (equally invalid of courfe) as is expreffed in $D_{\prime}$C'D'(xyZ).

An examination of the complex particular relation P=$I_{\prime}$+I' +$O_{\prime}$+O', whether by the diagram or by unaffifted thought, will fhow that when this relation exifts between X and Y, it alfo exifts between x and Y, X and y, x and y. Hence PC, CP, PD, DP, give P. Moreover, two complex particulars give no poffibility of any conclufion, all being equally poffible. Thus PP may give C_{ι} or C or C', or D_{ι} or D or D'.

Now combine one of the others, as D_{ι}, with P: examine PD_{ι} and $D_{\iota}P$. It will be found that the complex particular of a fubidentical may be either complex particular, fubidentical, or fupercontrary; or that PD_{ι} may be either P, D_{ι} or C'. Examine all the cafes, and the rules will be found in

$$(D_{\iota}C_{\iota})P \qquad P(D_{\iota}C')$$
$$(D'C')P \qquad P(C_{\iota}D')$$

thus interpreted. Either premife from between the parenthefes, with P, in order as written, may have either, and muft have one, of the three for its conclufion. That $D_{\iota}P$ muft give either D_{ι} C_{ι} or P, and fo muft $C_{\iota}P$: but PC_{ι} muft have either P, C_{ι}, or D'.

Before proceeding to the fimple fyllogifm, as I have called it, I will ftate that I much doubt the propriety of the terms *fimple* and *complex*. Undoubtedly the phrafes are hiftorically juft, for each of the fyllogifms which I propofe to call *complex* is, as we fhall fee, neceffarily compofed of three of thofe which are always called *fimple*. But in another point of view, the phrafeology ought to be reverfed; the fimple fyllogifm is the affirmation of the exiftence of one out of feveral of the complex ones. Thus X)Y+Y)Z=X)Z, or $A_{\iota}A_{\iota}A_{\iota}$, is really ($D_{\iota}$ or D, not known which) (D_{ι} or D, not known which) (D_{ι} or D, not known which) and afferts that there is either $D_{\iota}D_{\iota}D_{\iota}$ or $D_{\iota}DD_{\iota}$ or $DD_{\iota}D_{\iota}$ or DDD.

But it will be faid, furely the complex propofition requires the *conjunctive* exiftence of two fimple ones: $D_{\iota}=A_{\iota}+O'$; and is therefore *compound* at leaft. I anfwer that, on the other hand, the fimple propofition requires the *disjunctive* exiftence of two complex ones: as $A_{\iota}=D_{\iota}$ or D. Which is moft fimple, *both*, or *one or the other?* to me, I think, the firft. Certainly the fyllogifm $D_{\iota}D_{\iota}D_{\iota}$ is one which I more readily apprehend than $A_{\iota}A_{\iota}A_{\iota}$. Indeed, to moft minds, the latter is the former, if they are left to themfelves: and the cafes $D_{\iota}DD_{\iota}$, &c. are only admitted when produced and infifted on.

But further, is the fimple propofition properly called *fimple?* Is there in it but one affertion to deny or admit? Is but one queſtion anſwered? When I affirm 'Every X is Y,' I affirm 1. Compariſon of X and Y. 2. Coincidences. 3. The greateſt poſſible amount of them. 4. That *every* X has been uſed in obtaining them. In 'Some Xs are Ys the firſt two of the preceding are employed. In 'No X is Y,' we have, 1. Compariſon of Xs and Ys. 2. Excluſions. 3. The greateſt amount. 4. The compariſon of *every* X with every Y. And 'Some Xs are not Ys' omits the third, and ſubſtitutes Xs for every X in the fourth.

Now the ſubidentical, for inſtance, only contains, beſides what is in the ſubaffirmative, the notion that there are more Ys than Xs in exiſtence. The ſubcontrary conſiſts, over and above what is in the ſubnegative, in that Xs and Ys are not every thing that the propoſition might have applied to: and ſo on. On theſe conſiderations, I think it may be allowed to treat the words ſimple and complex as only of hiſtorical reference, and to conſider the firſt as disjunctively connected with the ſecond, the ſecond as conjunctively connected with the firſt, in the manner above noted. I think I ſhall make it clear enough, that the paſſage from the conjunctions to the disjunctions is better ſuited to a demonſtrative ſyſtem than the converſe. If the plan which I propoſe ſhould gain any reception, I ſhould imagine that *disjunctive* and *conjunctive* would be the names given to the claſſes which I have called ſimple and complex: the conjunctive compoſed of ſeveral of the disjunctive, the disjunctive conſiſting of one or the other out of ſeveral of the conjunctive.

When a propoſition R, is the neceſſary conſequence of two others, P and Q, it neceſſarily follows that the denial of R, muſt be the denial of one at leaſt of P and Q. For every propoſition admits but of affirmation or denial: and he who affirms *both* P and Q muſt *affirm* R. If then P be affirmed and R denied, the denial of Q muſt follow: if Q be affirmed and R denied, the denial of P muſt follow.

A *ſimple ſyllogiſm* is one, the two premiſes and concluſion of which are to be found among the ſimple propoſitions A_1, E_1, I_1, O_1, A^1, E^1, I^1, O^1. Thus we have $A_1E_1E_1$ or $X)Y + Y.Z = X.Z$, as an inſtance. The order of reference is always XY, YZ, XZ.

The following theorems will be neceſſary;—1: *A particular premiſe cannot be followed by a univerſal concluſion.*

If possible, let A_1I_1 for example, have a universal conclusion. Take the complex premises D_1P or $(A_1+O')(I_1+I'+O_1+O')$. All that can be inferred is that one of *three* conclusions (page 85) is valid, and neither D nor C: either D_1 or P or C_1. But if a universal be true, one of *two* conclusions must be valid (page 69) and one of them D or C. If then A_1 and I_1 alone yielded a universal conclusion, quite as much must D_1P: or a form which is indifferent to three conclusions, and not having D nor C, is necessarily productive of one of two conclusions, one of which is D or C. This contradiction cannot exist: or A_1I_1 cannot yield a universal conclusion.

2. *From two particular premises no conclusion can follow.*

If possible, let I_1I_1 yield a conclusion; which by the last theorem, must be only particular. Now PP or $(I_1+I'+O_1+O')(I_1+I'+O_1+O')$ is indifferent to all complex conclusions: quite as much is I_1I_1. But if these premises yield a particular conclusion, two complex conclusions are denied (page 69). This contradiction cannot exist: or particular premises can yield no conclusion.

Let a simple syllogism with premises and conclusion all universal, be called *universal:* and with either premise (and therefore the conclusion) particular, be called *particular.* Then every universal syllogism has two particular syllogisms deducible from it. Thus if $A_1E_1E_1$ be valid, then A_1 joined with the denial of E_1 gives the denial of E_1: or $A_1I_1I_1$ *seems to be valid.* But the alteration of the places of the propositions requires us to say that it is $A'I_1I_1$ which is valid: and this point requires close attention.

Take $A_1E_1E_1$ or $X)Y+Y.Z=X.Z$. Then $X)Y$ with the denial of $X.Z$ (or XZ) gives the denial of $Y.Z$ (or YZ); and we have

$$X)Y+XZ=YZ$$

This is valid, if the first be (as it is) valid: but its symbol is not $A_1I_1I_1$. For the middle term is, in our notation, made *middle* in the order of reference, which is therefore YX, XZ, YZ: and the syllogism is $A'I_1I_1$. Similarly we have

$$XZ+Y.Z=X:Y$$

produced by coupling the denial of $X.Z$ with $Y.Z$. But this is $I_1E_1O_1$: for the order of reference is now XZ, ZY, XY, and

E_1 is not changed by change of order. The rule is as follows. When the denials of the conclufion and of a premife are made to take the places of that premife and the conclufion, the order of reference remains undifturbed as to the tranfpofed terms, and is changed as to the ftanding term. This laft muft therefore have the prepofition of the inconvertible propofition changed; but not that of the convertible propofition.

Thus $E'A_1E'$, if valid, gives $E'I'O_1$ and $I'A'I'$. Again, in a fimilar way it may be fhown that from each particular fyllogifm follows a univerfal: thus $I_1E'O'$, if valid, fhows that denial of O', and E', give denial of I_1 or $A'E'E_1$. In this cafe neither is valid. And $E'I'O_1$, befides $E'A_1E'$, alfo gives $A_1I'I'$.

Such claffification of thefe *opponent forms* as is ufeful, will prefently be given.

Since there are eight forms of affertion, with reference to each of the orders XY YZ, it follows that there are fixty-four combinations of a pair of premifes each. But of thefe the only ones which have a chance of yielding a conclufion are, 1. fixteen with premifes both univerfal; 2. thirty-two with one univerfal and one particular. If, for a moment, *U* ftand for univerfal and *P* for particular, the form of a fyllogifm is either *UUU*, *PUP*, *UPP*, or *UUP*. Of thefe, the firft, fecond, and third are fo related that each form has the other two for its opponents: but the fourth has its own form in each of its opponents.

Now examine one of the complex affirmative fyllogifms, fay $D_1D_1D_1$, by the diagram in page 79. The premifes are A_1+O' and A_1+O', giving the four combinations A_1A_1, A_1O', $O'A_1$ and $O'O'$. The conclufion is A_1+O': but it is not merely twofold, but threefold: for the *à fortiori* character explained in page 81, fhows that O' is obtainable on two different grounds, and is the fum, as it were, of two different and neceffary parts of the conclufion. That every X is Z, follows from X)Y and Y)Z, or we have the fyllogifm.

$$A_1A_1A_1 \quad X)Y+Y)Z=X)Z$$

But as far as the Zs which are below (12) are concerned, it follows that they are not Xs becaufe they are the Ys which are not Xs: or we have

$$O'A_1O' \quad Y:X+Y)Z=Z:X$$

and as to the Zs below (23) they are not Xs becaufe they are not Ys, among which are all the Xs. Accordingly we have

$$A_{\prime}O'O' \quad X)Y+Z{:}Y=Z{:}X$$

or $D_{\prime}D_{\prime}D_{\prime}$ requires the coexiftence of $A_{\prime}A_{\prime}A_{\prime}$, $O'A_{\prime}O'$, $A_{\prime}O'O'$. Apply this reafoning to the contraries x, y, z, or elfe examine $D'D'D'$ in the fame way, and we find that $D'D'D'$ requires the coexiftence of $A'A'A'$, $O_{\prime}A'O_{\prime}$, $A'O_{\prime}O_{\prime}$.

By applying the preceding refults to x, Y, Z, &c. as in page 82, or, as is better at firft, by examining all the cafes of the diagram in page 79, we get the following table of derivations from the eight affirmatory complex fyllogifms. The firft column fhews the terms which muft be ufed, to deduce all from $D_{\prime}D_{\prime}D_{\prime}$

XYZ	$D_{\prime}D_{\prime}D_{\prime}$ . .	$A_{\prime}A_{\prime}A_{\prime}$	$X)Y+Y)Z=X)Z$	
		$O'A_{\prime}O'$	$Y{:}X+Y)Z=Z{:}X$	(12)
		$A_{\prime}O'O'$	$X)Y+Z{:}Y=Z{:}X$	(23)
x y z	$D'D'D'$ . .	$A'A'A'$	$Y)X+Z)Y=Z)X$	
		$O_{\prime}A'O_{\prime}$	$X{:}Y+Z)Y=X{:}Z$	(12)
		$A'O_{\prime}O_{\prime}$	$Y)X+Y{:}Z=X{:}Z$	(23)
x YZ	$C'D_{\prime}C'$ . .	$E'A_{\prime}E'$	$x.y+Y)Z=x.z$	
		$I_{\prime}A_{\prime}I_{\prime}$	$XY+Y)Z=XZ$	(12)
		$E'O'I_{\prime}$	$x.y+Z{:}Y=XZ$	(23)
X y z	$C_{\prime}D'C_{\prime}$ . .	$E_{\prime}A'E_{\prime}$	$X.Y+Z)Y=X.Z$	
		$I'A'I'$	$xy+Z)Y=xz$	(12)
		$E_{\prime}O_{\prime}I'$	$X.Y+Y{:}Z=xz$	(23)
XYz	$D_{\prime}C_{\prime}C_{\prime}$ . .	$A_{\prime}E_{\prime}E_{\prime}$	$X)Y+Y.Z=X.Z$	
		$O'E_{\prime}I'$	$Y{:}X+Y.Z=xz$	(12)
		$A_{\prime}I'I'$	$X)Y+yz=xz$	(23)
x y Z	$D'C'C'$ . .	$A'E'E'$	$Y)X+y.z=x.z$	
		$O_{\prime}E'I_{\prime}$	$X{:}Y+y.z=XZ$	(12)
		$A'I_{\prime}I_{\prime}$	$Y)X+YZ=XZ$	(23)
x Y z	$C'C_{\prime}D'$ . .	$E'E_{\prime}A'$	$x.y+Y.Z=Z)X$	
		$I_{\prime}E_{\prime}O_{\prime}$	$XY+Y.Z=X{:}Z$	(12)
		$E'I'O_{\prime}$	$x.y+yz=X{:}Z$	(23)
X y Z	$C_{\prime}C'D_{\prime}$ . .	$E_{\prime}E'A_{\prime}$	$X.Y+y.z=X)Z$	
		$I'E'O'$	$xy+y.z=Z{:}X$	(12)
		$E_{\prime}I_{\prime}O'$	$X.Y+YZ=Z{:}X$	(23)

Before forming any rule, or making any remark, I proceed to

collect the results of the remaining cases. And first, let a premise be brought to its limit, D or C: say that $D_iD_iD_i$ becomes DD_iD_i. In the diagram it immediately appears that one of the particular conclusions is lost; not contradicted, but nullified: for (12) disappears, because X and Y are identical names. That is, $A_iA_iA_i$ remains, and $A_iO'O'$: but the conclusion of $O'A_iO'$ is nullified. But this very circumstance creates, not a new conclusion, for it is only a part of one already existing, but a new form of deduction. The premises are now A_i+A' and A_i+O', and the conclusion is A_i+O'. The syllogisms $A_iA_iA_i$ and $A_iO'O'$ are as before, and for the same reasons: but there is now the combination $A'A_i$ among the premises, which produces the conclusion I_i, and we have

$$A'A_iI_i \quad Y)X+Y)Z=XZ$$

This syllogism, though new as far as $D_iD_iD_i$ is concerned, is only a strengthened form of $I_iA_iI_i$, a concomitant of $E'A_iE'$. For (page 65) I_i is true whenever A' is true, so that $A'A_i$ includes I_iA_i and its necessary consequence I_i. But if I_i had been strengthened into A_i instead of A', we should have had $A_iA_iI_i$ which though perfectly valid, yet admits of a stronger conclusion, as seen in $A_iA_iA_i$.

Of the two modes of strengthening a particular proposition (as I_i into A_i or A') there is one which strengthens the quantity of the first form of the proposition, and another that of the second. Thus XY or I_i becomes X)Y or A_i when the first form, and Y)X or A', when the second form, is strengthened. Similarly O_i or X:Y becomes X.Y or E_i, and y.x or E', according as the form strengthened is X:Y or y:x. The preposition remains the same, or changes, according as the first or second form is strengthened. If the *first* form of the *second* premise of a syllogism, or the *second* form of the *first* premise, be strengthened, no strength is added to the conclusion. Thus, as far as the syllogisms in this chapter are concerned, I_iA_i gives as much as $A'A_i$, and E_iO_i as E_iE_i. But if the first form of the first premiss, or the second form of the second, be strengthened, the conclusion has its first form strengthened.

A very simple and obvious theorem contains all these results. The concluding terms are, in our order of reference, the first term of the firft premife and the fecond term of the fecond. The conclufion is never ftrengthened by augmenting the quantity of the middle term, nor *only* weakened (it may be altogether deftroyed) by weakening the middle term. A wider field of comparifon does not by itfelf give more comparifons: nor can more comparifons arife except by augmenting the number of things compared in that field. Since the conclufion can obvioufly fpeak of no more than was in the premifes, no term of that conclufion can be augmented in quantity, until the fame thing has taken place in its premife. But no ftrengthening of a propofition ftrengthens both terms: confequently, to make fuch a thing effective, it muft be the concluding, and not the middle, term which is ftrengthened.

The following table is only worth inferting as a collection of exercifes. The fourth column fhows the eight *ftrengthened particular fyllogifms*, as I will call them, having univerfal premifes but only a particular conclufion, not ftronger than might have been inferred from the particular fyllogifm itfelf.

Alteration of	into	removes	and fubftitutes	ftrengthened from	occurring in
$D_{,}D_{,}D_{,}$	$DD_{,}D_{,}$	$O'A_{,}O'$	$A'A_{,}I_{,}$	$I_{,}A_{,}I_{,}$	$C'D_{,}C'$
$D'D'D'$	$D'DD'$	$A'O_{,}O_{,}$		$A'I_{,}I_{,}$	$D'C'C'$
$D_{,}D_{,}D_{,}$	$D_{,}DD_{,}$	$A_{,}O'O'$	$A_{,}A'I'$	$I'A'I'$	$C_{,}D'C_{,}$
$D'D'D'$	$DD'D'$	$O_{,}A'O_{,}$		$A_{,}I'I'$	$D_{,}C_{,}C_{,}$
$D_{,}C_{,}C_{,}$	$DC_{,}C_{,}$	$O'E_{,}I'$	$A'E_{,}O_{,}$	$I_{,}E_{,}O_{,}$	$C'C_{,}D'$
$D'C'C'$	$D'CC'$	$A'I_{,}I_{,}$		$A'O_{,}O_{,}$	$D'D'D'$
$D_{,}C_{,}C_{,}$	$D_{,}CC_{,}$	$A_{,}I'I'$	$A_{,}E'O'$	$I'E'O'$	$C_{,}C'D_{,}$
$D'C'C'$	$DC'C'$	$O_{,}E'I_{,}$		$A_{,}O'O'$	$D_{,}D_{,}D_{,}$
$C_{,}D'C_{,}$	$CD'C_{,}$	$I'A'I'$	$E'A'O_{,}$	$O_{,}A'O_{,}$	$D'D'D'$
$C'D_{,}C'$	$C'DC'$	$E'O'I_{,}$		$E'I'O_{,}$	$C'C_{,}D'$
$C_{,}D'C_{,}$	$C_{,}DC_{,}$	$E_{,}O_{,}I'$	$E_{,}A_{,}O'$	$O'A_{,}O'$	$D_{,}D_{,}D_{,}$
$C'D_{,}C'$	$CD_{,}C'$	$I_{,}A_{,}I_{,}$		$E_{,}I_{,}O'$	$C_{,}C'D_{,}$
$C_{,}C'D_{,}$	$CC'D_{,}$	$I'E'O'$	$E'E'I_{,}$	$O_{,}E'I_{,}$	$D'C'C'$
$C'C_{,}D'$	$C'CD'$	$E'I'O_{,}$		$E'O'I_{,}$	$C'D_{,}C'$
$C_{,}C'D_{,}$	$C_{,}CD_{,}$	$E_{,}I_{,}O'$	$E_{,}E_{,}I'$	$O'E_{,}I'$	$D_{,}C_{,}C_{,}$
$C'C_{,}D'$	$CC_{,}D'$	$I_{,}E_{,}O_{,}$		$E_{,}O_{,}I'$	$C_{,}D'C_{,}$

I will now examine the negatory complex fyllogifms, premifing however than we cannot get any new conclufions from them.

For we have now got all the ſixteen caſes in which both premiſes are univerſal: and we know that there can be no ſyllogiſm with a particular premiſe, except it have one of thoſe with univerſal premiſes for its opponents.

Take $D_\prime D' : C'$ or $A_\prime + O'$ and $A' + O_\prime$ together deny $E' + I_\prime$, that is, deny the coexiſtence of E' and $I_\prime$, that is, deny either E' or $I_\prime$, that is, aſſert either I' or $E_\prime$. This ſyllogiſm then may be written thus,

$$(A_\prime + O')\ (A' + O_\prime)\ (\text{either } E_\prime \text{ or } I')$$

Now the fact is that this disjunction is ſuperfluous; it is I' which is always aſſerted, and $E_\prime$ is never a neceſſary conſequence of $D_\prime D'$. For $A_\prime A'$ gives I' as already ſhown, and $A_\prime O_\prime$ and $O'A'$ are inconcluſive (and $O'O_\prime$ of courſe). And the rationale of the inference is as follows: ſince X is a ſubidentical of Y, and Y a ſuperidentical of Z, it follows that Y is ſuperidentical both of X and Z; conſequently, Y not filling the univerſe (our ſuppoſition throughout) it follows that there are things which are neither Xs nor Zs, namely, all which are not Ys. Again, in $C_\prime C_\prime : C'$, which the ſame reaſoning ſhows to be only $C_\prime C_\prime I'$, none either of X or of Z is in Y, therefore every inſtance in Y is both x and z. And thus it will appear that in every negatory complex concluſion *the whole middle term*, or the *whole of its contrary*, makes the ſubject matter of the ſtrengthened particular ſyllogiſm which is all that can be collected.

Our concluſion is that no negatory complex ſyllogiſm is of any more logical effect than the ſtrengthened particular derived from it. Thus we may ſay that, ſo far as the extent and character of the inference is concerned, the former is the latter.

I will now paſs to the general rules of the complete ſyſtem of ſyllogiſms;—

The reader muſt take pains to remember two rules of formation, perfect contraries of each other, for the dependence of the *accents* (or *prepoſitions*) on the *ſign* (affirmative or negative character) of the firſt premiſe. I expreſs them in the briefeſt way poſſible.

Direct Rule. Affirmation (in the firſt premiſe) makes the ſecond *premiſe* agree with both the other propoſitions, or *iſolates* nothing: negation makes the ſecond premiſe differ from both the others, or iſolates the ſecond premiſe. *Inverſe rule.* Affirmation *iſolates the firſt premiſe*, makes the firſt premiſe differ from both the others in prepoſition: negation *iſolates the concluſion*, makes the concluſion differ from both the others. Theſe rules might be expreſſed ſo as to make their contrariety more complete. Thus in the direct / inverſe rule, affirmative commencement ſhows like / unlike prepoſitions in the two premiſes, and the concluſion agreeing / differing with / from the firſt premiſe in prepoſition; but negative commencement ſhows unlike / like prepoſitions in the two premiſes, and the concluſion agreeing with / differing from the firſt premiſe in prepoſition.

The ſubjects of the following rules are,

1. The eight affirmatory complex ſyllogiſms.
2. The eight univerſal ſimple ſyllogiſms.
3. The eight ſtrengthened particular ſimple ſyllogiſms.
4. The ſixteen particular ſimple ſyllogiſms.

Omit the negatory complex ſyllogiſms, as fully contained in the third of this enumeration, and the complex ſyllogiſms which contain the unaccented D or C, as carrying a momentary accent for the rule, to be expunged when the formation is completed. Conſider $D_{,}$, D, D', $A_{,}$, A', $I_{,}$, I', as of the affirmative ſigns, and $C_{,}$, C, C', $E_{,}$, E', $O_{,}$, O', as negative.

Rule 1. In the complex ſyllogiſm all parts are complex; in the univerſal ſimple ſyllogiſm all parts are univerſal; in the ſtrengthened particular only the concluſion is particular; in the particular only a premiſe is univerſal.

Rule 2. Premiſes of like ſign have an affirmative concluſion; of unlike ſign, a negative.

Rule 3. The complex, the univerſal, the particulars which begin with a particular, follow the direct rule; the ſtrengthened particulars, and the particulars which begin with a univerſal (all that commence with a univerſal, and conclude with a particular) follow the inverſe rule. [Or thus; all which begin and end alike, follow the direct rule; all which begin and end differently, the inverſe.]

The complex ſyllogiſms and univerſals are eaſily remembered

by rule: the particulars almoſt as eaſily. The following ſub-rules may be noted, as far as theſe laſt are concerned.

Sub-rule 1. *Firſt and ſecond premiſes.* A and O in the firſt premiſe demand unlike prepoſitions in the two premiſes: E and I demand like prepoſitions. Thus A,O, muſt be inconcluſive: A,O' muſt be concluſive. But E,O, muſt be concluſive: and E,O' muſt be inconcluſive.

Sub-rule 2. *Firſt premiſe and concluſion.* A univerſal in the firſt premiſe demands an unlike prepoſition in the concluſion: a particular firſt premiſe, a like prepoſition in the concluſion.

Sub-rule 3. *Second premiſe and concluſion.* Every ſecond premiſe demands its own prepoſition in a concluſion of like ſign: and the other prepoſition in a concluſion of unlike ſign.

As far as the four ſpecies are concerned, every ſyllogiſm formed according to the three rules is valid; and every one not ſo formed is invalid. The following remarks are partly recapitulatory, partly new.

Remark 1. Every complex ſyllogiſm gives one univerſal ſyllogiſm* and two particular ones, its concomitants: and the concomitants are formed by changing one of the premiſes of the univerſal and the concluſion, into their particular concomitant propoſitions (page 63.)

Remark 2. Every ſyllogiſm has its *contranominal*, which aſſerts of the contraries in the ſame manner as the firſt does of the direct terms: and contranominals have all their accents different, as in O'A,O' and O,A'O, (page 62.)

Remark 3. Every ſyllogiſm has *two* opponents, made by interchanging the contradictories of one premiſe and of the concluſion, and altering the accent of the remaining premiſe, if inconvertible (A or O) (page 88.)

Remark 4. Every complex ſyllogiſm has two ſuch opponents formed in the ſame way, the Ds being the inconvertibles, the Cs the convertibles. Thus (:) meaning *denial of*, the opponents of C,D'C, are C,:C,:D' and :C,D,:C,. The firſt of theſe is

(E, + I') (I, *or* E') (O' or A,)

containing the valid ſyllogiſms E,E'A,, E,I,O', I'E'O'; being

* *Syllogiſm*, not preceded by *complex*, means ſimple ſyllogiſm.

$E_{,}E'A_{,}$ and its concomitants. And $:C_{,}D_{,}:C_{,}$ gives $E'A_{,}E'$ (the contranominal of $E_{,}A'E_{,}$) and its concomitants. And the ſame of the reſt.

Remark 5. Each univerſal ſyllogiſm has two weakened forms, made by weakening one premiſe and the concluſion. When the *firſt* premiſe is weakened, it is without change of prepoſition: but when the ſecond, with change. Thus the weakened forms of $E_{,}A'E_{,}$ are $O_{,}A'O_{,}$ and $E_{,}I_{,}O'$.

Remark 6. Each particular ſyllogiſm has two ſtrengthened forms, one of which is a univerſal, the other only a ſtrengthened particular. Thus the ſtrengthened forms of $O_{,}A'O_{,}$ are $E_{,}A'E_{,}$ and $E'A'O_{,}$.

Remark 7. In every ſyllogiſm except the ſtrengthened particular, the middle term is univerſal in one premiſe, and particular in the other: and its contrary is therefore the ſame. But in the ſtrengthened particular, the middle term is univerſal in both premiſes, or particular in both. This affords a complete criterion of ſyllogiſm, as will be noticed hereafter: in fact, the completeneſs of this ſyſtem crowds us with relations, from many of which general rules might be deduced, though they need only appear here by caſual remark.

In $O'A_{,}O'$, $A'O_{,}O_{,}$, $I_{,}A_{,}I_{,}$, $E_{,}O_{,}I'$, $O'E_{,}I'$, $A'I_{,}I_{,}$, $I_{,}E_{,}O_{,}$, $E_{,}I_{,}O'$, the middle term enters univerſally in the univerſal, and particularly in the particular. In all the others it enters particularly in the univerſal, and univerſally in the particular. In the firſt ſet, the convertible premiſes are all *ſubs*, the inconvertibles are *ſubs* in the ſecond premiſe, and *ſupers* in the firſt. In the ſecond ſet, theſe rules are inverted.

Remark 8. Of the twelve poſſible pairs of premiſes AA, AE, AI, AO, EA, EE, EI, EO, IA, IE, OA, OE, which *can* give a concluſion, each one *will*, in two ways, which two ways are inverted in their accents. Thus EO appears in $E'O'I_{,}$ and $E_{,}O_{,}I'$. The two premiſe-letters and one accent dictate all the reſt: thus $I'A$ can belong to nothing but $I'A'I'$. When the ſyſtem is well learnt, it will be found unneceſſary to write more than $I'A$, for the ſymbol of $I'A'I'$. I now ſpeak only of fundamental ſyllogiſms: the ſtrengthened ſyllogiſm $A_{,}A'I'$ might be ſignified by $A'A'$.

Remark 9. The ſyllogiſms of the three firſt claſſes are all really ſpecimens of one, thoſe of the fourth of two, among them, with the eight variations XYZ, xYZ, XYz, xYz, XyZ, xyZ, Xyz, xyz. The rules for conducting theſe changes are

> Change of ſubject is change of both accent and letter.
> Change of predicate is change of letter.
> Change of both is change of accent.

thus to paſs from E'E,A' to A,E,E, we note in XY change of ſubject, in YZ change of neither, in XZ change of ſubject: therefore xYZ is the ſet of terms into which XYZ muſt be changed: and the E'E,A' ſyllogiſm of either ſet is the A,E,E, ſyllogiſm of the other.

The 24 ſyllogiſms, which are 24 with reference to the order XY, YZ, XZ, are only 12 if the order ZY, YX, ZX, be allowed. Thus A,I'I' of the firſt is the I'A'I' of the ſecond. Theſe ſyllogiſms are eſſentially the ſame in the mode of inference they afford. To change a ſyllogiſm into another of the ſame mode of inference, invert the premiſes and change the prepoſition of all the inconvertibles. Thus A'O,O, and O'A,O' are of the ſame inference. The pairs which in this point of view are identical are

A,A,A, = A'A'A'	E'A,E' = A'E'E'	E,A'E, = A,E,E,	E'E,A' = E,E'A,
O'A,O' = A'O,O,	I,A,I, = A'I,I,	I'A,I' = A,I'I'	I,E,O, = E,I,O'
A,O'O' = O,A'O,	E'O'I, = O,E'I,	E,O,I' = O'E,I'	E'I'O, = I'E'O'

The ninth remark admits of conſiderable extenſion. The '*ſome*' of a logical propoſition may have a much more definite character in ſome caſes than in others. It may be a ſelected, or at leaſt a diſtinguiſhable *ſome*, which want nothing but a nominal diſtinction to make the particular propoſition eaſily and uſefully univerſal. Whether it can be done more or leſs eaſily, and more or leſs uſefully, is no queſtion of formal logic. If it be ſuppoſed done, the particular is converted into a univerſal. In 'ſome Xs are Ys,' if we make a name for every X which is Y, ſay M, we have then 'Every M is Y'. This propoſition may be purely identical, or it may not. If we call every X which is Y by the name M merely becauſe it is Y, then our univerſal is only 'Every *X which is Y* is Y'. But if the name M be conferred from any other circumſtance, which diſtinguiſhes the Xs that are Ys from other Xs, then the change from the particular to the univerſal by means of the new reftriction impofed by the new name, is the expreffion of new knowledge.

The quantities in the conclufion are of two kinds. There are thofe which are brought in with the terms, and which continue in the conclufion fuch as they were introduced in the premifes: and there are thofe which depend on the union of the premifes, and which are what they are only in virtue of the joint exiftence of the premifes. For example, in $I_1A_1I_1$ we have 'fome Xs are Ys, but every Y is Z, therefore fome Xs are Zs': if we afk, what Xs are Zs, the anfwer is, thofe which are Ys, and no others, fo far as this conclufion affirms. But when we look at $O'A_1O'$ or 'fome Ys are not Xs, and every Y is Z; therefore fome Zs are not Xs: and if we then afk *what* Zs are not Xs; the anfwer is, that this quantity does not enter with Z, but depends upon the other premife, namely, upon the number of Ys which are not Xs. In a particular fyllogifm, let us call the quantity of the fubject in the conclufion *intrinfic* or *extrinfic* according as it is that of the premife which introduces that fubject, or of the other premife. Examination will fhow that in every particular fyllogifm which concludes in I_1 or I', in which both terms are particular, the quantities of the terms are, of the one intrinfic, of the other extrinfic: but that where the conclufion is in O_1 or O', either the quantity of the fubject is intrinfic and that of the contrary of the predicate extrinfic, or *vice verfâ*.

When the quantity of a particular term in the conclufion is intrinfic, the invention of a name will convert the fyllogifm into a univerfal. Thus $I_1A_1A_1$ or $XY+Y)Z=XZ$, if M be taken to reprefent all thofe Xs which are Ys, and nothing elfe, becomes $M)Y+Y)Z=M)Y$, of the form $A_1A_1A_1$. Again, $O'A_1O'$ or $Y:X+Y)Z=Z:X$, thrown into the form $x:y+z)y=x:z$, becomes $m.y+z)y=m.z$, of the form $E_1A_1E_1$, when the xs which are ys are diftinguifhed from the reft of the univerfe by the name m. There is nothing either illegitimate or uncommon in diftinguifhing by a peculiar name *certain fome* (or even *uncertain fome*, if *certainly always the fame fome*) of another name. Again, fince we know that every univerfal fyllogifm is reducible to the form $A_1A_1A_1$ by ufe of contraries, we have now reafon to know that there is no fundamental inference, of the kind treated in this chapter, which is any other than that in $A_1A_1A_1$, or, the *contained of the contained is contained.* And there is no better exercise than learning to read off each of the syllogisms, universal and particular, into this one form, by perception, and without use of rules. Take as an instance $X{:}Y+y.z=XZ$: what is the container, what is the contained, and what is the middle container of one and contained of the other. It is a parcel of Xs which are contained in y, all y in Z, and therefore *that parcel* of Xs in Z.

This general principle suggests a notation for all the complex, universal, and fundamental particular, syllogisms. If we abbreviate $X)Y+Y)Z=X)Z$ into XYZ), and if we denote by XYZ, without), that it is only a parcel of Xs (all or some, defined or undefined, but always the same), we have the following,

For $A_,A_,A_,$ read	XYZ) or zyx)	For $A'A'A'$ read	xyz) or ZYX)
— $O'A_,O'$ —	xYZ	— $O_,A'O_,$ —	Xyz
— $A_,O'O'$ —	Zyx	— $A'O_,O_,$ —	zYX
For $E'A_,E'$ read	xYZ) or zyX)	For $E_,A'E_,$ read	Xyz) or ZYx)
— $I_,A_,I_,$ —	XYZ	— $I'A'I'$ —	xyz
— $E'O'I_,$ —	ZyX	— $E_,O_,I'$ —	zYx
For $A_,E_,E_,$ read	XYz) or Zyx)	For $A'E'E'$ read	xyZ) or zYX)
— $O'E_,I'$ —	xYz	— $O_,E'I_,$ —	XyZ
— $A_,I'I'$ —	zyx	— $A'I_,I_,$ —	ZYX
For $E'E_,A'$ read	xYz) or ZyX)	For $E_,E'A_,$ read	XyZ) or zYx)
— $I_,E_,O_,$ —	XYz	— $I'E'O'$ —	xyZ
— $E'I'O_,$ —	zyX	— $E_,I_,O'$ —	ZYx

Here, using P,Q,R, as general terms, PQR) denotes that all Ps are Qs, and all Qs are Rs, whence all Ps are Rs: while PQR only denotes that there is a parcel of Ps among the Qs, and all Qs are among the Rs, whence that parcel of Ps is among the Rs.

The rules for the connection of these systems are not complicated, considering the extent of the cases they are to include. Let the letters A,E, &c. be called *proponents* ; X,Y,Z, *nominals:* and by the *order* of the nominals we always mean that X is first, &c. both in XYZ, and ZYX. The nominals being *direct* (X,Y,Z) and *contrary* (x,y,z), remember that, *first*,

An affirmative { *firſt* / *ſecond* / *third* } proponent denotes that the { *firſt* and ſecond / *ſecond* and third / *third* and firſt } nominals agree (are both direct or both contrary).

A negative { *firſt* / *ſecond* / *third* } proponent denotes that the { *firſt* and ſecond / *ſecond* and third / *third* and firſt } nominals differ (are one direct, one contrary).

Thus EIO muſt give Xyz or xYZ or zyX or ZYx
IEO muſt give XYz or xyZ or zYX or Zyx

Secondly, whether the middle term be Y or y depends only on the accent of the middle proponent: a *ſub*-accent gives Y, a *ſuper*-accent gives y. In the univerſal ſyllogiſm however, either gives either.

Thirdly, the XYZ ſyllogiſms are the particulars which begin with a particular: and the ZYX ſyllogiſms are the particulars which begin with a univerſal.

For example, required O,E'I,. Seeing the particular O, at the beginning, take the order XYZ, ſeeing the ſuperaccent in E' make it XyZ. Seeing the negative O,, let the exiſting diſagreement of the firſt and ſecond nominals continue: and the ſame of the ſecond and third from the negative E,. Conſequently XyZ is the ſyllogiſm expreſſed in nominals. Or the rationale of the inference in O,E'I, is that a parcel of Xs are among the Zs becauſe among the ys which are all among the Zs.

Again, required the nominal mode of expreſſing E'I'O,. Seeing the univerſal E at the beginning, write down ZYX; for the ſuperaccent in I', write down ZyX; for the negative in E', continue yX; for the affirmative in I', write zy: hence zyX is the nominal form of E'I'O,.

Required the proponent mode of expreſſing xYz. Here xY, Yz, ſhow us that the premiſes are negative forms, and the direction of the order x, Y, z, that the firſt premiſe is particular. Then OE are the premiſes, and I the concluſion. And Y tells us that the middle proponent has a ſubaccent. Whence OE,I is, ſo far as it goes, the proponent expreſſion. And, by the laws of form, the other accents muſt be as in O'E,I', ſince the ſyllogiſm follows the direct rule (page 93).

Required the proponent mode of expreſſing ZYx. Here we note in ſucceſſion—univerſal commencement—firſt premiſe negative—ſecond, affirmative—middle accent ſub. This gives EI,O of the inverſe rule, or E,I,O'.

Required the proponent notation for the univerſal xYZ) or zyX). We ſee at once EA,E, or E'A,E'.

The concomitants of a univerſal are found by changing the firſt nominal into the contrary, in each of the forms, and throwing away the ſign of univerſality [)]. Thus the concomitants of XyZ) or zYx) are xyZ and ZYx.

The weakened forms of a univerſal are found by merely throwing away the ſymbol of univerſality [)] from the two forms of the univerſal. Thus the weakened forms of XYZ) which is alſo zyx) are XYZ and zyx.

But we have not yet reached the climax of ſymbolic ſimplicity in the mere repreſentation of ſyllogiſms. An algebraiſt would ſay that the ſtructure of the inference, as now conſidered, does not depend upon the names; but only upon their reference to the names in the fundamental form XYZ). He would therefore propoſe a ſimple ſymbol to repreſent *letting alone*, and another to repreſent *changing into the contrary*. Theſe, with a ſign of complete univerſality, and another of inverſion of order, are all that he would find neceſſary. Let 0 and 1 ſignify letting alone and changing into the contrary: let the terminal parentheſis denote complete univerſality, as before, and let inverſion of order be denoted by a negative ſign prefixed. Thus XYZ or I,A,I,, would be denoted by 000; Zyx or A,O'O' by—011; A,E,E, or XYz) by 001) or its equivalent—011. Thus—011 tells us that ſome of the Zs are ys, all the ys are xs, whence ſome of the Zs are xs. To write its proponent form, obſerve that — inſtructs us to write a univerſal firſt; 11 to make it affirmative; 1 in the middle to ſuperaccent the middle propoſition; 01 to make the ſecond premiſe negative. We have then A,O'O' or X)Y+Z:Y=Z:X which is Zy+y)x=Zx, as aſſerted.

All that relates to univerſals in the preceding, applies to the complex ſyllogiſms. Let a couple of parentheſes imply a complex ſyllogiſm: thus D,D,D, may be (XYZ) or (000). Then in (010) or (XyZ), we are to ſee that X is a ſubidentical of y, and y of Z, whence X is the ſame of Z. But Xy and yZ warn us to write *contraries* for the firſt and ſecond premiſes and y to ſuperaccent the middle letter: whence $C_{\prime}C'D_{\prime}$ is the ſyllogiſm expreſſed by the names XYZ. The equivalent forms—(101) and (zYx) expreſs it by ſaying that z is a ſubidentical of Y and Y of x, whence z is a ſubidentical of x.

I now look at the ſtrengthened particular ſyllogiſms. All inference which is fundamental, that is, which will come from nothing weaker than the premiſes given, has been reduced to the one eaſy caſe of 'the contained of the contained is contained.' The ſtrengthened particular, the type of which is $A'A_{\prime}I_{\prime}$, obeying the inverſe rule of formation, and written at more length in Y)X + Y)Z=XZ, may be ſtated thus 'all names are common as to what they contain in common.' If we denote this ſtrengthened ſyllogiſm by XYZ|, a ſymbol intended to imply ſomething between XYZ and XYZ) in the amounts of quantity introduced, we ſhall find that the eight ſtrengthened ſyllogiſms muſt be repreſented by

$A'A_{\prime}I_{\prime}$ = XYZ|	$A_{\prime}A'I'$ = xyz|
$A'E_{\prime}O_{\prime}$ = XYz|	$A_{\prime}E'O'$ = xyZ|
$E'A'O_{\prime}$ = Xyz|	$E_{\prime}A_{\prime}O'$ = xYZ|
$E'E'I_{\prime}$ = XyZ|	$E_{\prime}E_{\prime}I'$ = xYz|

The rules of connexion are preciſely thoſe for the particular ſyllogiſms: and inverſion is abſolutely ineffective. Thus XYZ| =ZYX|.

A few words will ſerve to diſpoſe of the *mixed complex ſyllogiſms* in which a complex premiſe is combined with a ſimple one, univerſal or particular. Firſt, when a complex and a univerſal are premiſed, and ſigns and accents are as in the *direct rule* (page 92), the concluſion is as it would be if the A were heightened into D, or E into C. Thus $E_{\prime}D'$ gives $C_{\prime}$, the ſame as $C_{\prime}D'$. For $E_{\prime}$ is C or $C_{\prime}$, and both CD' and $C_{\prime}D'$ give $C_{\prime}$, but with different quantities. But if the premiſes be conſtructed on the inverſe rule, there is no more inference than can be obtained when the complex premiſe is lowered into a univerſal: or we have only a ſtrengthened particular. Thus in $D_{\prime}E'$ or ($A_{\prime}$ + O')E', $A_{\prime}E'$ gives the ſtrengthened particular $A_{\prime}E'O'$, and O'E' is inconcluſive. And when the complex premiſe is combined with a particular, we have only what would follow if the complex premiſe were lowered into a univerſal. Thus $D_{\prime}I'$, or

$(A_{,} + O')I'$ can only give $A_{,}I'I'$; and $D'I'$ or $(A' + O_{,})I'$ gives no conclusion, for $A'I'$ is inconclusive.

The classification of opponent forms may be thus treated. We know that opponent forms of AEE, for instance, be it $A_{,}E_{,}E_{,}$ or $A'E'E'$, must be IEO and AII. Now whether $A_{,}E_{,}E_{,}$ shall have $I_{,}E_{,}O_{,}$ or $E_{,}I_{,}O'$, whether $A'I_{,}I_{,}$ or $I_{,}A_{,}I_{,}$, depends upon the introduction of a new and arbitrary notion of the order to be adopted. Our first syllogism being described by XY, YZ, XZ, the opponent which ends in the contradiction of the first premise is in XZ, YZ, XY; which, keeping Z middle, is either to be described with reference to XZ, ZY, XY, or to YZ, ZX, YX. Now in adopting the first of these three orders, there is nothing which compels us therefore to prefer the second to the third, or *vice versâ.*

The effect of the change of order which consists in the interchange of Z and X is as follows. The premises change places; A and O with altered accents, altered also in the conclusion, E and I with unaltered accents. Thus $A_{,}I'I'$ becomes $I'A'I'$; $E'O'I_{,}$ becomes $O_{,}E'I_{,}$. Accordingly, it is matter of new arrangement whether for instance, $I_{,}E_{,}O_{,}$ or $E_{,}I_{,}O'$ shall be called the opponent of $A_{,}E_{,}E_{,}$; and I prefer to give the name to both. The consequence is, the following distribution of opponents;—

AA	AO	OA	AE	EA	AI	IA	EI	IE.
EE	EO	OE						

The three sets represent letters combined in representation of premises: the first two containing six syllogisms each, the third twelve. The third must be divided into two sets of six each, in one of which the subaccents are in greater number, in the other the superaccents. There are then four sets in all. Pick any two out of a set, which only differ in change of order: these two have the same opponent forms, namely, the other four of the set. For instance, $A'I_{,}I_{,}$ and $I_{,}A'I_{,}$, in which subaccents predominate. Take AE, EA, EI, IE, and complete syllogisms in such manner as to make subaccents predominate: giving $A_{,}E_{,}E_{,}$, $E_{,}A'E_{,}$, $E_{,}I_{,}O'$ $I_{,}E_{,}O_{,}$. The last four are the opponents of the first two.

In the set of strengthened particulars the opponent forms will be found to be universals weakened in the conclusion without being weakened in the premiſes. Thus $A_{\prime}A'I'$ has $A'E'O'$ for one of its opponents: but $A'E'$ may produce the univerſal concluſion E' as well as its weaker form O'.

Some readers, particularly thoſe who have a tincture of algebra, are more helped by ſymbolic notation than by language: with others it is the converſe. To ſuit the latter, obſerve that the language of page 78 may eaſily be adapted to ſimple ſyllogiſms. Thus $A_{\prime}$ being ſubaffirmation, $I_{\prime}$ may be ſome ſubaffirmation, O' may be ſome ſupernegation; and ſo on. Thus inſtead of $E'I'O_{\prime}$ we may ſay that 'ſupernegation of ſome ſuperaffirmation gives ſome ſubnegation.' Practice in this language would make the phraſe ſuggeſt ſomething more than the notation it is derived from. The phraſe refers to Z: there is a term partially ſuperaffirmed of Z, namely Y; and a complete ſubnegative of Y, namely X. The partial ſubaffirmation declares ſome things neither Y nor Z; the complete ſupernegation declares that whatever is not Y is X. Conſequently there are ſome Xs which are not Zs: or X is a partial ſubnegative of Z. This ſubject will be reſumed.

In what precedes are two views of the deduction of all the varieties of ſyllogiſm. The firſt, taking the complex ſyllogiſm as the ſource, connects the ſtrengthened ſyllogiſms and the particular ones with the univerſals, and thus in fact reduces every thing to the conſtituents of $D_{\prime}D_{\prime}D_{\prime}$ or $DD_{\prime}D_{\prime}$. The ſecond proceeds from $A_{\prime}A_{\prime}A_{\prime}$, $A'A_{\prime}I_{\prime}$, $A_{\prime}I'I'$, and $I_{\prime}A_{\prime}I_{\prime}$, and forms the claſſes of univerſal, ſtrengthened, and particular, ſyllogiſms by ſubſtituting contraries in every way in which it can be done. Theſe two ſyſtems have cloſe connexion, but not ſo cloſe as might perhaps be thought: for $I_{\prime}A_{\prime}I_{\prime}$ is not one of thoſe which are connected with $A_{\prime}A_{\prime}A_{\prime}$ in the formation of a complex ſyllogiſm.

The two new views which I now proceed to give are alſo cloſely connected, and different from the former ones, in which we held it equally admiſſible to refer one of the concluding terms to the middle, as in X)Y, or the middle to one of the concluding terms, as in Y)X. But now I aſk whether it be not poſſible ſo to conſtruct the ſyſtem, that we may firſt lay down the middle term and its contrary, as conſtituting the univerſe of the ſyllogiſm, and then complete the premiſes and their concluſion, by properly laying down the concluding terms in their places. We may ſucceed, if, in the firſt inſtance, we conſider none but convertible propoſitions. And this we can do; for univerſal excluſion and particular incluſion comprehend all aſſertion. Thus univerſal incluſion is only univerſal excluſion from the contrary, and particular excluſion is only particular incluſion in the contrary.

Setting out then with the middle term and its contrary, and reſtricting ourſelves to E and I, let E ſignify (univerſal) excluſion from the middle term, and e from its contrary; let I ſignify (particular) incluſion in the middle term, and i in its contrary. Chooſing a pair of concluding terms, we reject II, Ii, and ii on grounds already demonſtrated, and very eaſily ſeen in this view, and proceed to conſider Ee, EE and ee, EI and ei, Ei and eI.

Ee. From this a univerſal concluſion muſt follow. If one term be completely excluded from the middle and the other from its contrary, the terms are completely excluded each from the other. The fundamental forms are,

$$E_{\iota}A'E_{\iota},\ X.Y+Z.y=X.Z\ ;\ A_{\iota}E_{\iota}E_{\iota},\ X.y+Z.Y=X.Z$$

and by uſe of XZ, Xz, xZ, xz, we thus bring out the eight univerſal ſyllogiſms.

EE and ee. From theſe a particular incluſion muſt follow. Excluſion of both terms from a third, gives partial incluſion of their contraries in each other: for all that third term belongs to the contraries of the other two. The fundamental forms are,

$$E_{\iota}E_{\iota}I',\ X:Y+Z.Y=xz\ ;\ A_{\iota}A'I'\ X.y+Z.y=xz$$

from which, as before, the eight ſtrengthened ſyllogiſms are deduced.

EI and ei. From theſe a particular incluſion muſt follow. The excluſion of one term from a third, and the incluſion of part of a ſecond term in that third, tell us that part of the particularized term is in the contrary of the univerſalized term. The fundamental forms are,

$$E_1 I_1 O', \; X.Y + ZY = Zx \;;\; A_1 O' O', \; X.y + Zy = Zx$$
$$I_1 E_1 O_1, \; XY + Z.Y = Xz \;;\; O_1 A' O_1, \; Xy + Z.y = Xz$$

from which the sixteen particular syllogisms are deduced.

Ei and eI. From these no conclusion can be drawn. All that is signified is that one concluding term is wholly excluded from a third, and the second partially excluded (or included in the contrary).

It thus appears that a syllogism with one particular premise is valid when the premises reduced to convertible forms, show the middle term in both or the contrary of it in both; otherwise, invalid. Also, that the conclusion in its convertible form, takes directly from the particular premise and contrariwise from the universal.

It also appears that a syllogism with both premises universal is always valid; with a universal conclusion when the premises (made convertible) show one the middle term and the other its contrary; with a particular conclusion when both show the middle term or both its contrary. And the convertible form of the conclusion takes directly from both in the first case, and contrariwise from both in the second.

The other view which I here propose is really a different mode of looking at that just given. By the time we have made every name carry its contrary, as a matter of course, we become prepared to take the following view of the nature of a proposition. A name by itself is a sound or a symbol: its relation to things (be they objects or ideas) is twofold. There may be *in rerum natura* that to which the name applies, or there may not. I do not here speak of how many things there may be to which a name applies: it is not essential to know whether they be more or fewer, either absolutely or relatively. The introduction of *contraries* may be made the expulsion of *quantity*. With reference to application, then, let a name be called *possible* or *impossible* according as the thing to which it applies can be found or not.

A name may be compounded of others; the compound name being that of everything to which all the components apply. Thus *wild animal* is the name of all things to which both the names *wild* and *animal* apply. To call this compound name impoffible is to fay that there is not fuch a thing as a wild animal: to call it poffible is to fay that there is fuch a thing.

X and Y being two names, the compound name may be reprefented by XY when poffible, and by XY) when impoffible. This does not alter the meaning of our fymbol XY, as hitherto ufed: as yet it has been 'there are Xs which are Ys' and now it is 'XY, the name of that which is both X and Y, is the name of fome thing or things;' and thefe two are the fame in meaning, fo far as their ufe in inference is concerned. Nor need XY), as juft defined, be treated as a departure from, otherwife than as an extenfion of, the ufe of X)Y. In X)Y, we affert that X is fomething, namely Y: in X) we affert that X is *nothing whatever.* The proper notation, however, for indicating that the name X has no application, is X)u, u being the contrary of U, which laft includes everything in the univerfe fpoken of; fo that u may denote nonexiftence.

The propofition 'Every X is Y' afferts that Xy is the name of nothing, or X)Y=Xy). Similarly 'No X is Y' afferts that XY is the name of nothing, or X.Y=XY). But 'Some Xs are Ys' and 'Some Xs are not Ys' merely affert the poffibility of the names XY and Xy.

A fyllogifm, then, is the affertion that from the poffibility or impoffibility of the names produced by compounding X or x, Z or z, each with Y or y, may be inferred the poffibility or impoffibility of a name compounded of X or x with Z or z. The rules of the laft fyftem are now fo eafily changed into the language of the prefent one, that it is hardly worth while to ftate more than one for example. Thus, if X compounded with Y, and Z compounded with y, both give impoffible names, then X compounded with Z gives an impoffible name. This is XY)+Zy)=ZX) or X.Y+Z.y=Z.X, or $E_{I}A'E_{I}$

The view here taken of compound names will be extended in the next chapter.

CHAPTER VI.

On the Syllogifm.

WHEN the premifes of a fyllogifm are true, the conclufion is alfo true, and when the conclufion is falfe, one or both of the premifes are falfe. There are two kinds of modifications which it may be ufeful to confider: thofe which concern the entrance of the propofition into the argument; and thofe which affect the connexion of the fubject and predicate.

As to the propofition itfelf, it may be true or falfe abfolutely, or it may have any degree of truth, credibility, or probability. This relation will be hereafter confidered; and, according to the principles of Chapter IX. fo far as the propofition is probable it is credible, and fo far as it is credible, it is true. But as to other modes of looking at the fyllogifm, are we entitled to fay that every thing which can be announced as to the premifes may be announced in the fame fenfe as to the conclufion? The anfwer is, that we cannot make fuch announcement abfolutely; but of the premifes *as derived from that conclufion* we can make it. In what manner foever two premifes are applicable, their conclufion as from thofe premifes is alfo applicable: becaufe the conclufion is in the premifes. For inftance, in the fyllogifm 'all men are trees, all trees are rational, therefore all men are rational,' the premifes are abfurd and falfe, and the conclufion taken independently is rational and true: but that conclufion, as from thofe premifes, is as abfurd as the premifes themfelves. Again, in 'all pirates are convicted, all convicts are punifhed, therefore all pirates are punifhed,' the premifes are *defirable*, and fo is the conclufion with thofe premifes. But the conclufion is not defirable in itfelf: as that pirates fhould be punifhed with or without trial. Neither may we fay 'X ought to be Y and Y ought to be Z, therefore X ought to be Z' except in this manner, that we affirm X ought to be Z in a particular way. We may not even fay that when 'X ought to be Y, and Y *is* Z' it follows that 'X ought to be Z,' for it may be that Y ought not to be Z. Thus a royalift, in 1655, would fay that the hundred excluded members of Cromwell's parliament ought to be allowed to take their ſeats, and alſo that all who took any ſeats in that parliament were rebels; but he would not infer that the hundred members ought to be rebels. There is nothing which, being the property of the premiſes, is neceſſarily the independent property of the concluſion, except *abſolute truth*. It ſhould be noted that in common language and writing, the uſual meaning of concluſions is that they are ſtated as of their premiſes and to ſtand or fall with them, even as to truth. Though a concluſion may be true when its premiſes are falſe, the proponent does not mean, for the moſt part, to claim more than his premiſes will give, nor that any thing ſhould ſtand longer than the premiſes ſtand.

Next, we are not to argue from what we may ſay of a propoſition to what we may ſay of the inſtances it contains, except as to what concerns the truth of thoſe inſtances, or elſe to what concerns the inſtances as parts of a whole. If I ſay 'Every X is Y' I aſſert, no doubt, of each X independently of the reſt: that is, the truth of 'Every X is Y' involves the truth of 'this X is Y.' But if, to take ſomething elſe, I maintain 'Every X is Y' to be a deſirable rule, I do not therefore aſſert 'this X is Y' to be a deſirable caſe, except upon an implied neceſſity that there ſhould be a rule. And if I ſay that 'every X is Y' is unintelligible, I do not ſay that 'this X is Y' is unintelligible; and ſo on. Thus, where there muſt be a rule, as in law, 'every man's houſe is his caſtle' is deſirable, becauſe there is but one alternative 'no man's houſe, &c.' But the propoſition, by itſelf, may not be deſirable as to the inſtance of a generally reputed thief or receiver.

There is one caſe, however, in which a term cannot be applied to the general propoſition, unleſs it can be applied in a higher degree to the inſtances. The propoſition 'Every X is Y' cannot be announced as of any degree of probability, unleſs each inſtance has a much higher degree of probability. If μ, ν, ρ, &c. be the probabilities of the ſeveral inſtances, ſuppoſed independent, that of the propoſition (Chapter IX.) is $\mu\nu\rho\ldots$ which product muſt be leſs than that of any one of the fractions of which it is formed.

I now come to the conſideration of circumſtances which modify the internal ſtructure of the premiſes themſelves. And firſt of conditions.

A *conditional* propofition is only a grammatical variation of the ordinary one; as in 'If it be X, then it is Y.' The common form of this, 'Every X is Y,' is called *categorical*, or *predicative*. Of the two forms, categorical and conditional, either may always be reduced to the other; as follows,

'Every X is Y' or 'If X, then it is Y'
'No X is Y' or 'If X, then it is not Y'

The particular propofitions might be given conditionally in various ways, but the transformation is not fo common. Thus 'fome Xs are Ys' might be 'if X, then it may be Y' or 'if X, then Y muft not therefore be denied of it,' &c.

Of the two common fubject-matters of names, ideas and propofitions, it is moft common to apply the categorical form to the firft, and the conditional form to the fecond: in truth we might call the conditional form a grammatical convenience for the expreffion of dependence of propofitions on one another, and of names which require complicated forms of expreffion. Thus in pages 2 and 3, the conditional forms, containing *if*, are more fimple than the correfponding categorical forms.

A condition may be either *neceffary*, or *fufficient*, or both. A neceffary condition is that without which the thing cannot be; a fufficient condition is one with which the thing muft be. In pages 73, 74, I have fufficiently pointed out the completenefs of the connexion between the conditional and the categorical forms. In any one cafe the fufficient muft contain all that is neceffary, and may contain more.

After what is faid in page 23, it is not neceffary to dwell on the reduction of a conditional* fyllogifm to a categorical one. The premifes contain the conclufion: whatever gives us the premifes, gives us the conclufion. But I think that the reduction of conditional to categorical forms, though juft, and, for inference, complete, is not the reprefentation of the whole of what paffes in our minds.

As an example of what I mean, look forward to the numerical fyftem of Chapter VIII. Precedent to all propofitions,

* Wallis, as far as I know, was the firft who afferted that all fyllogifms are, or can be made, categorical. He did this in the fecond thefis attached to his logic, headed *Syllogifmi Hypothetici, aliique Compofiti, referendi funt omnes ad Ariftotelicos Categoricorum Modos.*

there are the numerical conditions which prefcribe the limits of the univerfe under confideration. Say there are 250 inftances in that univerfe: this is the firft condition. Of thefe 100 are Xs and 200 are Ys; giving a fecond and third condition. If we take a propofition, as 20XY, and afk whether it be fpurious or not, we have reference to the three conditions underftood. But this is not neceffary: for it would be poffible categorically to exprefs thefe conditions by '20Xs out of 100 in a univerfe of 250 inftances containing 200Ys are to be found among thofe 200 Ys? It is of courfe the rule of brevity not to drag about thefe conditions with every propofition which is employed, but rather to ftate them once for all. There is however fomething more. The conditions are a reftriction upon the arguments intended to be introduced, and a reftriction throughout. The attachment of them to each individual propofition does not exprefs this: if they be feen in twenty confecutive propofitions, there is no more than a prefumption that they are to be feen in the twenty-firft. It is better that the limits allowed fhould be marked out by one boundary than that the feveral arguments fhould each have a defcription of the boundary to itfelf.

Juft as a univerfe of names is defined by fpecifying one or more names to conftitute collectively the *fummum genus*, or *univerfe*, fo one of propofitions may be defined by ftating propofitions which are to be true, or which are not to be contradicted, as the cafe may be. Thefe propofitions may be conditions preceding all, or fome only, of the premifes which are ufed in argument; or fome may precede fome, and others others. In analyfing arguments, it would be found that many propofitions which enter as premifes, enter each with a condition underftood, and well underftood, to be granted. Whatever the conditions may be, fo long as the confequent propofitions act logically together to produce the final refult, then that fame refult depends at laft only on the conditions, and muft be affirmed when the conditions, and their connexion with their confequents, are affirmed. But then it muft be underftood that the refult alfo ftands upon the conditions, and may fall with them. Let us now examine the common fyllogifm, and fee whether there be any preceding conditions, on which the refult depends.

On looking into any writer on logic, we fhall fee that *exiftence* is claimed for the fignifications of all the names. Never, in the ftatement of a propofition, do we find any room left for the alternative, *fuppofe there fhould be no fuch things*. Exiftence as objects, or exiftence as ideas, is tacitly claimed for the terms of every fyllogifm. The exiftence of an idea we muft grant whenever it is diftinctly apprehended, and (therefore) not felf-contradictory: we cannot for inftance admit the notion of a lamp which is both metal and not metal; but, as an idea, we are at liberty to figure to ourfelves fuch a lamp as that with which Aladdin made his fortune. An attempt at a felf-contradicting idea is no idea; we have not that apprehenfion of it in which an idea confifts: but in no other way can we fay that the attempt to produce an idea fails. It may then be more convenient here to dwell on *objective* definition of terms, as more eafily conceived with relation to exiftence and non-exiftence. Accordingly, let us take the propofitions X)Y and X.Y, of the character of which the particulars muft partake, as to the point before us. By the meaning of y, in relation to Y, it follows that every thing is either Y or y: if we fay that Y does not exift, then every thing is y. If then X exift, and Y do not, the propofition X)Y, or X.y is falfe, and X)y, or X.Y is true. If neither X nor Y exift, I will not fo far imitate fome of the queftions of the fchools as to attempt to fettle what nonexifting things agree or difagree. If Y exift, but not X, then y)x is certainly true, but not thence X)Y, for when x is, as here, the whole univerfe, the proof of y)x=X)Y fails to prefent intelligible ideas, that is, fails to be a proof. But Y)x or Y.X is true.

If all my readers were mathematicians, I might purfue thefe extreme cafes, as having intereft on account of their analogy with the extreme cafes which the entrance of zero and of infinite magnitude oblige him to confider. But as thofe who are not mathematicians would not be interefted in the analogy, and thofe who are can purfue the fubject for themfelves, I will go on to fay that the preceding order is not the natural one. We cannot, to ufeful purpofe, laying down the truth of the propofition, *firft*, then proceed to enquire how the non-exiftence of one or both terms affects the propofition. The exiftence of the terms muft be firft fettled, and then the truth or falfehood of the propofition. The affirmative propofition requires the exiftence of both terms:

the negative propofition, of one; being neceffarily true if the other term do not exift, and depending upon the matter, as ufual, if it do exift.

Let us make the exiftence of the terms to be preceding conditions of the propofitions. The fyllogifm $A_1A_1A_1$ is then as follows,

	If X and Y both exift,	Every X is Y
	If Z alfo exift	Every Y is Z
Therefore	If X, Y, Z all exift	Every X is Z.

As to the concluding terms, X and Z, they remain, as it were, to tell their own ftory. Whatever conditions accompany their introduction unto the premifes, thefe fame conditions may be conceived to accompany them in the conclufion. But the middle term difappears: and, not fhowing itfelf in the conclufion, the conditions which accompany it muft be exprefsly preferved. The conclufion then is 'every X is Z, if Y exift' which may be thrown into theform of a dilemma, 'Either every X is Z, or Y does not exift'.

But taking X and Z to exift, let us confider the following fyllogifm, *as it appears to be*,

Every X is (Y, if Y exift)
Every (Y, if Y exift) is Z
Therefore Every X is Z.

If this be not a valid fyllogifm, what *expreffed* law of the ordinary treatifes does it break? The middle term, a curious one, is ftrictly middle: but there is no rule for excluding middle terms of a certain degree of fingularity. That it does break, and very obvioufly, an implied rule, I grant. And as to this work, the rule laid down in Chapter III. is broken in its fecond condition (page 50). The two ufes of the word *is* do not amount to one fuch ufe as is made in the conclufion. That X is (conditionally) Y which is (on the fame condition) Z, gives that X is (on the fame condition) Z. Accordingly, the abfolute conclufion is only true upon fuch conditions as give the middle term abfolute exiftence.

But it muft be particularly noted that it is enough if this exiſtence be given to the middle term by the fulfilment of the conditions which precede the entrance of one of the concluding terms. The condition of the act of inference is, that the compariſon muſt be really made, if the terms to be compared with the middle term really exiſt, or, which is the ſame, if the conditions under which they are to enter be ſatisfied. The other terms being ready, there muſt *then* be a real middle term: and there will be, if the mere entrance of one of the concluding terms be proof of the exiſtence of a middle term; while, if the other terms cannot be brought in, from nonexiſtence, there is no occaſion to inquire about a middle term, for it is otherwiſe known that the compariſon cannot be completed. I will take two concrete inſtances, in the firſt of which one of the concluding terms, if exiſting, is held to furniſh a middle term as real as itſelf, and in the ſecond of which no ſuch ſuppoſition occurs. Of courſe I have nothing here to do with the truth of the premiſes.

Philip Francis, (if the author of Junius), was an accuſer whoſe ſilence was ſimultaneous with a government appointment: an accuſer &c. reflects diſgrace upon the government (if they knew that their nominee *was* the accuſer): therefore Francis (if &c.) reflects diſgrace upon government (if &c.).

Homer (if there were ſuch a perſon) was a perfect poet (if ever there were one): a perfect poet (if &c.) is faultleſs in morals: therefore Homer (if &c.) was faultleſs in morals.

The firſt inference is good, even though we grant that our only poſſible mode of knowing of the exiſtence of an accuſer &c. is by eſtabliſhing that Francis was Junius: it is even good againſt one who ſhould aſſert that the accuſer &c. is a contradiction in terms in every actual and imaginable caſe except that of Junius.

In the ſecond caſe, we put it that the man Homer (if he ever exiſted; ſome critics having contended for the contrary) was a perfect poet, if ever there were one. There may never have been one; and then Homer (exiſtent or nonexiſtent) was not a perfect poet. There is no condition here, which being fulfilled, is held to amount to an aſſertion that the middle term muſt have exiſted: but the condition of the exiſtence of the middle term is independent. Accordingly, the ſecond inference is not good: it ſhould be Homer (if &c.) was a perfect poet, if ever there were one: that is, or elſe there never was a perfect poet.

These points refer to the matter of a syllogism, and not to the form; or rather, perhaps, hold a kind of intermediate relation.

There is another process which is often necessary, in the formation of the premises of a syllogism, involving a transformation which is neither done by syllogism, nor immediately reducible to it. It is the substitution, in a compound phrase, of the name of the genus for that of the species, when the use of the name is particular. For example, 'man is animal, therefore the head of a man is the head of an animal' is inference, but not syllogism. And it is not mere substitution of identity, as would be 'the head of a man is the head of a *rational animal*' but a substitution of a larger term in a particular sense.

Perhaps some readers may think they can reduce the above to a syllogism. If *man* and *head* were connected in a manner which could be made subject and predicate, something of the sort might be done, but in appearance only. For example, 'Every man is an animal, therefore he who kills a man kills an animal.' It may be said that this is equivalent to a statement that in 'Every man is an animal; some one kills a man; therefore some one kills an animal,' the first premise, and the second premise *conditionally*, involve the conclusion as *conditionally*. This I admit: but the last is not a syllogism: and involves the very difficulty in question. 'Every man *is* an animal; some one *is* the killer of a man': here is no middle term. To bring the first premise into 'Every killer of a man is the killer of an animal' is just the thing wanted. By the principles of chapter III, undoubtedly the copula *is* might in certain inferences be combined with the copula *kills*, or with any verb. But so simple a case as the preceding is not the whole difficulty. If any one should think he can syllogize as to the instances I have yet given, let him try the following. 'Certain *men*, upon the report of certain other *men* to a third set of *men*, put a fourth set of *men* at variance with a fifth set of *men*.' Now every man is an animal: and therefore 'Certain *animals*, upon the report of certain other *animals*, &c.' Let the first description be turned into the second, by any number of syllogisms, and by help of 'Every man is an animal.'

The truth is, that in the formation of premises, as well as in their use, there is a postulate which is constantly applied, and therefore of course constantly demanded. And it should be demanded openly. It contains the *dictum de omni et nullo* (see the next chapter), and it is as follows. For every term ufed univerfally *lefs* may be fubftituted, and for every term ufed particularly, *more.* The fpecies may take the place of the genus, when all the genus is fpoken of: the genus may take the place of the fpecies when fome of the fpecies is mentioned, or the genus, ufed particularly, may take the place of the fpecies ufed univerfally. Not only in fyllogifms, but in all the ramifications of the defcription of a complex term. Thus for 'men who are not Europeans' may be fubftituted 'animals who are not Englifh.' If this poftulate be applied to the unftrengthened forms of the Ariftotelian Syllogifm, (page 17) it will be feen that all which contain A are immediate applications of it, and all the others eafily derived.

I now pafs to the confideration of the invention of names, and of the diftinctions which are made to exift for the want of it.

Any one may invent a name, that is, may choofe a found or fymbol which is to apply to any clafs of ideas or of objects. The clafs fhould, no doubt, be well defined : but fmall caution is here neceffary, for invented words are generally much more definite than thofe which have undergone public ufage. They come from the coiner's hand as fharp at the edge as a new halfpenny : and in procefs of time we look in vain for any edge at all. The right of invention being unlimited, and the actual ftock having been got together without any uniform rule of formation, *there can be no reafon why we fhould admit any diftinction which can be abrogated by the invention of a name*, fo far as inference is concerned. I do not difpute that the modes of fupplying the want of names may be of importance in many points of view : what I deny is, that they create any peculiar modes of inference.

The invention of names muft either be by actually pointing out objects named, or by defcription in terms of other names. With the former mode of invention, as 'let this, that, &c (fhowing them) be called X' we can have nothing to do. As to the latter, we may make a fymbolic defcription of the procefs by joining together the names to be ufed, with a fymbol indicative of the mode of ufing them, in extenfion of the fyftem in page 106. Thus, P, Q, R, being certain names, if we wifh to give a name to everything which is all three, we may join them thus, PQR : if we wifh to give a name to every thing which is either of the three (one or more of them) we may write P,Q,R : if we want to fignify any thing that is either both P and Q, or R, we have

PQ,R. The contrary of PQR is p,q,r; that of P,Q,R is pqr; that of PQ,R is (p,q)r: in contraries, conjunction and disjunction change places. This notation would enable us to expreſs any complication of the preceding conditions: thus, to name that which is one and one only of the three, we have Pqr, Qrp, Rpq; for that which is two and two only, PQr, QRp, RPq. Thus, XY includes the inſtances common to X and Y; but X,Y includes all X and all Y: accordingly X,Y is a wider term than XY, except when X and Y are identical. As in page 106, XY, the term, ſuppoſed to exiſt, is XY, the propoſition of chapter IV; if we wiſh to diſtinguiſh, we may make X-Y the term, and XY the propoſition, the hyphen having its common grammatical uſe. Thus, X-Y P-Q tells us the ſame as XYP-Q, both meaning, *for inference*, no more than that there exiſt objects or ideas to which the four names are applicable. But the firſt tells it thus, ſome XYs are PQs; and the ſecond thus, ſome things are Xs, Ys, and PQs.

With reſpect to this and other caſes of notation, repulſive as they may appear, the reader who refuſes them is in one of two circumſtances. Either he wants to give his aſſent or diſſent to what is ſaid of the form by means of the matter, which is eaſing the difficulty by avoiding it, and ſtepping out of logic: or elſe he deſires to have it in a ſhape in which he may get that moſt futile of all acquiſitions, called a *general idea*,* which is truly, to uſe the contrary adjective term as colloquially, *nothing particular*, a whole without parts.

If the difficulty of abſtract aſſertion be to be got over, the eaſieſt way is by firſt conquering that of abſtract expreſſion, to the extent of becoming able to make a little uſe of it.

Suppoſe we aſk for the alternative of the following ſuppoſition, 'Both X, and either P, or Q and one of the two R or S.' This is no impoſſible complication: for inſtance, 'He was rich, and if not abſolutely mad, was weakneſs itſelf ſubjected either to bad advice or to moſt unfavourable circumſtances.' The repreſentation of the complex term is X {P, Q(R,S)}; of the contrary,

* "Je vous avoue, dit . . . ., que j'ai cru en deviner quelque choſe, et que je n'ai pas entendu le reſte. L'abbé de . . . . a ce diſcours, fit réflexion que c'était ainſi que lui-même avait toujours lu, et que la plupart des hommes ne liſaient guere autrement."

x, p(q,rs) or x,pq,prs. If not the above, he was either not rich, or both not mad and not very weak, or neither mad nor badly advifed, nor unfavourably circumftanced.

When a name thus formed, whether conjunctively or disjunctively, enters a fimple inference, it gives rife to what have been called the *copulative* fyllogifm, the *disjunctive* fyllogifm, and the *dilemma*. The two laft are not well diftinguifhed by their definitions as given: the disjunctive fyllogifm feems to be that in which *names* are confidered disjunctively, the dilemma that in which *propofitions* are fo ufed. But a propofition entering as part of a propofition, enters merely as a name, the predicates being ufually only *true* or *falfe*, or fome equivalent terms. A propofition may only enter for its matter, or it may enter in fuch a way that its truth is the matter: in this laft cafe it is only as a name that it is the fubject of inference. Thus, 'It is true that he was fired at' is 'the affertion (that he was fired at) is a true affertion.' I believe the beft way would be to apply the term *disjunctive* argument fo as to include the dilemma, marking by the latter word (as a term rather of rhetoric than of logic) every argument in which the disjunctive propofition is meant to be a difficulty for the opponent on every cafe, or *horn*, of it.

Whatever has right to the name P, and alfo to the name Q, has right to the compound name PQ. This is an abfolute identity, for by the name PQ we fignify nothing but what has right to both names. According X)P + X)Q = X)PQ is not a fyllogifm, nor even an inference, but only the affertion of our right to ufe at our pleafure either one of two ways of faying the fame thing inftead of the other. But can we not effect the reduction fyllogiftically? Let Y be identical with PQ; we have then PQ)Y and Y)PQ, and alfo Y)P and Y)Q. Add to thefe X)P and X)Q, and we have all the propofitions afferted. But we cannot deduce from them alone X)Y, the refult wanted, by any fyllogiftic combination of the fix. Nor muft it be thought furprifing that we cannot, by a train of argument, arrive at demonftration of it being allowable to give to anything which has right to two names, a third name invented exprefsly to fignify that which has fuch right. We might as well attempt to fyllogize into the refult, that a perfon who fells the meat he has killed is a butcher.

I lay ſtreſs upon this, to an extent which may for a moment appear like diligently grinding nothing in a mill which might be better employed, for two reaſons. Firſt, the young mathematician is very apt to try, in algebra, to make one principle deduce another by mere force of ſymbols: and the above attempt may ſhow him what he is liable to. Secondly, I am inclined to ſuppoſe that the diſtinction drawn between the claſſes of ſyllogiſms to which I preſently come, and the ordinary categorical ones, is due to what muſt be deſcribed in my language as a want of perception of the abſolute, *leſs than inferential* (ſo to ſpeak) identity of X)P+X)Q and X)PQ. But all other propoſitions of the kind, however ſimple, may be made deductions. For inſtance, 'if X be both P and Q, and if P be R, and Q be S, then X is both Q and S' is thus deduced: X)P+P)R=X)R, and X)Q+Q)S=X)S, and X)R+X)S is X)RS. Even P)R+Q)R=P,Q)R is deducible; being P)R+Q)R=r)p+r)q=r)pq=P,Q)R. Thus it is ſeen that, as ſoon as the *conjunctive* poſtulate is laid down, the identity of the correſponding disjunctive poſtulate with it may be ſhown. Next, if X muſt be either P or Q, or X)P,Q, and if P be always R, and Q be always S, then X)R,S may be *deduced* from the preceding.

Firſt, that X)P and Y)Q give XY)PQ can be deduced; evident as it may be, it is a ſucceſſion of applications. XY)X+X)P gives XY)P, and XY)Y+Y)Q gives XY)Q, and XY)P+XY)Q is XY)PQ by the poſtulate. Next, X)P,Q is pq)x, and P)R is r)p, and Q)S is s)q, whence, as juſt proved rs)pq. Now, rs)pq+pq)x=rs)x, which is X)R,S. It will be a good exerciſe for the reader to tranſlate this proof into ordinary language.

I may now proceed to extend this idea and notation relative to propoſitions of complex terms. The complexity conſiſts in the terms being conjunctively or disjunctively formed from other terms, as in PQ, that to which both the names P and Q belong conjunctively; and as in P,Q that to which one (or both) of the names P and Q belong disjunctively. The contrary of PQ is p,q; that of P,Q is pq. *Not both* is either not one or not the other, or not either. *Not either P nor Q* (which we might denote by :P,Q or .P,Q) is logically '*not* P *and not* Q' or pq: and this is then the contrary of P,Q.

The disjunctive name is of two very different characters, according as it appears in the univerſal or particular form: ſo very different that it has really different names in the two caſes, *copulative* and *disjunctive*. This diſtinction I here throw away: oppoſing *disjunctive*, (having one or more of the names) to *conjunctive*, (having all the names). The disjunctive particle *or* has the ſame meaning with the diſtributive copulative *and*, when uſed in a univerſal. Thus, 'Every thing which is P or Q is R or S' means 'Every P *and* every Q is R *or* S.' But PQ is always 'both P and Q in one.' Accordingly

Conjunctive	PQR uſes *and* collectively.
Disjunctive	P,Q,R in a univerſal uſes *and* diſtributively,
	P,Q,R in a particular uſes *or* disjunctively, in the common ſenſe of that word.

'Either P or Q is true,' is an ambiguous phraſe, which is P,Q)T or T)P,Q according to the context.

The manner in which the component of a name enters, whether conjunctively or disjunctively, is to paſs as it were for a part of the quality of the name itſelf. Thus the contrary of P (conjunctive, as indicated by the abſence of the comma) is ,p (disjunctive, as indicated by the comma). To teſt this aſſertion about the mode of making contraries, let us aſk what is that of 'one only of the two P or Q?" We know it of courſe to be 'both or neither.' The name propoſed is Pq, Qp and its contrary is (p,Q)(q,P), that is, one of the two p,Q, *and* one of the two q,P. It is then either pq, pP, qQ, or PQ: the ſecond and third cannot exiſt, therefore it is pq, PQ, as already ſeen. I need hardly have remarked that (P,Q)(R,S) is PR, PS, QR, QS.

Obſerve that though X)PQ gives X)P, and that XPQ gives XP, we may not ſay that XY)P gives X)P, nor that X)P,Q gives X)P. But any disjunctive element may be rejected from a univerſal term, and any conjunctive element from a particular one. Thus P)QR gives P)Q and P,Q)R gives P)R. Alſo P.Q,R gives P.Q and PQ:R,S gives P:R. All theſe rules are really one, namely that PQ is of the ſame extent at leaſt as PQR. This will appear from our rules of tranſpoſition preſently given.

Let change from one member of the propoſition to the other be called *tranſpoſition.* I proceed to inquire how many tranſpoſitions the various forms will bear, and what they are. It will however be neceſſary to complete our forms by the recognition, as a propoſition, of the ſimple aſſertion of exiſtence or non-exiſtence. By XU we mean that there are in the univerſe things to which the name X applies, and we ſpeak only of ſuch things under the name. Accordingly X)U and XU do not differ in meaning. By u, the contrary of U, we can only denote non-exiſtence; thus X.U or X)u throws the name X out of conſideration. Thus Y)X=U)X,y; Y.X=YX)u, &c. To ſignify, for inſtance, that X and Y are complements (contraries or ſubcontraries, page 75) we have U)X,Y, which our rules will tranſpoſe into xy)u, or x.y.

Having to conſider ſubject and predicate, conjunctive and diſjunctive, affirmative and negative, univerſal and particular, we muſt think of ſixteen different forms. Thus the four forms of the univerſal affirmative are

XY)PQ; X,Y)PQ; XY)P,Q; X,Y)P,Q

It will be beſt here to neglect the contranominal converſes of A and O equally with the ſimple converſes of E and I: thus XY)PQ may be read as identical with p,q)x,y. There is alſo one obvious tranſpoſition which we muſt not merely neglect but throw out; ſince it does not give a reſult identical with its predeceſſor. I mean the tranſpoſition of M)PQ into MP)Q: the ſecond follows from the firſt but not the firſt from the ſecond. Alſo the correſponding change of M.P,Q into Mp.Q, for the ſame reaſon.

This being premiſed, the following are the rules;—

Direct tranſpoſition is the change from one member to the other without alteration of name or junction: *contrary*, with alteration of both.

The convertibles (E,I) allow direct tranſpoſition of conjunctive elements either way, from ſubject to predicate, or from predicate to ſubject: and theſe are the only direct tranſpoſitions. Thus X.YZ=XY.Z, and X-YZ=XY-Z.

The inconvertibles (A,O) allow contrary tranſpoſition of conjunctive elements from ſubject to predicate, and of disjunctive elements from predicate to ſubject: beſt remembered by allowing SP to ſtand for *conjunctive* and PS for *disjunctive*.. And theſe are the only contrary tranſpoſitions. Thus XY)M=X)M,y and M)X,Y=My)X.

An element that can be rejected cannot be tranſpoſed, and *vice verſâ*. Thus X,Y)M gives X)M, and Y cannot be tranſpoſed.

The following table exhibits the varieties of the forms A and E, equivalents being written under one another, and converſions, contranominal or ſimple, oppoſite.

XY)P,Q	pq)x,y	XY.PQ	PQ.XY
Xp)Q,y	Yq)P,x	XP.QY	QY.XP
Xq)P,y	Yp)Q,x	XQ.PY	PY.XQ
X)P,Q,y	pqY)x	X.PQY	PQY.X
Y)P,Q,x	pqX)y	Y.PQX	PQX.Y
p)Q,x,y	XYq)P	P.QXY	QXY.P
q)P,x,y	XYp)Q	Q.PXY	PXY.Q
XYpq)u	U)P,Q,x,y	XYPQ.U	U.XYPQ

XY)PQ	p,q)x,y	XY.P,Q	P,Q.XY
X)PQ,y	[p,q]Y)x	X.[P,Q]Y	[P,Q]Y X
Y)PQ,x	[p,q]X)y	Y.[P,Q]X	[P,Q]X.Y
XY[p,q])u	U)[x,y],PQ	XY[P,Q].U	U.XY[P,Q]

X,Y)P,Q	pq)xy	X,Y.PQ	PQ.X,Y
[X,Y]p)Q	q)xy,P	[X,Y]P.Q	Q.[X,Y]P
[X,Y]q)P	p)xy,Q	[X,Y]Q.P	P.[X,Y]Q
[X,Y]pq)u	U)xy,P,Q	[X,Y]PQ.U	U.[X,Y]PQ

X,Y)PQ	p,q)xy	X,Y.P,Q	P,Q.X,Y
[x,y][p,q])u	U)xy,PQ	[X,Y][P,Q].U	U.[X,Y][P,Q]

If for) we write (:) in the left hand diviſions, and eraſe the (.) and uſe the hyphens of page 115, on the right, we have the tranſpoſitions of O and I. And if we write p and q for P and Q on the left, and change the form X)Y into X.y, we thereby change the forms of A into thoſe of E. If more than two elements were uſed, the tranſpoſitions would now be perfectly eaſy.

It appears that there are no leſs than ſixteen A forms into

which XY)P,Q may be varied: the reaſon is that both ſubject and predicate are tranſpoſibly conſtructed. But XY)PQ ſhows only a tranſpoſible ſubject; X,Y)P,Q only a tranſpoſible predicate: and theſe have only four forms each. Laſtly, X,Y)PQ, having neither tranſpoſible, has only two forms. By tranſpoſibly conſtructed, I mean capable of having the elements ſeparated by tranſpoſition. The whole term is always tranſpoſible: that is, the complete ſubject, or the complete predicate, may be looked on as conjunctive or disjunctive, at pleaſure. Thus in X)Y, if we conſider this as XU)Y,u, we may make this yU)x,u or y)x. So that the ordinary contranominal converſion may be conſidered as a caſe of the more general rule. Juſt as, in arithmetic, a number, 5, may be made to obey the laws of $a+b$ as $0+5$, or of ab as 1×5.

Syllogiſms of complex terms might be widely varied, even if we choſe to conſider only each firſt caſe of the preceding table as fundamental. Thus

XY)P,Q+VW)P,Q=(x,y)-(v,w)   A₍A'I'

would give ſixty-four varieties of premiſes. I now proceed to ſhow that the ordinary disjunctive and dilemmatic forms are really common ſyllogiſms with complex terms, reducible to ordinary ſyllogiſms by invention of names.

Example 1. Every S is either P, Q, R; no P is S; no Q is S; therefore every S is R. Let S repreſent 'the true propoſition' (ſingular), and let P, Q, R be names of propoſitions, and this then repreſents a very common form, which would be expreſſed thus 'either A is B, or C is D, or E is F; but A is not B, C is not D; therefore E is F.' I ſay that, where the neceſſary names exiſt, the final ſtep of this could not be diſtinguiſhed from a common ſyllogiſm; which accordingly it becomes by invention of names.

We have S)P,Q,R, whence Spq)R. But S.P and S.Q or S)p and S)q give S)pq, with which S)S combined gives S)pqS. And S)pqS+pqS)R=S)R. Let M be the name of what is S and not P and not Q, and the thing required is done. Here then is a ſyllogiſm of the ordinary kind, to one premiſe of which we are led by a uſe of the *conjunctive poſtulate* (page 116): the neceſſity for which is the diſtinction between the claſs we are conſidering and others. It happens here that two of the terms of our final ſyllogiſm are identical: for Spq is of no greater extent than S. But the use made of S)S is perfectly legitimate.

Example 2. 'If A be B, E is F; and if C be D, E is F; but either A is B or C is D; therefore E is F.' This can be reduced to

$$P)R+Q)R+S)P,Q=S)R$$

which is immediately made a common syllogism by changing P)R+Q)R into P,Q)R.

Example 3. 'From P follows Q; and from R follows S; but Q and S cannot both be true; therefore P and R cannot both be true.' This may be reduced to

$$P)Q+R)S+T.QS=T.PR$$
$$\text{or } PR)QS \quad +T.QS=T.PR$$

Example 4. 'Every X is either P, Q, or R; but every P is M, every Q is M, every R is M; therefore every X is M.' This is a common form of the dilemma; it is obviously reducible to P,Q,R)M+X)P,Q,R=X)M.

Example 5. 'Every X is either P or Q, and every Q is X.' This is wholly inconclusive, and leads to an identical result, as follows; X)P,Q gives Xp)Q, which with Q)X gives Xp)X, a necessary proposition.

Example 6. If we throw X)R into the form X)R,R, we have Xr)R, or 'Every X which *is not* R *is* R,' a contradiction in terms. But it evidently implies that there can be no Xs which are not Rs; and thus also we return to X)R. Take 'every X is either P, Q, or R; every P is M; every Q is M; and every M is R.' Here X)P,Q,R=Xr)P,Q, which with P,Q)M gives Xr)M, which with M)R gives Xr)R or X)R.

Example 7. 'Every X is either P or Q, and only one.' This gives two propositions, X)P,Q+X.PQ. Now X)XP,XQ is identical with X)P,Q, and this may be looked on as an extreme case of

$$X)P,Q+X)Y=X)PY,QY$$

but X.PQ gives XP)q and XQ)p, from which we can obtain

$$X)XP,XQ+XP)q+XQ)p=X)p,q$$

$$\text{Hence } X)P,Q+X)p,q=X)[P,Q,][p,q.]$$
$$=X)Pp,Pq,Qp,Qq=X)Pq,Qp$$

since Pp and Qq are subject to X.Pp and X.Qq. All this being worked out in ſyllogiſtic detail, ſhows us that the tranſition from 'Every X is P or Q, and no X is both' to 'Every X is either P and not Q, or Q and not P' is capable of being made ſyllogiſtically. The ſtudent of logic may thus acquire the idea, which ſo ſoon becomes familiar to the ſtudent of mathematics, of perfectly ſelf-evident propoſitions which are deducible from one another, as diſtinguiſhed from thoſe which are not.

Example 8. 'Every X is one only of the two, P or Q; every Y is both P and Q, except when P is M, and then it is neither; therefore no X is Y.' Here is a caſe in which it is the fact of the exception and not its nature which determines the inference: M may be anything. This ought to appear in our reduction: and it does appear in this way. From X)P,Q it is obvious that X)P,Q,R,S, and ſyllogiſtically demonſtrable from X)P,Q, and Xrs)X. Now in the ſecond premiſe we have

$$\mathrm{Y)PQm,pqM,\ or\ [p,q,M][P,Q,m])y}$$
$$\mathrm{or\ pQ,Pq,PM,QM,pm,qm)y}$$

from which, by rejection, follows pQ,Pq)y. And the firſt premiſe is X)Pq,Qp. Whence X)y or X.Y.

It is not neceſſary to multiply examples: I will conclude this part of the ſubject by pointing out that the ordinary propoſitions X)Y, &c. are, with reference to their inſtances, disjunctively compoſed: the difference between the univerſal and particular lying in the latter being indefinite in the number of its inſtances. Thus, if there be three Xs and four Ys, the four propoſitions are, applying the name to each inſtance, as ſeen written at length in X,X,X)Y,Y,Y,Y; X,X,X.Y,Y,Y,Y; (X,X,X)(Y,Y,Y,Y); and (X,X,X):Y,Y,Y,Y.

The propoſition in page 25, is a caſe of the preceding method. I leave the reader to ſhow it, and alſo that the hypotheſis is ſlightly overſtated.

I now come to the *ſorites*, the *heap* or chain of ſyllogiſms, in which the concluſion of the firſt is a premiſe of the ſecond, and ſo on. Take a ſet of terms, P, Q, R, S, &c. and let the order of reference be PQ, QR, RS, &c. Then $A_1A_1A_1A_1$ &c. is a ſorites, and the only one uſually conſidered: thus,

$$\mathrm{P)Q + Q)R + R)S + S)T = P)T}$$

The firſt two links give P)R, which with the third gives P)S, which with the fourth gives P)T. Thus we have *links*, *intermediate concluſions*, and a *final concluſion*.

A great number of different ſorites may be formed, under the following conditions,

The firſt particular propoſition which occurs, be it link or concluſion, prevents any future link from being particular: for all the concluſions thence become particular.

Examine the caſes of ſyllogiſm which proceed by the firſt rule of accentuation (page 92), that is, which have beginning and ending both univerſal, or both particular: theſe only can occur in a ſorites, except at the end, or in the place where a particular propoſition firſt enters. It will be found that the concluſion, when the argument goes on, muſt come after ſomething connected with that which comes after it by the firſt rule of accentuation: except at the place where a particular concluſion comes in for the firſt time. For inſtance, $E_,E'$ gives $A_,$, which, ſtill keeping concluſions univerſal, muſt be followed by $A_,$ or $E_,$, which follow E' by the firſt rule. Again, take $O_,E'$, which gives $I_,$; this muſt be followed either by $A_,$ or $E_,$, which follow E' by the ſame rule: and ſo on. Accordingly,

Any chain of univerſals, in which affirmation is followed by a like prepoſition, and negation by a different one, as $A_,A_,E_,A'E'A_,E_,E'$, &c. may be part of the chain of a ſorites. And the chain muſt be either of this kind wholly, or once only broken in one of two ways: either by the direct entrance of a particular propoſition, or by a breach of the rule. In a chain of this kind, unbroken, the concluſions are affirmative or negative, according as an even or odd number of negatives goes to the formation of them. All the concluſions have the ſame accent as the firſt link.

Let a particular premiſe be introduced, as in $A_,E_,E'I'$ &c. The accent of the particular introduced muſt be the ſame as or contrary to that of the firſt link, according as the preceding number of negatives is odd or even. For the accent of the firſt link remains as long as the concluſion is univerſal, and a ſyllogiſm with the ſecond premiſe particular follows the ſecond rule. Thus, inſerting the intermediate concluſions, the above is $A_,E_,(E_,)E'(A_,)I'(I')$. And after (I') muſt come A' or E', ſo that the firſt rule ſtill continues. But the accent of the concluſions changes.

Now let the rule of accentuation be broken. The *accent* of the conclufion ftill requires the firft rule to be refumed. Thus, $E_{,}E'$(rule unbroken) gives $A_{,}$, and $E_{,}E_{,}$ (rule broken) gives I', and $A_{,}$ requires $A_{,}$ or $E_{,}$ to follow E', while I' requires A' or E' to follow $E_{,}$. This one breach of rule only changes the conclufion from univerfal to particular. The accent of the conclufion changes as before.

The links of a forites, then, are either a chain of univerfals following the firft rule of accentuation, or fuch a chain with *one* breach of the rule, or fuch a chain with one particular inferted, of the fame or contrary accent to the firft link, according as the preceding negatives are odd or even, and made the commencement of the refumption of the rule (if broken). In all the cafes the conclufion is affirmative or negative according as the preceding negatives are even or odd in number: the unbroken chain has a univerfal conclufion with the accent of the firft link, and the broken one a particular with the contrary accent.

$A'E'E_{,}A'E'$	$E_{,}A'A_{,}E_{,}E'A_{,}$	$A_{,}E_{,}A'O_{,}A'E'$
$E'A'A'E'$	$E_{,}O'I'O'O'$	$E_{,}E_{,}I'I'O'$

Here are examples of the three kinds. The chain is in the firft row, the intermediate and final conclufions in the fecond. Thus the fecond example prefents the fyllogifms $E_{,}A'E_{,}$, $E_{,}A_{,}O'$, $O'E_{,}I'$, $I'E'O'$, $O'A_{,}O'$; and at length is

$$P.Q+R)Q+R)S+S.T+t.u+U(V=V:P$$

The forites ufually confidered are only $A_{,}A_{,}A_{,}\ldots$ and $A'A'A'\ldots$ To thefe might be added without abandoning the Ariftotelian fyllogifm, fuch as $A_{,}E_{,}A'A'A'\ldots$, $A_{,}E_{,}A'A_{,}A_{,}\ldots$ But it would not be very eafy to follow the chain in thought without introducing the intermediate conclufions, and thus deftroying the fpecific character of the procefs.

And juft as the ordinary univerfal fyllogifm can be reduced to $A_{,}A_{,}A_{,}$, fo the univerfal forites can always be reduced to a chain of $A_{,}$. Thus $A'E'E_{,}A'E'$ or

$$Q)P+q.r+R.S+T)S+t.u=p.u$$

is

$$u)T+T)S+S)r+r)Q+Q)P=u)P$$

CHAPTER VII.

On the Ariſtotelian Syllogiſm.

FROM the time of Ariſtotle until now, the formal inference has been a matter of ſtudy. In the writings of the great philoſopher, and in a ſomewhat ſcattered manner, are found the materials out of which was conſtructed the ſyſtem of ſyllogiſm now and always prevalent: and two diſtinct principles of excluſion appear to be acted on. Perhaps it would be more correct to ſay that the followers collected two diſtinct principles of excluſion from the writings of the maſter, by help of the aſſumption that everything not uſed by the teacher was forbidden to the learner. I cannot find that Ariſtotle either limits his reader in this manner, or that he anywhere implies that he has exhauſted all poſſible modes of ſyllogizing. But whether theſe excluſions are to be attributed to the followers alone, or whether thoſe who have more knowledge of his writings than myſelf can fix them upon the leader, this much is certain, that they were adopted, and have in all time dictated the limits of the ſyllogiſm. Of all men, Ariſtotle is the one of whom his followers have worſhipped his defects as well as his excellencies: which is what he himſelf never did to any man living or dead; indeed, he has been accuſed of the contrary fault.

The firſt of theſe excluſions is connected with the celebrated *dictum de omni et nullo*, namely, that what is diſtributively affirmed or denied of all, is diſtributively affirmed or denied of every ſome which that all contains. It is there ſaid that in every ſyllogiſm the middle term muſt be univerſal in one of the premiſes, in order that we may be ſure that the affirmation or denial in the other premiſe may be made of ſome or all of the things about which affirmation or denial has been made in the firſt. This law, as we ſhall ſee, is only a particular caſe of the truth: it is enough that the two premiſes together affirm or deny of more than all the inſtances of the middle term. If there be a hundred boxes, into which a hundred *and one* articles of two different kinds are to be put, not more than one of each kind into any one box, ſome one box, if not more, will have two articles, one of each kind, put into it. The common doctrine has it, that an article of one particular kind muſt be put into every box, and then ſome one or more of another kind into one or more of the boxes, before it may be affirmed that one or more of different kinds are found together. This excluſion is a ſimple miſtake, the mere ſubſtitution of the aſſertion that none but a certain law of inference *can* exiſt, for the determination that no other *ſhall* exiſt. Any one is at liberty to limit the inferences he will uſe, in any manner he pleaſes: but he may err if he declare his own arbitrary boundary to be a natural limit impoſed by the laws of thought.

The other excluſion may involve, on the ſame terms, an error of the ſame kind; or may equally be the expreſſion of arbitrary will: but there is what is more reaſonably matter of opinion about it. Ariſtotle will have no contrary terms: not-man, he ſays, is not the name of anything. He afterwards calls it an indefinite or *aoriſt* name, becauſe, as he aſſerts, it is both the name of exiſting and non-exiſting things. If he had here made the diſtinction between ideal and objective, he would have ſeen that *man* and *not-man* equally belong to both (objectively) exiſting and non-exiſting things: *man*, for example, belongs as a name to Achilles and the ſeven champions of Chriſtendom, whether they ever exiſted in objective reality or not: and *not-man* belongs, in either caſe, to their horſes. I think, however, that the excluſion was probably dictated by the want of a definite notion of the extent of the field of argument, which I have called the *univerſe* of the propoſitions. Adopt ſuch a definite notion, and, as ſufficiently ſhown, there is no more reaſon to attach the mere idea of negation to the contrary, than to the direct term.

The excluſion of contraries throws out the propoſitions E' and I', or $x.y$ and xy, which cannot be expreſſed without either contraries, as in $x.y = x)Y = y)X$, and $xy = x:Y = y:X$, or reference to things not named by X and Y, as in 'Every *thing* is either X or Y' and 'Some things are neither Xs nor Ys,' the moſt natural readings of 'No not-Xs are not-Ys, and 'Some not-Xs are not-Ys.' There remain then ſix modes of connexion of X and Y, namely $X)Y$ and $Y)X$, $X{:}Y$ and $Y{:}X$, and $XY(=$

YX) and X.Y(=Y.X). Theſe ſix are made eight; for in the common ſyſtem, XY and YX are conſidered as diſtinct in form, and alſo X.Y and Y.X. But theſe eight are only treated as four: for reference to order is not made in the ſimple propoſition. Thus X)Y and Y)X are both denoted by A, XY and YX by I, X.Y and Y.X by E, and X:Y and Y:X by O. But the ſtandard of order which is neglected as to the propoſition by itſelf, is adopted in the ſyllogiſm in the following manner.

The *predicate* of the concluſion is called the *major* term, and the ſubject of the concluſion the *minor* term. This language is faſhioned upon the idea of an affirmative propoſition, in which major and minor have reference to *magnitude*. In 'every X is Z' Z is a name which entirely contains X and is therefore *at leaſt as great* as X, *greater than or equal to* X. Here is, before it was introduced into mathematics, the idea now ſo familiar to the mathematician, of allowing his language to include the extreme limit of its meaning. When the ſame terms are applied to negative propoſitions, the notion of magnitudinal incluſion is loſt; and major and minor, being ſtill retained, muſt be preſumed to refer to real or ſuppoſed importance. The premiſes are called major and minor, according as they contain the major or minor term of the concluſion: and the major premiſe is always written firſt. Accordingly, Z and X being the major and minor terms, there are four poſſible arrangements, which are called the four figures. Ariſtotle gives three, and tradition has it that Galen ſupplied the fourth in number and order.

1.	2.	3.	4.
YZ	ZY	YZ	ZY
XY	XY	YX	YX
XZ	XZ	XZ	XZ

To me, the moſt ſimple arrangement is that which takes up what was left off with, as in the fourth figure: and 'X is in Y, Y is in Z, therefore X is in Z' is more natural than 'Y is in Z, X is in Y, therefore X is in Z.'

It is now plain, that whenever one only of the three propoſitions is convertible, there are two diſtinct ways in which the ſyllogiſm may be written: when two only, four: and when all three (if there were ſuch a thing), eight.

The ſyſtem rejects all concluſions which may be made ſtronger: thus when X.Z follows, it does not allow X:Z to make a diſtinct form. But when X)Z is the concluſion, it does not reject ZX, for, not conſidering ZX as identical with XZ, it does not conſider X)Z as a ſtrengthened form of ZX. But it does not reject ſyllogiſms in which as ſtrong a concluſion can be deduced from a weaker premiſe: accordingly, we muſt ſearch for Ariſtotelian forms among the ſtrengthened ſyllogiſms of chapter V, as well as among the fundamental ones. Now, taking all the forms which ſhow neither E' or I', let us write down the ſymbols of them, and the number of caſes we may expect from each. Moreover, ſince transformation of order makes no difference here, I put the ſyllogiſms together as in page 96, into twelve pairs.

Fundamental $A_1A_1A_1$, $A'A'A'$, 1; $O'A_1O'$, $A'O_1O_1$, 1; $A_1O'O'$, $O_1A'O_1$, 1; $E'A_1E'$, $A'E'E'$, rejected; $I_1A_1I_1$, $A'I_1I_1$, 4; $E'O'I_1$, $O_1E'I_1$, rejected; $E_1A'E_1$, $A_1E_1E_1$, 4; $I'A'I'$, $A_1I'I'$, rejected; E_1O_1I', $O'E_1I'$, rejected; $E'E_1A'$, $E_1E'A_1$, rejected; $I_1E_1O_1$, E_1I_1O', 4; $E'I'O_1$, $I'E'O'$, rejected.

Weakened $A_1A_1I_1$, 1.

Strengthened $A'A_1I_1$, 1; $A_1A'I'$, rejected; $A'E_1O_1$, E_1A_1O', 2; $A_1E'O'$, $E'A'O_1$, rejected; $E'E'I_1$, rejected; E_1E_1I', rejected.

There are then fifteen fundamental, one weakened, and three ſtrengthened, forms of ſyllogiſm in the received ſyſtem. I now put them down, with their derivations, forms of expreſſion in full, ordinary ſymbols, figures into which they fall, and the magic words by which they have been denoted for many centuries, words which I take to be more full of meaning than any that ever were made.

Fundamental.

$A_1A_1A_1$	$A'A'A'$	Y)Z + X)Y = X)Z	AAA	I	*Barbara*
$O'A_1O'$	$A'O_1O_1$	Y:Z + Y)X = X:Z	OAO	III	*Bokardo*
$A_1O'O'$	$O_1A'O_1$	Z)Y + X:Y = X:Z	AOO	II	*Baroko*
$I_1A_1I_1$	$A'I_1I_1$	Y)Z + XY = XZ	AII	I	*Darii*
—	—	Y)Z + YX = XZ	AII	III	*Datiſi*
—	—	ZY + Y)X = XZ	IAI	IV	*Dimaris*
—	—	YZ + Y)X = XZ	IAI	III	*Diſamis*

Fundamental.

$E_{\prime}A^{\prime}E_{\prime}$	$A_{\prime}E_{\prime}E_{\prime}$	Y.Z+X)Y=X.Z	EAE	I	*Celarent*
—	—	Z.Y+X)Y=X.Z	EAE	II	*Ceſare*
—	—	Z)Y+Y.X=X.Z	AEE	IV	*Camenes*
—	—	Z)Y+X.Y=X.Z	AEE	II	*Cameſtres*
$E_{\prime}I_{\prime}O^{\prime}$	$I_{\prime}E_{\prime}O_{\prime}$	Y.Z+XY =X:Z	EIO	I	*Ferio*
—	—	Z.Y+XY =X:Z	EIO	II	*Feſtino*
—	—	Y.Z+YX =X:Z	EIO	III	*Feriſon*
—	—	Z.Y+YX =X:Z	EIO	IV	*Freſiſon*

Weakened.

$A_{\prime}A_{\prime}I_{\prime}$	$A^{\prime}A^{\prime}I_{\prime}$	Z)Y+Y)X=XZ	AAI	IV	*Bramantip*

Strengthened.

$A^{\prime}A_{\prime}I_{\prime}$	$A^{\prime}A_{\prime}I_{\prime}$	Y)Z+Y)X=XZ	AAI	III	*Darapti*
$A^{\prime}E_{\prime}O_{\prime}$	$E_{\prime}A_{\prime}O^{\prime}$	Y.Z+Y)X=X:Z	EAO	III	*Felapton*
—	—	Z.Y+Y)X=X:Z	EAO	IV	*Feſapo*

The words which repreſent the different *moods* (as they are called) are uſually collected under their figures in the following lines.

Barbara, Celarent, Darii, Ferioque prioris.
Ceſare, Cameſtres, Feſtino, Baroko, ſecundæ.
Tertia Darapti, Diſamis, Datiſi, Felapton,
Bokardo, Feriſon habet. Quarta inſuper addit
Bramantip, Camenes, Dimaris, Feſapo, Freſiſon.

The vowels of the different words give the ſymbol of the ſyllogiſm; thus A,A,A, are ſeen in *Barbara.* The conſonants in the firſt figure have no ſpecial meaning: but in the other figures every conſonant except T and N (which are only euphonic) has its meaning as follows;—every mood of every figure can (with two exceptions) in one way or another, be reduced to a mood of the firſt figure: and the letters ſhow the way of doing it. The initial tells to which mood the reduction brings us: thus Ceſare is reduced to Celarent, and alſo Cameſtres; Feſtino is reduced to Ferio, and ſo on. The two exceptions are denoted by the letter K (as in Baroko and Bokardo); we ſhall preſently notice them further. And S means that the preceding premiſe is to be ſimply converted. P, that what was called converſion *per acci-*

dens is to be made, ZX for X)Z, or X)Z for ZX: accordingly, P only occurs in the weakened or ſtrengthened ſyllogiſms. M means that the premiſes are to be tranſpoſed. Thus the meaning of the word *Diſamis* is nothing leſs than what follows. 'There is a ſyllogiſm in which the middle term is the ſubject of both premiſes, and when reduced to the firſt figure it becomes *Darii*: the major premiſe, which muſt be converted in reduction, is a particular affirmative: the minor premiſe, which muſt become the major one in reduction, is a univerſal affirmative: and the concluſion, which muſt be converted in reduction, is a particular affirmative.' Thus,

$$YZ + Y)X = XZ \quad \textit{Diſamis}$$
becomes $$Y)X + ZY = ZX \quad \textit{Darii}$$

The moods *Baroko* and *Bokardo* do not admit of reduction to the firſt figure, by any fair uſe of the phraſe: but the logicians were determined they ſhould do ſo, and they accordingly hit upon the following plan, which they called reduction *per impoſſibile.* AOO and OAO being the opponent forms (pages 88, and 102) of AAA, the two moods in queſtion were connected with *Barbara* (whence their letter B) by ſhowing that the latter would make the denial of their concluſions force one premiſe to contradict the other. Thus, *Baroko*, or if Z)Y and X:Y then X:Z was proved in the firſt figure as follows. If under theſe premiſes, X:Z be not true, then X)Z is true; but Z)Y is true: and Z)Y + X)Z, by *Barbara*, gives X)Y. But X:Y: therefore, if *Baroko* be not a legitimate form, X)Y and X:Y are both true at once, which is abſurd. Had contraries been uſed, Z)Y + X:Y = X:Z would have been thrown into the firſt figure as y)z + Xy = Xz, *Darii*, or y.Z + Xy = X:Z, *Ferio*. And Y:Z + Y)X = X:Z, *Bokardo*, is ſeen reduced to the firſt figure in Y)X + zY = zX, *Darii*.

Ariſtotle did not uſe the fourth figure, conſidering it, as is ſaid, to be only an inverſion of the firſt. The introduction of it among the figures is attributed to Galen, and it does not often appear in ordinary works of logic before the beginning of the laſt century. If the order of the premiſes be inverted, ſo as to make the firſt figure appear, the major and minor terms will appear wrongly placed in the concluſion. The words uſed for theſe indirect moods of the firſt figure were uſually the fifth and following ones in

Barbara, *Celarent*, *Darii*, *Ferio*, Baralip-*ton*
Celantes, Dabitis, Fapeſmo, Friſeſom-*orum*

the final ſyllables in Italics being only euphonic (Friſeſmo-orum would have been more correct). Some uſed the words *Fareſmo* and *Fireſmo*.

In calling the moods of the fourth figure by the name of indirect moods of the firſt figure, notice was taken of the circumſtance that a tranſpoſition of the premiſes would give the arrangement of the firſt figure, in every thing but the proper arrangement of major and minor terms, which is inverted. A little conſideration will ſhow the reader that the earlier Ariſtotelians were wiſer than the later ones in this matter. Conſider the fourth and firſt figures as coincident, and the arbitrary notion of arrangement by major and minor vaniſhes. It was not till this mere matter of diſcipline was made an article of faith that the fourth figure had any ground of ſeceſſion from the firſt.

It might ſeem as if the union of the firſt and fourth figures would demand that of the ſecond and third: the firſt pair containing all the moods in which the middle term occupies different places in the two premiſes, the ſecond pair thoſe in which it has the ſame place in both. If this were done, each of the two main ſubdiviſions muſt be itſelf ſubdivided into two. And this would perhaps have been the more ſkilful mode of diviſion.

The diſtinction of figures has been condemned by many, and particularly by Kant. Whether attacked or defended, it is eſſential that the true grounds of the ſide taken ſhould be more explicitly ſtated than is often done. The root of the diſtinction of figure is undoubtedly the diſtinction between the two forms XY and YX, X.Y and Y.X. It would be equally abſurd, either to deny the identity of XY and YX, conſidered as material of inference, or to deny their difference in many other points of view. In this work I am concerned only with what can be inferred, and to what extent of quantity, and accordingly the diſtinction is to me immaterial. But if I had not merely to ſtudy the way of uſing premiſes, but alſo that of arriving at them, it might very well happen that the aſpects under which the ſame inference is seen in different figures would give it very different shades of character. A simple instance will show that though the comparison, and its extent, are all that can be attended to in forming the conclusion, these points of meaning are not the only ones. A person who wished to contest the old use of the word *green*, as applied to unripe fruit, would say that 'some green fruits are ripe,' if he wanted specially to show the misapplication of the word. But if he rather wanted to show the badness of the method of denying ripeness, he would say 'some ripe fruits are green.' The propositions are endless in which, X and Y being the terms, it is at one time X which is brought to Y for comparison, and at another Y to X. The subject of a proposition is always the object of examination; whether the form be X)Y, X.Y, XY, or X:Y, we examine and report upon the Xs.

If we arrange the four figures separately, we shall better see their several peculiarities.

First Figure.

Barbara	Y)Z+X)Y=X)Z	*Celarent*	Y.Z+X)Y=X.Z
Darii	Y)Z+XY =XZ	*Ferio*	Y.Z+XY =X:Z

What is here declared, is in every case the *dictum de omni et nullo* in its simplest form, in a manner which justifies the preference given to this figure. The middle term being completely contained in, or completely excluded from, the major term; such inclusion or exclusion then follows of all such part of the minor term as is declared in the second premise to be in the middle term. The inference then is in this sentence 'What is true of the whole middle term, is true of its part.' And it is obvious that in this figure the major premise must be universal, the minor premise affirmative. The four forms are all found among the conclusions. I think that the inversion of the premises which the system of chapter V. employs will be found to give the forms which are most easily translated into language independent of the middle term. The sentence 'All (or some) of the Xs are what must be Zs, therefore all (or some) of the Xs are Zs' includes *Barbara* and *Darii:* and 'All (or some) of the Xs are what cannot be Zs, and therefore cannot be Zs,' contains *Celarent* and *Ferio.*

Second Figure.

Ceſare	$Z.Y + X)Y = X.Z$	*Cameſtres*	$Z)Y + X.Y = X.Z$
Feſtino	$Z.Y + XY = X:Z$	*Baroko*	$Z)Y + X:Y = X:Z$

In this figure (in which only negatives can be proved) the appearance of the dictum is not ſo direct. The terms of the concluſion are both objects of examination, and one is wholly included, and the whole or part of the other excluded (Ceſare, Cameſtres, and Baroko) or one is wholly excluded, and the whole or part of the other included (Ceſare, Cameſtres, and Feſtino). Or rather, to juſtify the diſtinction, we ſhould ſay that the whole of the major term is $\begin{matrix}\text{included}\\ \text{excluded}\end{matrix}$ and the whole of the minor $\begin{matrix}\text{excluded}\\ \text{included}\end{matrix}$ which gives $\begin{matrix}\text{Cameſtres}\\ \text{Ceſare}\end{matrix}$ in which the whole of the minor is therefore excluded from the major; or elſe the whole of the major is $\begin{matrix}\text{included}\\ \text{excluded}\end{matrix}$ and part of the minor $\begin{matrix}\text{excluded}\\ \text{included}\end{matrix}$ which gives $\begin{matrix}\text{Baroko}\\ \text{Feſtino}\end{matrix}$ in which that part of the minor is excluded from the major. And it is evident enough why the premiſes muſt be of different ſigns.

In the firſt figure, though all the forms be eſſentially one, (page 98,) the reduction of either to the form *Barbara* requires either the explicit uſe of contraries, or invention of a name ſub-identical to X. Accordingly, no mood of that figure is reducible to any other by the uſually admitted reductions. But this cannot be ſaid of any of the other figures. In the one before us, Ceſare and Cameſtres are identical, even without changing the figure. That which is *Ceſare* when X is major and Z minor, is *Cameſtres* when X is minor and Z major. In the firſt figure, the ſame attempt made on *Celarent* or *Darii*, removes them into another figure.

Third Figure.

Darapti	$Y)Z + Y)X = XZ$	*Felapton*	$Y.Z + Y)X = X:Z$
Diſamis	$YZ + Y)X = XZ$	*Bokardo*	$Y:Z + Y)X = X:Z$
Datiſi	$Y)Z + YX = XZ$	*Feriſon*	$Y.Z + YX = X:Z$

The firſt and ſecond figures contain a pair of univerſals each, with one particular derived from each, by a legitimate weakening of one premiſe and the concluſion at the ſame time: but in no inſtance is the quantity of the middle term weakened. And all the ſyllogiſms in theſe two figures are fundamental (page 77). In the caſe now before us, both the leading ſyllogiſms are not fundamental, but ſtrengthened, and capable of being weakened in two different ways. The middle term is here examined in both premiſes: if it be wholly included in, or excluded from, one of the concluding terms, and wholly or partly included in, or excluded from, the other (but not ſo that there ſhall be excluſion from both) we have it that the whole or part mentioned in one caſe is included in, or excluded from, that which the whole is included in, or excluded from, in the other. There can be none but particular concluſions.

Fourth Figure.

Bramantip	Z)Y + Y)X = XZ	*Camenes*	Z)Y + Y.X = X.Z
Dimaris	ZY + Y)X = XZ	. . . .	

Feſapo Z.Y + Y)X = X:Z
Freſiſon Z.Y + YX = X:Z

We have now one univerſal ſyllogiſm in a form which does not admit of being weakened *in this figure*, and two ſtrengthened ſyllogiſms, each of which has one weakened form, one of them, *Bramantip*, admitting a ſtronger concluſion in another figure. Every concluſion except A appears. The mode of inference of the three firſt ſyllogiſms has been deſcribed in the other figures. In Feſapo and Freſiſon, the perfect excluſion of the major term from the middle, accompanied by the total or partial incluſion of the middle in the minor, ſecures the excluſion from the major, of as much of the minor as it has in common with the middle.

I ſhall now proceed to the rules uſually given, and to ſome remarks on the degree in which they apply to the more general ſyſtem in chapter V. Aldrich gives them as follows—

> Diſtribuas medium: nec quartus terminus adſit:
> Utraque nec præmiſſa negans, nec particularis:
> Sectetur partem concluſio deteriorem;
> Et non diſtribuat, niſi cum præmiſſa, negetve.

These rules, I need hardly say, are perfectly correct, when the contraries of the terms are excluded, and also all notion of quantity except all, or the indefinite some. Taking them in the natural order, which versification has a little disturbed, we have;—

1. There are to be but three terms, of which it is understood two only appear in the conclusion, the excluded or middle term appearing in both of the premises. This is true in my system, when by terms are understood also contraries of terms. I should suppose that there can be no objection to the admission of contraries, unless there be one to the conception of a contrary. Any one may, with Aristotle, object to the word not-man, as not the name of anything: on the grounds which immediately induced him to call it an aorist, or indefinite, name. But it can hardly be affirmed that any one admitting not-man as a name, should thereupon refuse to recognise the identity of 'horse *is not* man,' with 'horse *is* not-man.'

2. The middle term is to be *distributed* in one or the other of the premises. By distributed is here meant *universally spoken of.* I do not use this term in the present work, because I do not see why, in any deducible meaning of the word *distributed*, it can be applied to universal as distinguished from particular. In using a name, it seems to me that we always distribute: that is, scatter as it were, the general name over the instances to which it is to apply. When I say some horses are animals, I distribute certain horses among the animals; and when all, all. Leaving the word, the principle is one which clearly must be true whenever we are restricted in quantity to all or some (indefinite), and when contraries are not admitted. In the former case we have, in one form or another, to make $m+n$ greater than η (chapter VIII.) when we cannot know what relation either m or n has to η, unless one of them, or both, be equal to η. We have no alternative then, but to require that m or n shall be η. The cases in which there is apparently no dependence on η will be discussed in the next chapter.

But when contraries are introduced, this rule is not universally true. The exception is seen in

$$A_1A'I' \text{ or } X)Y+Z)Y=xz.$$

If all the Xs be Ys, and also all the Zs, it follows that there are things which are neither Xs nor Zs, namely, all which are not

Ys. It is here, as elsewhere, implied that the middle term is not the universe of the proposition.

When we come, then, to use contraries, the simple rule of the middle term is no longer universally true. What other rule are we to put in its place? We know, of course, that every syllogism can be reduced to an Aristotelian syllogism, and even to one or other of two among them, $A_1A_1A_1$ or $I_1A_1I_1$, or to the first of these, if we be at liberty to use invention of names (page 97). Again, each term, or its contrary, is mentioned universally in every proposition: so that there is certainly one way in which every pair of premises may be made to exhibit a middle term universally used in one of them. The rule to be substituted for the *distribuas medium* is, that all pairs of universals are conclusive, but a universal and a particular require that the middle term should also be a universal and a particular, that is, universal in one and particular in the other. Thus, in X)Y + Z)Y, as it stands, the middle is particular in both; transpose into y)x + y)z and the middle is now universal in both, by which we see the Aristotelian conclusion. Again, in X)Y + ZY, which is of the same kind, the transposition gives y)x + Z:y, which is faulty, because, though there be a particular premise, there is not anywhere a particular middle term. The cases in which the middle is of the same name in both places (universal in four, particular in four), are the strengthened syllogisms only. There is nothing to be surprised at in its thus appearing that the particularity of the middle term is just as much a test of a good syllogism as its universality: of every name and its contrary, one enters universally, and one particularly, in every proposition which contains it; and the system in chapter V. is as much concerned with contrary as with direct terms. It is thence visible beforehand, to the mathematician at least, that any test must be defective, unless universal and particular enter into it in the same manner.

The above contains a complete canon of validity, as soon as the law of the three terms is understood, which is only a law of definition. We may state it as follows: Two premises conclude —when both are universal, always; when one only is universal, so often as it happens that the middle term (be it Y or y) is once only universal; when neither is universal, never. By this rule alone the thirty-two conclusive cases can be distinguished from the thirty-two inconclusive ones.

3. When both premiſes are negative, there is no Ariſtotelian ſyllogiſm. In the ſyſtem completed by contraries, there are eight ſuch ſyllogiſms, as many in fact, as there are with premiſes both affirmative. But a pair of negative premiſes never conclude with both terms of the premiſes, but with the contrary of one or both: and this muſt be ſubſtituted, as a rule of concluſion, for the one juſt named.

4. Both premiſes muſt not be particular. This rule, which relates wholly to quantity, muſt be preſerved in every ſyſtem which admits no definite ratio, except that of one to one, or *all* (pages 56, 57). I cannot learn that any writer on logic ever propounded even the very ſimple caſe of 'Moſt Ys are Xs, moſt Ys are Zs, therefore ſome Xs are Zs,' as a legitimate inference. And this, though it is certain that the quantitative prefix *moſt (plurimi)* has before now excited diſcuſſion as to whether it belonged to a univerſal or a particular.

5. By *ſectetur partem concluſio deteriorem* it is underſtood that the negative is called weaker or lower *(deterior)* than the affirmative, and the particular than the univerſal; and that the concluſion is to be as weak as negative, or as particular, if there be a premiſe which is negative or particular. This rule muſt be preſerved, when contraries are introduced, ſo far as relates to particulars. But ſo far as negatives are concerned, the rule muſt be that *one* negative premiſe gives a *negative* concluſion, and *two* an affirmative one.

7. The laſt line, *et non diſtribuat, niſi cum premiſſa, negetve,* ſpoils the ſymmetry to procure a verſe. The concluſion is not to be negative without a negative premiſe: that is, affirmative premiſes give an affirmative concluſion. Alſo, no term is to be diſtributively, (*i. e.* univerſally) taken in the concluſion, unleſs it were ſo taken in its premiſe. A breach of this rule would be equivalent to drawing a concluſion about what was not (or about more than was) introduced into the premiſes.

When contraries are introduced, the diſtinction between poſitive and negative is made to appear, what it really is, one of language, or rather one of choice of names. But the diſtinction of form is not aboliſhed, but is exactly what it was before. We cannot lay down any rules for the formation of the concluſion unleſs, in our eight ſtandard forms, we preſerve the mode of writing which belongs to the fundamental derivation of the forms (page 61). Thus, the order being XY, A' is x)y and not Y)X, and O' is x:y and not Y:X. This method of writing being reſtored, when neceſſary, in pages 89 and 91, it follows immediately that the rule of accentuation in the notation gives the rule by which we determine whether the concluſion takes the terms from the premiſes, or prefers contraries. According as the prepoſition of the concluſion agrees with or differs from that of a premiſe, ſo does the concluſion take a term from that premiſe, or its contrary. Thus, $A_{'}A_{'}A_{'}$ takes both terms from the premiſes, but $A_{'}A'I'$ takes a contrary from the firſt premiſe only. This laſt we ſee if we write the ſyllogiſm as X)Y + y)z = xz. Accordingly, we have—

Syllogiſms taking both concluding terms direct from the premiſes. Univerſals which begin with A; particulars which begin with I: eight in number; being all which iſolate no accent.

Taking the firſt term only from the premiſe. Univerſals beginning with E; particulars beginning with O: eight in number; being all which iſolate the middle accent.

Taking the ſecond term only from the premiſe. Strengthened forms and particulars which begin with A: eight in number, being all which iſolate the firſt accent.

Taking neither term from the premiſes. Strengthened forms and particulars which begin with E: eight in number, being all which iſolate the third accent.

This is a new mode of ſtating the law of accentuation (pages 92-3) which I have preferred to place here, for fear of overloading chapter V. with rules. I have not ſtated one half of thoſe which ſuggeſted themſelves. This multiplicity of relations is a preſumption of the completeneſs of the ſyſtem.

In the Ariſtotelian ſyſtem, there is multiplication of the ſame modes of inference, under the varieties of figure. In that which I propoſe, there is a reduplication of moſt of the eſſential caſes; for whatever caſe is found, the ſame is alſo found with X and Z. interchanged, and alſo the order of the premiſes. Again, whatever caſe is found, it is found contranominally; or with all the accents (or prepoſitions) altered. There are other ways (and many of them) in which the ſyſtem is only in one half a duplicate of what it is in the other. If all theſe modes of dividing the ſyſtem into two correlative parts divided it into the ſame two parts, there can be no queſtion that one alone of thoſe parts ſhould have been preſented as the object of conſideration. But this does not happen in any inſtance: ſo that it is impoſſible to diſpenſe with the whole of the thirty-two caſes. The Ariſtotelian caſes do not form or include any half whatever of this ſyſtem.

CHAPTER VIII.

On the numerically definite Syllogiſm.

IN the laſt chapter I conſidered no other quantity in names except all and ſome: the latter meaning 'one or more, it may be all.' To this extent of quantity we are limited in moſt kinds of reaſoning, by want of knowledge of the definite extent of our propoſitions: and the few phraſes (page 58), as 'moſt,' 'a good many,' &c. by which we endeavour to eſtabliſh differences of extent in ordinary converſation, have been hitherto held inadmiſſible into logic. In this ſcience it ſeems to have been always intended that the baſes on which its forms are conſtructed ſhall be nothing but the ſuppoſition of the moſt imperfect and inaccurate knowledge. Though in geometry we are permitted to aſſume as the object of reaſoning the ideal ſtraight line, the 'length without breadth' of Euclid, which has no objective prototype, and though we ſee the advantage of reaſoning upon ideas, and allowing the eſſential inaccuracies of material application to produce no effect except in material application,—yet in the conſideration of the pure forms of thought, the learner has always been denied the advantage of ſtudying the more perfect ſyſtem of which his inferences are the imperfect imitation.

The ordinary univerſal propoſitions are of a certain approach to definite character, both of them with reſpect to their ſubjects, and the negative one with reſpect to its predicate alſo. In X)Y for example, what is known is as much known of any one X as of any other. Perfect definiteneſs would conſiſt in a more exact degree of deſcription, and would require a higher degree of knowledge. But in this chapter I ſpeak only of *numerical* definitenefs, of the fuppofition that we know *how many* things we are talking about. We may be well content to examine what we fhould do if we were a ftep or two higher in the fcale of creation, if by fo doing we can manage to add fomething to our methods of inference in the higheft to which we have as yet attained.

A numerically definite propofition is of this kind. Suppofe the whole number of Xs and Ys to be known: fay there are 100 Xs and 200 Ys in exiftence. Then an affirmative propofition of the fort in queftion is feen in '45 Xs (or more*) are each of them one of 70 Ys': and a negative propofition in '45 Xs (or more) are no one of them to be found among 70 Ys.'

But it muft be particularly noticed that in fpeaking of a number of Xs, as 45 Xs, I do not mean *certain* 45 Xs which can be diftinguifhed from all the reft, fo that of any X it is poffible to be known whether it belong to the 45 of the propofition, or to the remaining 55. This degree of definitenefs is one ftep higher than that which I here propofe to confider, and which is defcribed by 'there are 45 Xs which are contained among 70 Ys, it not being known which Xs are the 45 Xs, nor which Ys are the 70 Ys:' or elfe by 'there are 45 Xs which are not any of them identical with any one of 70 Ys, the precife Xs and Ys in queftion being unknown.'

It cannot of courfe be difputed that if any thing fhould neceffarily follow from *any* 45 Xs being found among *any* 70 Ys, it will not the lefs follow from our knowing which are the Xs and which are the Ys. But this laft fuppofition only brings us to really univerfal propofitions. If, there being 100 Xs, 45 of them can be fpecifically feparated from the reft, fo as to be known, the procefs of feparation is equivalent to putting them

* Thefe words (or more) fhow that the word *definite* has reference only to the *lower* boundary. Of courfe nothing can be fhown in right of "45 or more, *perhaps*" except what is true in right of the 45. It is defirable that as the premifes, fo fhould be the conclufion, of a fyllogifm: this would not be the cafe if we ufed premifes definite both ways. For example, there being 100 Ys in exiftence, it will prefently appear that '*Exactly* 55 Ys are Xs and *exactly* 60 Ys are Zs,' though it enable us to fay that '15 Xs are Zs' does not allow us to fay 'Exactly 15 Xs are Zs,' but only '15 Xs (or more) are Zs.'

under a ſeparate name, ſubidentical to X, and the reſt, which are equally diſtinguiſhable, under another name, alſo ſubidentical to X, and contrary of the firſt name, when the univerſe is X. Whether the name be long or ſhort, does not matter, nor whether it carry the ſeparating diſtinction in its etymology or not. To ſeparate in any way inſtance from inſtance by language, is to name.

If then 45 definite Xs were known to be contained among 70 definite Ys, and if theſe Xs were each named M, and thoſe Ys each N, and if the reſt of the Xs and Ys were named P and Q, we ſhould have the following propoſitions,

M)X, P)X, N)Y, Q)Y, M)N, M.P, N.Q,

and all inferences. Moreover, in each caſe, we ſhould have the total number of inſtances which are contained under each name; the numbers carrying with them evidence that every X is either M or P, and every Y either N or Q. Subſtitute M.N for M)N and we have the correſponding negative propoſition.

But if 45 unſeparated and inſeparable Xs be ſuppoſed known each to be among 70 ſimilarly ſituated Ys, there is no immediate method of making any other propoſition out of the terms X and Y except its converſe, that 45 of theſe 70 Ys are 45 Xs, and (if the whole number of Ys be known, ſay 200) that there are 45 Xs which are not any one among 200 — 70, or 130 Ys. This is then a ſimple propoſition, which becomes of a highly complex character, when the Xs and Ys named in it are taken as definitely ſeparable from the reſt. I ſhall call it the *ſimple numerical* propoſition.

The diſtinction may be eaſily illuſtrated by example. "All the planets but one' is a particular propoſition; it is 'ſome planets:' there is no one planet of right included in it. But 'all the planets except Neptune' is a univerſal propoſition: 'a-planet-not-Neptune' is a name of Mercury, of Venus, &c.; and of every planet it can be ſtated whether it be in the name or not. That which is true inferentially of 'all the planets but one' left particular, is true of 'all the planets but Neptune:' but that which is true of the latter is not neceſſarily true of the former.

Taking X, Y, Z as the terms of the ſyllogiſm, ξ the number of Xs in exiſtence, η the number of Ys, and ζ the number of Zs, and υ the number of inſtances in the univerſe, there are of courſe ſixteen poſſible caſes of knowledge, more or leſs, of theſe primary quantities, from all unknown to all known. Of theſe ſixteen caſes, it will be requiſite to conſider two only. Firſt, when the extent η of the middle term is known, and all the reſt unknown; ſecondly, when all are known. The *algebraical* formulæ of the latter caſe will enable us to point out how the ſuppoſition of any leſs degree of knowledge would affect our power of inference.

I propoſe the following notation. Let mXY denote either of the equivalent propoſitions, that m Xs are to be found among the Ys, or that m Ys are to be found among the Xs. Let mX : nY denote either of the equivalent propoſitions, that there are m Xs which are not any one among n Ys, or n Ys which are not any one among m Xs.

The ſymbol 10X is the algebraical ſymbol for ten equal Xs added together, X being a magnitude: it is then a *collective* ſymbol. In this work, X being a name, it implies every one out of ten inſtances of that name, *diſtributively*, but not *collectively*. This diſtinction is very material, not only in this chapter, but throughout every part of logic. 'Every X is Y' is diſtributively true, when, by 'Every X' we mean each one X: ſo that the propoſition is 'The firſt X is Y, *and* the ſecond X is Y, *and* the third X is Y, &c.' In this caſe the ſubject is X, and the word *every* belongs to the quantity of the propoſition. But 'every X is Y' is *collectively* true, when we do not mean that any one X is a Y, nor that any number of Xs are Ys, but that all the Xs make a Y. In this caſe the propoſition is *ſingular:* there is but one inſtance of the ſubject mentioned, that ſubject being, not X, but the collection 'all the Xs.' Thus 'the ten men are members of a committee' is diſtributive: 'the ten men are a committee' is collective.

If, in ſuch a propoſition as 10XY, we were to ſuppoſe the 10 Xs ſpecifically ſeparated from the reſt, being certain aſſignable ten individuals from among all the Xs, then 10X *becomes a name* for each of the ten, as much as X, and may be conſidered as a univerſal term. And now 10XY and {10X})Y mean the ſame things.

Let η be known, and η only of the four, υ, ξ, η, ζ. The only collections of premiſes which it is neceſſary to conſider are

$$mXY + nYZ$$
$$mXY + nZ:sY$$
$$mX:rY + nZ:sY$$

Without ſome knowledge of the number of ys, of which by ſuppoſition we have none, it would be uſeleſs to attempt to draw an inference from a pair in which Y and y enter together, partially quantified, as in $mXY + nZ:ry$. And nZy merely amounts to $nZ:\eta Y$.

The above three are all we need conſider: and even of theſe the third is incapable of inference, ſince both premiſes are negative, and moreover, not reducible to a poſitive form by uſe of contraries, the only way in which negative premiſes really acquire a concluſion in chapter V.

Let us firſt conſider the premiſes $mXY + nYZ$. They tell us that among the η Ys we find m Xs and n Zs: accordingly, neither m nor n exceeds η. If m and n together fall ſhort of η, nothing can be inferred: Y is extenſive enough (that is, there are inſtances enough of Y) to hold the m Xs and the n Zs without any coincidence of an X with a Z. As to other Xs or Zs, we do not know whether they exiſt; or, if they exiſt, we do not know that any one of them is a Y. But if m and n together exceed η, it is impoſſible that m Xs and n Zs can find place among η Ys, except by putting either two Xs or two Zs, or an X and a Z, with *one* of the Ys. Now as by the nature of the ſuppoſitions, there cannot be two Xs, nor two Zs, to one Y, we muſt have the inference 1XZ as often as there are units in the exceſs of $m+n$ over η. That is,

$$mXY + nYZ = (m + n - \eta)\ XZ$$

Next, let us take $mXY + nZ:sY$. There may be two inferences, perfectly diſtinct from each other, the connexion of which can only be explained in the more general ſyſtem to which we ſhall preſently come. Firſt, let m and s together exceed η. Then $m + s - \eta$ of the Ys have the common property of being Xs, and of being clear of the n Zs. Accordingly, we have

$$mXY + nZ:sY = (m + s - \eta)X:nZ$$

Next, let $n+s$ be greater than η. Take the s Ys among which no one of the n Zs is found. Becauſe $n+s$ is greater than η, n is greater than $\eta-s$, the number of Ys left. Accordingly, $n-(\eta-s)$ of the n Zs cannot be *any* Ys, and therefore cannot be any of the m Xs which are Ys. Hence we have

$$mXY + nZ : sY = mX : (n+s-\eta)Z$$

In the appendix to this chapter (at the end of the work) will be ſeen the manner in which all the Ariſtotelian ſyllogiſms can be brought under the firſt caſe, and the firſt* inference of the ſecond caſe. No Ariſtotelian ſyllogiſm can be deduced from the ſecond inference except when $s=\eta$, in which caſe it agrees with the firſt. For, when s is not η, we muſt, to make ſuch a ſyllogiſm, have $m=\eta$, and then, to make $nZ:sY$ Ariſtotelian, s not being η, we muſt have all the Zs in n, or $n=\zeta$. We thus get $Y)X+sY:Z$, the premiſes of *Bokardo*. But the concluſion is $\eta X:(\zeta+s-\eta)Z$, that of *Bokardo* being $sX:Z$. And this will be found to be the only Ariſtotelian ſyllogiſm which has this ſecond and numerically quantified inference, depending upon the number of Zs exceeding the number of Ys unnamed in the particular premiſe.

I now proceed to ſuppoſe that all the quantities are taken into account. Some preliminary conſiderations will be uſeful, as follows.

Let two propoſitions be called identical, when, either of them being true, the other muſt be true alſo: ſo that nothing can be inferred from the one, which does not equally follow from the other. Such propoſitions are X.Y and Y.X, ſuch are X)Y and y)x, and ſo on. Again, two propoſitions may be identical relatively to a third: thus, P being true, Q and R may either follow from the other; accordingly, as long as it is underſtood that P is true, Q and R may, relatively to that ſuppoſition, be treated as identical.

The word *identical*, as applied to *propoſitions*, is here made to mean more than uſual, but not with more licenſe than when the word is applied to *names*. Thus, *man* and *rational animal* are

* I was not in poſſeſſion of the ſecond inference till I had written what is in page 157.

not identical names, *quà* names, for they neither ſpell nor ſound alike: the identity underſtood is that of meaning; where one applies, there ſhall the other apply alſo. Similarly, as to propoſitions (of which ſubject, predicate, and copula are the material parts, juſt as ſpelling and ſound are thoſe of names), identity does not conſiſt in ſameneſs of parts, nor in reducibility to ſameneſs, but in ſimultaneous truth or falſehood, ſo that what either is, be it true or falſe, the other is alſo, in every caſe. Thus two propoſitions, one of which ſignifies that an end has been gained, and the other that the ſole and ſufficient means of gaining it have been uſed, are identical.

All the theory of *names*, their *application* or *non-application*, may be applied to *propoſitions*, their *truth* or *falſehood*. To ſay that a propoſition is true in a certain caſe, is to ſay that a certain name applies to a certain caſe: to ſay that it is falſe, is to ſay that a certain name does not apply, but that its *contrary* does. That contrary is what logicians uſually call *contradictory*: and the *name* is not ſimply *true* or *falſe*, but the adjective attached to the propoſition. The conditions under which we are to ſpeak limit us to a number of caſes which conſtitute what we may now call, not the *univerſe* of the *names in the propoſitions*, but the univerſe of the *truth or falſehood* of the propoſitions. Thus we ſhall ſuppoſe ourſelves now to be ſpeaking, not of all inſtances to which the *name* U applies, but of all in which the *propoſition* U is *true*, or in which the name 'true U' applies. A caſe in which a propoſition P is true may be marked P, one in which it is falſe, p. We may now apply the names ſubidentical, &c. and the ſymbols, together with all the ſyllogiſms, complex and ſimple; but on each a remark may be neceſſary.

Subidentical, *identical*, and *ſuperidentical*. If P be a propoſition ſubidentical of Q, that is, if every caſe in which P is true be one in which Q is true, but ſo that Q is ſometimes true when P is not, the propoſition Q is uſually mentioned as *eſſential* to P, and as a *neceſſary conſequence* of it. Whenever P is true, Q is true; Q neceſſarily follows from P; if Q be falſe, P cannot be true; Q is eſſential to P; are all mere ſynonymes. Accordingly '*neceſſary conſequent*' and '*ſuperidentical or identical*' are ſynonymous terms: that is (page 68), *neceſſary conſequent* and *ſuperaffirmative*. Identity of courſe conſiſts in *each* propoſition being true when the other is true. I think that, according to general notions, it would be held more juſt to ſay that a propoſition *contains* its neceſſary conſequence than that it *is contained:* but a moment's conſideration will ſhow that the latter analogy is at leaſt as ſound. If the ſecond be true whenever the firſt is true, it may be true in other caſes alſo: ſo that we only ſay the ſecond contains the firſt, and it may be more.

Subcontrary, *contrary*, and *ſupercontrary*. It is uſual to call 'No X is Y' and 'Every X is Y' by the name of contraries, and to ſay that 'contraries may be both falſe, but cannot be both true.' This is a technical uſe of the word: in common language we ſhould ſay that either a propoſition or its contrary muſt be true; 'have you any thing to ſay to the *contrary*' generally means what a logician would expreſs by putting the word *contradictory* in the place of *contrary*. I am compelled to uſe the words contrary and contradictory as ſynonymous: at which compulſion I am well pleaſed, never having ſeen any good reaſon why, in the ſcience which conſiders the relations of *dicta*, the *contraria* ſhould be any thing but the *contra dicta*. The proper word for contrary, commonly uſed to expreſs the relation of X)Y and X.Y, is *ſubcontrary*. Here are two propoſitions P and Q which cannot both be true, but may both be falſe: here is a pair which can never be aſſerted of the ſame inſtance, and of which, in many inſtances, neither can apply. In the ſame manner, the propoſitions XY and X:Y, uſually called *ſubcontrary* (for no reaſon that I can find except that they are written *under* the ſo called *contraries* in a ſcheme or diagram very common in books of logic) ſhould be called *ſupercontrary:* they are never both falſe, and may be both true. This is a complete inverſion of the uſual propoſitions: an inverſion which ſeems to me to be imperatively required, if only my uſe of *ſub* and *ſuper* in Chapter IV. be allowed.

In applying theſe names to propoſitions, it muſt be remembered that we make the ſame ſort of aſcent which we make in paſſing from ſpecific to univerſal arithmetic, in uſing a ſymbol to ſtand for any number at pleaſure. For inſtance;—Perhaps it may be thought that XY and X:Y may ſometimes be only contraries, and not ſupercontraries, becauſe there may be names which make one only true and not both. But this is not correct:

for we are confidering the propofition itfelf *as an inftance among propofitions*, not the propofition as fubdivifible into inftances, in which name is compared with name. In fpeaking of propofitions, it is change from ufe of one name to ufe of another, or from ufe of one number to ufe of another, which is change of inftance: not change from one inftance of name to another. And juft as in a univerfe of names, every name introduced is fuppofed to belong, or not to belong, to every inftance in that univerfe: fo in a univerfe of propofitions, I fuppofe every propofition, or its contrary, to apply (whether it be or be not known which applies) in every inftance. We have never confidered fuch a thing as the univerfe U, in which there are cafes in which neither X nor x applies: we fuppofe there is always a power of declaring that the name X muft either belong or not belong to each inftance. In like manner, all the propofitions in each univerfe now confidered, are fuppofed to be connected with all the names in queftion: fo that X, Y, being two of them in their order of reference, A_1 or O_1 is true in each cafe, and A' or O', E_1 or I_1, and E' or I'. We might, if we pleafed, enter upon a wider fyftem. For though we cannot imagine of any object of thought, but that it is either X or not X, be X what *name* it may, yet we can imagine of *propofitions* that they may be wholly inapplicable, as being neither true nor falfe. The firft affertion is all the more true, that it could hardly be exemplified without exciting laughter: as I fhould do if I reminded the reader that a book is either a cornfield or not a cornfield. We have never confidered names under more predicaments than two; never, for inftance, as if we were to fuppofe three names X_1, X_2, X_3, of which everything muft be one or the other, and nothing can be more than one. But we fhould be led to extend our fyftem if we confidered propofitions under three points of view, as true, falfe, or inapplicable. We may confine ourfelves to fingle alternatives either by introducing not-true (including both falfe and inapplicable) as the recognized contrary of true: or elfe by confining our refults to univerfes in which there is always applicability, fo that true or falfe holds in every cafe. The latter hypothefis will beft fuit my prefent purpofe.

This digreffion is fomewhat out of place here, but I have preferred to retain the matter of it until I had occafion to ufe it.

I now proceed to aſſert that the ſimple numerical propoſition has no occaſion for a numerically definite predicate. Let us conſider firſt an affirmative propoſition, ſay 'Of 10 Xs, each is to be found among ſome 15 Ys.' Of courſe it is ſuppoſed there are 15 or more Ys in exiſtence. With this let us compare '10 Xs are to be found among the Ys.' Theſe two propoſitions are identical: if 10 Xs be among 15 Ys, there are 10 Xs among the Ys: and if 10 Xs be among the Ys they are certainly 10 Ys; put on 5 more Ys at pleaſure, and they can be ſaid to be among 15 Ys in juſt as many ways as we can chooſe 5 more Ys to make up the 15. Note, that if the 10 Xs were among certain ſpecified 15 Ys, then, though the firſt propoſition would give the ſecond, the ſecond would not neceſſarily give the firſt. But we are now ſuppoſing that numerical ſelection is only *numerically* definite: definite as to the number, not as to the inſtances which make up that number. When therefore we ſay '10 Xs are among 15 Ys' we ſay neither more nor leſs than when we ſay 10 Xs are among the Ys.' It is in fact '10 of the Xs are 10 of the Ys' and the converſe '10 of the Ys are 10 of the Xs' is the ſame propoſition.

Now let us take a negative propoſition, '10 of the Xs are not to be found, *any one of them*, among ſome 15 Ys,' abbreviated into '10 Xs are not in 15 Ys.' If there be 25 Ys in exiſtence this propoſition muſt be true; mean X and Y what they may. It is as true as that the X which is one Y is not any other Y. Say there are 25 or more Ys: take any 10 Xs you chooſe, and put them down on any 10 Ys you chooſe. Then certainly there are 15 Ys left, no one of which is any of thoſe 10 Xs. Again, if there be 25 Xs in exiſtence, ſtill the propoſition muſt be true. For if the 15 Ys were all there are, and they were all Xs, there ſtill remain 10 Xs which are not any one in the 15 Ys. Accordingly, the propoſition 'm Xs are all clear of n Ys,' whenever either the whole number of Xs, or the whole number of Ys, exceeds $m+n$, ſays no more than is conveyed in our permanent underſtanding that no object of thought can be more than one X or one Y. But let it be otherwiſe; let neither Xs nor Ys be as many as $m+n$ in number. Say there are 20 Xs and 23 Ys and let 10 Xs be clear of 15 Ys. There muſt now be at leaſt $15+10-20$, or 5 Ys which are *no* Xs *at all*, and at leaſt $15+10-23$, or 2 Xs which are *no* Ys *at all*. Firſt, it is plain that there are no 10 Xs among thoſe Ys which are clear of 15 Ys: for there are but 23 Ys in all. Therefore, 2 at leaſt of theſe 10 Xs muſt be Xs which are not Ys: which with 8 Xs that may be Ys, will be clear of the remaining 15 Ys. Therefore 2 Xs at leaſt are not Ys. Again, there are no 15 Ys among thoſe Xs which are clear of 10 Xs, for there are but 20 Xs in all. Five Ys which are not Xs muſt exiſt, which with 10 that may be Xs, will be clear of the remaining 10 Xs. Accordingly, if the whole number of Xs be ξ, and the whole number of Ys be η, the propoſition ‘there are m Xs which are no one to be found among n Ys’ is eſſentially true of every caſe of that univerſe, whenever $m+n$ is leſs than either ξ or η. But when $m+n$ is greater than both ξ and η, there are two propoſitions, neceſſarily involved, which are not eſſentials of all caſes of that univerſe: namely, that there are $m+n-\xi$ Ys which are not any Xs, and $m+n-\eta$ Xs which are not any Ys.

But, it may be aſked, if η ſhould be leſs than ξ, and $m+n$ greater than η, but ſtill leſs than ξ, may we not affirm that $m+n-\eta$ Xs are not Ys? Undoubtedly we may, but then we do not affirm ſo much as already belongs to every caſe of the univerſe. For if ξ be greater than η, no more than η Xs can be Ys, and there are left $\xi-\eta$ Xs which cannot be Ys: and $\xi-\eta$ is, in the caſe ſuppoſed, more than $m+n-\eta$.

Let υ be the number of inſtances in the univerſe, ξ and η being the number of Xs and of Ys. The following uſes of the notation will be readily ſeen to expreſs preceding reſults, or others immediately deducible.

ξ greater than η	$(\xi-\eta)X:\eta Y$	or $(\xi-\eta)Xy$
η greater than ξ	$(\eta-\xi)Y:\xi X$	or $(\eta-\xi)Yx$

$m+n$ greater than ξ and than η gives

$$mX:nY=(m+n-\eta)X:\eta Y=(m+n-\xi)Y:\xi X$$

A_1	$X)Y=\xi XY$	$=(\upsilon-\eta)yx$	$=\xi Y:(\upsilon-\xi)x$	
O_1	$X:Y=mX:nY$	$=mXy$	$=my:(\upsilon-\xi)x$	
A'	$Y)X=\eta XY$	$=(\upsilon-\xi)xy$	$=\eta X:(\upsilon-\eta)y$	
O'	$Y:X=mY:\xi X$	$=mYx$	$=mx:(\upsilon-\eta)y$	
E_1	$Y.X=\xi Xy$	$=\eta Yx$	$=\xi y:(\upsilon-\xi)x$	$=\eta x:(\upsilon-\eta)y$
I_1	$XY=mXY$	$=mX:(\upsilon-\eta)y$	$=mY:(\upsilon-\xi)x$	
E'	$x.y=(\upsilon-\xi)xY$	$=(\upsilon-\eta)yX$	$=(\upsilon-\xi)Y:\xi X$	$=(\upsilon-\eta)X:\eta Y$
I'	$xy=mxy$	$=mx:\eta Y$	$=my:\xi X$	

I now examine the modes of contradicting mXY and $mX:nY$. As to the firſt, it is obvious that (m always meaning that m are, but that more may be) either m or more Xs are Ys, or elſe $\xi-m+1$ or more Xs are not Ys. The contradiction then is either of the equivalents

$$(\xi-m+1)X:\eta Y \text{ and } (\eta-m+1)Y:\xi X$$

It will be ſatisfactory to evolve the contradiction of $mX:nY$ by a method which will again demonſtrate the caſes in which no contradiction exiſts; or in which the propoſition is always true. Let us put the two names in the leaſt favourable poſition for making $mX:nY$ true. Let p then be the number of Xs which are not Ys, all the reſt being Ys. Take the p Xs which are not Ys (p muſt not be ſo great as m, for then the propoſition is made good by the Xs which are not any Ys) and $m-p$ from thoſe which are Ys. All the m Xs thus obtained are clear of $\eta-(m-p)$ or $\eta-m+p$ Ys. Let this juſt be n: that is, let $p=m+n-\eta$. Then $\xi-p$, the number of Xs which *are* Ys, is $\xi-(m+n-\eta)$ or $\xi+\eta-m-n$. Let but one more X be Y, and the propoſition begins to be contradicted: for now $m+n-\eta-1$ Xs are not Ys, we muſt take up $\eta+1-n$ of thoſe which are Ys to make m Xs, and there only remain $\eta-(\eta+1-n)$ or $n-1$ Ys clear of the m Xs. And it is plain that if we cannot do it by uſing firſt all the Xs which are not Ys at all, ſtill leſs can it be done by uſing thoſe which are. Accordingly the contradiction of $mX:nY$ is

$$(\xi+\eta-m-n+1)XY$$

Then, in order to have a propoſition which can be contradicted, $m+n$ muſt be greater than ξ, or equal to $\xi+1$ at leaſt, for otherwiſe $\xi+\eta-m-n+1$ would be greater than η, or more Ys than η muſt be Xs, which is abſurd: and ſimilarly $m+n$ muſt be greater than η. Otherwiſe, all contradiction is abſurd, or $mX:nY$ is always true.

Aſſuming theſe laſt conditions, however, the contradiction of $mX:nY$ is made eaſier. To be capable of contradiction, it muſt amount to $(m+n-\eta)X:\eta Y$. Now when $m+n-\eta$ Xs are not Ys, and no more, $\xi+\eta-m-n$ Xs *are* Ys. One or more above this, or let $(\xi + \eta - m - n + 1)XY$, and $mX : nY$ cannot be true.

Thus much for contradictory or contrary propoſitions. I ſhall preſently conſider the *contranominal* propoſitions.

We muſt guard ourſelves from preſcribing the uſe of any premiſe which neceſſarily belongs to all caſes in the univerſe (of propoſitions). Let P be a propoſition which may or may not be true, laid down as a premiſe, and Q a propoſition which is true in every caſe. Let R be their neceſſary conſequence, or legitimate inference: then it is not 'whenever P and Q are true, R is true,' but 'whenever P is true, R is true.' So far as R is a conſequence of Q, ſo far it is a conſequence of every thing which neceſſarily gives Q; and thus it is a conſequence of the ſuppoſed conſtitution of the univerſe from which the propoſitions are taken. Now this conſtitution is always underſtood; it may be a convenience that R ſhould be deduced by firſt deducing Q, but it cannot be a neceſſity. And R is a conſequence of P and this conſtitution, not of P and Q.

For example, let the univerſe of propoſitions be all that can be formed out of the ſuppoſitions of the exiſtence of 20 Xs, and 30 Ys, and 40 Zs, in one univerſe of names. Let us join together 15XY and 10Z : 20Y. Our rules of inference will preſently ſhow us, that 5X : 10Z is the neceſſary conſequence of theſe premiſes: but this reſult is not only true when 15XY is true, without anything elſe, but even without that; becauſe 5 + 10 falls ſhort of 40.

Again, we muſt guard ourſelves from adopting the concluſion which follows from premiſes, when that concluſion is true in all caſes by the conſtitution of the univerſe: it is then a ſort of *ſpurious** concluſion, legitimate enough as an inference, but of a perfectly diſtinct character from inferences which would bear

* To this word, as here uſed, I have heard much objection; and when I firſt took it, it was unwillingly, and for want of a better. But on further conſideration I am well ſatisfied with it. The objection ariſes from the idea of falſe or worthleſs being generally attached to the word. But, though it may be uſual for ſpurious things to be worthleſs, it is not neceſſary. If a London maker of razors ſhould put the name of a great Sheffield houſe upon them, thoſe razors would be ſpurious. Suppoſe them as good as thoſe of the Sheffield maker, or better, they are ſtill ſpurious: though it may be true enough

doubt but for the premifes, or would bear contradiction under other premifes. Say that in the above univerfe we join the propofitions 15XY and 30Z : 20Y. Both thefe propofitions are capable of contradiction: the fecond is 20Z : ηY (η means 30, but the fymbol reminds the reader that 30 is *all*) or 10Y : ζZ (ζ being 40). Now, by laws of inference, 15XY + 30Z : 20Y yields 5X : 30Z, which is always true in that univerfe.

Here is a cafe in which premifes capable of contradiction give a conclufion which is not.

The rule of inference is obvioufly as follows. We cannot fhow that Xs are Zs by comparifon of both with a third name, unlefs we can affign a number of inftances of that third name, *more than filled up* by Xs and Zs: that is to fay, fuch that the very leaft number of Xs and Zs which it can contain are together more in number than there are feparate places to put them in. If our premifes, for example, feparate fome 30 Ys, and dictate that among thofe 30 Ys there muft be 20 Xs and 15 Zs, it is clear that there muft be at leaft 5 Zs which are Xs. For if we put down the 20 Xs which are to go in, and try to put the Zs into feparate places, we are ftopped as foon as we have filled up the 10 remaining out of the 30 Ys, and muft put the other 5 Zs among the Ys which have been made Xs. Accordingly, fo many Xs at leaft muft be Zs as there are units in the number by which the Xs and Zs to be placed, together exceed the number of places for them. All the other rules of inference are modifications of this. For example, to prove that 10 Xs are not Zs, we muft fhow fome number of inftances (be they Ys or ys, or part one and part the other) *overfull* (in the above fenfe) of Xs and zs, to the amount of 10 at leaft; fo that 10 Xs are zs, or are not Zs. To prove that fome xs are ys, we muft fhow a number of inftances in which the *leaft* numbers of xs and zs

that the chances are rather in favour of their refembling the ware of Peter Pindar's hero. In this work, a fpurious inference is that which paffes for the confequence of certain premifes, but does not in reality follow from thofe premifes any more than from an infinity of others: being true by the conftitution of the univerfe. It is made to have the mark of thofe premifes, when in truth we cannot know whether thofe premifes be poffible or not, until we have firft examined a conftitution which virtually contains our conclufion.

which it can contain, *overfill* it, or in which the *greateſt* number of Xs and Zs which it can contain *underfill* it, or do not fill it, though made completely ſeparate.

In examining the fundamental laws of ſyllogiſtic inference, it is not neceſſary to conſider any thing but the poſitive forms. For $mX:nY$, when not ſpurious (and we ſhall ſee that the ſpurious caſes may be rejected) is $(m+n-\eta)X:nY$, which is $(m+n-\eta)Xy$ or $(m+n-\xi)xY$. There are, then, but two fundamental caſes: one in which the predicates are the ſame, one in which they are contraries. We ſhall accordingly have to conſider

$$mXY+nZY \qquad \text{and} \qquad mXY+nZy:$$

and it will preſently appear that not more than one, even of theſe, is abſolutely neceſſary. In each caſe we muſt aſk, what collective inſtances of Y or of y, or partly of one and partly of the other, receive any dictation as to how they are to be filled with Xs, with xs, with Zs, or with zs: and what is the leaſt number of each which can be allowed to every ſuch collection. But there is yet ſomething to do, ſuggeſted by the preceding remarks. Let us take one propoſition, a type of all we ſhall have to conſider, ſay mXY. This means that XY is true to at leaſt m inſtances. Now, this propoſition may involve Xy, or xY, or xy. Firſt, as to Xy. To get the leaſt number of Xs among the ys, we muſt put the greateſt number among the Ys. If all the Xs will go among the Ys (or if η be greater than or equal to ξ) there need be no Xs among the ys: but if not (or if η be leſs than ξ) then $\xi-\eta$ Xs muſt be among the ys, in every caſe. Accordingly

$$mXY \text{ gives } (\xi-\eta)Xy$$

where by $\xi-\eta$ underſtand 0, not only when ξ is equal to η, but when it is leſs. This reſult is ſpurious, ſince it is true or falſe, by the mere conſtitution of the univerſe, independently of mXY. Secondly, as to xY. Since mXY is equally mYX, the ſame reaſoning ſhows that

$$mXY \text{ gives } (\eta-\xi)xY$$

where $\eta-\xi$ is to be underſtood in the ſame way. This reſult is alſo ſpurious for a like reaſon.

Thirdly, as to xy. Since there muſt be m Xs among the Ys, the greateſt poſſible number of xs is $\eta-m$. If this be as great as $\upsilon-\xi$, the whole number of xs, there need be no xs among the ys: but if $\eta-m$ be leſs than $\upsilon-\xi$, there muſt then be at leaſt $\upsilon-\xi-(\eta-m)$ xs among the ys, or $\upsilon+m-\xi-\eta$. Conſequently

$$m\mathrm{XY}=(\upsilon+m-\eta-\xi)\ \mathrm{xy}.$$

I here put the ſign $=$ becauſe theſe propoſitions are really equivalents. Treat the ſecond in the ſame way as that which deduced it from the firſt, and we have

$$(\upsilon+m-\eta-\xi)\mathrm{xy}=(\upsilon+\overline{\upsilon+m-\eta-\xi}-\overline{\upsilon-\eta}-\overline{\upsilon-\xi})\mathrm{XY}$$
$$=m\mathrm{XY}$$

If $\upsilon+m$ be not greater than $\eta+\xi$, the equivalent does not exiſt. We are already well acquainted with one caſe of this propoſition. Let $m=\xi$: then mXY is X)Y and the equivalent becomes $(\upsilon-\eta)$xy, which, as $\upsilon-\eta$ is the whole number of ys, is y)x.

The rule is, if two names have a certain number of inſtances at leaſt in common, to the whole number in the univerſe add that number of inſtances, and ſee if the ſum exceed the whole number of inſtances of both names together. If it do ſo, the exceſs ſhows the leaſt number of inſtances which the contraries of theſe two names muſt have in common. Follow this rule, and we have

$$m\mathrm{XY}=(\upsilon+m-\eta-\xi)\,\mathrm{xy}$$
$$m\mathrm{xY}=(\xi+m-\eta)\mathrm{Xy}$$
$$m\mathrm{Xy}=(\eta+m-\xi)\mathrm{xY}$$
$$m\mathrm{xy}=(\xi+\eta+m-\upsilon)\mathrm{XY}$$

Here are exhibited the equivalent *contranominal* forms. The following reſults may now be deduced.

Firſt, theſe contranominals being formed in the ſame way, each from the other, in any one pair, whatever we prove of the firſt from the ſecond, we alſo prove of the ſecond from the firſt. The mathematician would call them *conjugate* pairs. Next, ſince all the four pairs are but verſions of the firſt, with difference of names, whatever we prove univerſally of the firſt pair, we prove of all. Now, taking the firſt of any pair and making it poſſible, which is done by allowing m not to exceed the number of either of the names mentioned, the ſecond may be poſſible or impoſſible, according as the ſubtraction indicated can be done or not. *But whenever the ſecond is impoſſible, the firſt is ſpurious.* Take mXY, and let $(\upsilon+m-\xi-\eta)$xy be impoſſible, or $\upsilon+m$ (and ſtill more υ) leſs than $\xi+\eta$. Now as all the ξ Xs and η Ys muſt find place in the υ inſtances of the univerſe, and $\xi+\eta$ exceeds υ, we muſt, in every caſe of the univerſe of propoſitions, have at leaſt $(\xi+\eta-\upsilon)$XY. But $\upsilon+m$ is leſs than $\xi+\eta$ or $\xi+\eta-\upsilon$ greater than m: conſequently, mXY is ſpurious, a larger propoſition being always true.

As we are not to admit ſpurious propoſitions among our premiſes, we had better write all premiſes double, putting down each of the forms, and making double forms of inference. The preſence of the ſymbols of all neceſſary ſubtractions will remind the reader of the ſuppoſitions which muſt be made, to inſure a legitimate ſyllogiſm. I now take the ſeveral forms.

$$\frac{m \quad\quad \text{XY}}{(\upsilon+m-\xi-\eta)\ \text{xy}} + \frac{n \quad\quad \text{ZY}}{(\upsilon+n-\zeta-\eta)\ \text{zy}} = \frac{(m+n-\eta)\text{XZ}}{(\upsilon+m+n-\eta-\xi-\zeta)\text{xz}}$$

The law of inference here tells us (page 154,) that $m+n$ being greater than η, $(m+n-\eta)$XZ, be it ſpurious or not, follows from the upper premiſes. The lower premiſes alſo give their inference if

$$(\upsilon+m-\xi-\eta)+(\upsilon+n-\zeta-\eta) \text{ be greater than } \upsilon-\eta$$

$\upsilon-\eta$ being the number of the ys. This laſt is equivalent to ſaying that $\upsilon+m+n$ is greater than $\xi+\eta+\zeta$. Firſt, remark that one ſpurious premiſe neceſſarily gives a ſpurious concluſion. Say that $\upsilon+m$ is leſs than $\xi+\eta$, or that mXY is ſpurious. Then, ſince $\upsilon+m$ is leſs than $\xi+\eta$, and n does not exceed ζ, it follows that $\upsilon+m+n$ is leſs than $\xi+\eta+\zeta$; whence the contranominal of the concluſion does not exiſt, or the concluſion is ſpurious, as aſſerted.

Next, obſerve that the concluſion may be ſpurious, though neither of the premiſes be ſo. For though $\upsilon+m$ be greater than $\xi+\eta$, and $\upsilon+n$ than $\zeta+\eta$, and therefore $2\upsilon+m+n$ greater than $\xi+\zeta+2\eta$, or $\upsilon+m+n+(\upsilon-\eta)$ greater than $\eta+\xi+\zeta$, it by no means follows that $\upsilon+m+n$ alone is greater than $\eta+\xi+\zeta$. It is alſo viſible in the mode of formation of the ſecond inference, that to ſay $\upsilon+m$ exceeds $\xi+\eta$, and $\upsilon+n$ exceeds $\zeta+\eta$, only gives *existence* to the premises: to give them conclusion, the sum of the two excesses must itself exceed $\upsilon-\eta$, the whole number of ys.

Thirdly, we must not omit to examine the possible case in which a premise is *partially spurious*. For example, there are 10 Xs and 20 Ys in a universe of 25 instances; accordingly, 10+20−25, or 5, of the Xs *must* be Ys. Let one of the premises be 8XY: this is not then all contingent, and capable of contradiction; we only learn something about 3 out of the 8 Xs. And I call this proposition partially spurious. But it will give no trouble: for we must deal with the premises and their contranominal equivalents before we can pronounce for a conclusion; and of two propositions which are contranominal equivalents of each other, one *must* be partially spurious. To show this, observe that if mXY be *not* partially spurious, it is because υ is greater than $\xi+\eta$; or 2υ than $\xi+\eta+\upsilon$; or $(\upsilon-\xi)+(\upsilon-\eta)$ than υ. But then the numbers of xs and ys together exceed the whole number of instances in the universe; whence some xs must be ys, or the contranominal equivalent of mXY *is* partially spurious.

Now, to write down the various forms of inference. There are sixteen ways of trying for an inference: we may combine a proposition in XY, or xy, or xY, or Xy, with one in XZ, or xz, or xZ, or Xz. But these sixteen cases really combine four and four into only four distinct cases. Thus the one we have been considering, really contains the combinations of XY and YZ, XY and yz, xy and YZ, and xy and yz. It is in our power to make either pair the principal pair, and to give the other pair as contranominals of the first pair.

Thus, we may write the case of inference we have been considering, as in the first of the following list, the others being obtained from the first, by changing X into x, or Z into z, or both. The sign + placed in the middle implies the coexistence of the four propositions: and independent numeral letters are introduced as seen, which will presently be connected with the others by equations, instead of being expressed in terms of them.

$$\left.\begin{matrix} m\mathrm{XY} \\ m'\mathrm{xy} \end{matrix} + \begin{matrix} n\mathrm{YZ} \\ n'\mathrm{yz} \end{matrix}\right\} = \left\{\begin{matrix} p\mathrm{XZ} \\ p'\mathrm{xz} \end{matrix}\right.$$

The equations presently given for this case apply with certain changes to the other cases.

$$\begin{matrix} m\mathrm{xY} \\ m'\mathrm{Xy}\end{matrix} + \left.\begin{matrix} n\mathrm{YZ} \\ n'\mathrm{yz}\end{matrix}\right\} = \left\{\begin{matrix} p\mathrm{xZ} \\ p'\mathrm{Xz}\end{matrix}\right.$$

Here X and x are made to change their former places: in the equations, ξ and ξ' muſt change places.

$$\begin{matrix} m\mathrm{XY} \\ m'\mathrm{xy}\end{matrix} + \left.\begin{matrix} n\mathrm{Yz} \\ n'\mathrm{yZ}\end{matrix}\right\} = \left\{\begin{matrix} p\mathrm{Xz} \\ p'\mathrm{xZ}\end{matrix}\right.$$

Here Z and z change places: as muſt ζ and ζ' in the equations.

$$\begin{matrix} m\mathrm{xY} \\ m'\mathrm{Xy}\end{matrix} + \left.\begin{matrix} n\mathrm{Yz} \\ n'\mathrm{yZ}\end{matrix}\right\} = \left\{\begin{matrix} p\mathrm{xz} \\ p'\mathrm{XZ}\end{matrix}\right.$$

Here X and x, and alſo Z and z, change places; as muſt ξ and ξ', and ζ and ζ', in the equations.

In the new manner of writing the form we have already conſidered, being the firſt of the four, we have juſt written

$$\begin{matrix} m' \text{ for } \upsilon+m-\xi-\eta & . & p \text{ for } m+n-\eta \\ n' \text{ for } \upsilon+n-\zeta-\eta & . & p' \text{ for } \upsilon+m+n-\eta-\xi-\zeta \end{matrix}$$

Let us write ξ', η', ζ', for $\upsilon-\xi$, $\upsilon-\eta$, $\upsilon-\zeta$, the numbers of xs, ys, and zs: and then, $\xi+\xi'$, $\eta+\eta'$, $\zeta+\zeta'$, being all the ſame, (for each is υ) we may write $\eta-\xi'$ for $\xi-\eta'$, $\zeta'-\xi$ for $\xi'-\zeta$, and ſo on. That is, in the difference of two, one of which is accented, we may interchange the letters if we pleaſe. The equations of connection for the firſt or ſtandard caſe, are then

$$\begin{matrix} m'=m+\xi'-\eta=m+\eta'-\xi \\ n'=n+\zeta'-\eta=n+\eta'-\zeta \end{matrix} \quad\Bigg|\quad \begin{matrix} m=m'+\xi-\eta'=m'+\eta-\xi' \\ n=n'+\zeta-\eta'=n'+\eta-\zeta' \end{matrix}$$

$$\left.\begin{matrix} p=m+n-\eta=m'+n'+\eta-\zeta'-\xi' \\ \text{or } m'+n'+\xi-\eta'-\zeta' \\ \text{or } m'+n'+\zeta-\xi'-\eta' \end{matrix}\right\} \begin{matrix} =m+n'-\zeta' \\ =n+m'-\xi' \\ \ \end{matrix}$$

$$\left.\begin{matrix} p'=m'+n'-\eta'=m+n+\eta'-\zeta-\xi \\ \text{or } m+n+\xi'-\eta-\zeta \\ \text{or } m+n+\zeta'-\xi-\eta \end{matrix}\right\} \begin{matrix} =m'+n-\zeta \\ =n'+m-\xi \\ \ \end{matrix}$$

For the ſecond caſe we muſt write $m'=m+\xi-\eta=m+\eta'-\xi'$, and ſo on. I now proceed to the ſeveral diviſions into which our uſual modes of thinking make it convenient to ſeparate the caſes of this moſt general form.

Firſt, when every thing is numerically definite. In this caſe, as ſeen, every form requires an examination of the premiſes and concluſion, as to whether they are or are not ſpurious.

Secondly, when υ, the number of inſtances in the whole univerſe of names, is wholly unknown. In this caſe ξ' is indefinite when ξ is definite, and *vice verſâ*; and ſimilarly one at leaſt of each two, η or η', ζ or ζ', is indefinite. There are then no ſpurious concluſions; or, which is the ſame thing, none which are known to be ſuch: for the ſpurioufneſs of a premiſe or concluſion conſiſts in our *knowing* that it muſt be true of its two terms, independently of all compariſon of thoſe terms with a third.

Thirdly, when ξ, η, ζ, are all indefinite, as well as υ. In this caſe, as here ſtated, there is no poſſibility of inference. We cannot tell whether $m+n$ be or be not greater than η, if we do not know what η is, in any manner, or to any extent.

But here we introduce that degree of definiteneſs by which we diſtinguiſh the *univerſal* from the *particular* (or *poſſible* particular, ſee page 56) propoſition. If we can know that either of the two, m and n, is the ſame as η (greater neither can be) then we know that $m+n$ is greater than η. And at the ſame time we make Y univerſal, in one or the other of the premiſes. And the ſame if we can know that either m' or n' is η'.

The following are the forms which may all be derived from the firſt, by uſing all the varieties of contrary names and contranominal equivalents. If we want, for inſtance, to ſhow the connection of the fourteenth with the firſt, we throw the firſt into the form

$$(m+\eta'-\xi)\mathrm{xy}+n\mathrm{YZ}=(m+n+\eta'-\xi-\zeta)\mathrm{xz}$$

We then change x into X, and Z into z, changing at the ſame time ξ into ξ' and ζ into ζ': and thus we get

$$(m+\eta'-\xi')\mathrm{Xy}+n\mathrm{Yz}=(m+n+\eta'-\xi'-\zeta')\mathrm{XZ}$$

Now, for $m+\eta'-\xi'$ write m', that is, for m write $m'-\eta'+\xi'$ and we have

$$m'\mathrm{Xy}+n\mathrm{Yz}=(m'+n-\zeta')\mathrm{XZ}$$

which is one of the forms of the fourteenth. And $(n+m'-\xi)\mathrm{xz}$ is only the contranominal of $(m'+n-\zeta')\mathrm{XZ}$.

1. $mXY+nYZ=(m+n-\eta)XZ=(m+n+\eta'-\xi-\zeta)xz$
2. $m'xy+nYZ=(n+m'-\xi')XZ=(m'+n-\zeta)xz$
3. $mXY+n'yz=(m+n'-\zeta')XZ=(n'+m-\xi)xz$
4. $m'xy+n'yz=(m'+n'+\eta-\zeta'-\xi')XZ=(m'+n'-\eta')xz$
5. $mxY+nYZ=(m+n-\eta)xZ=(m+n+\eta'-\xi'-\zeta)Xz$
6. $m'Xy+nYZ=(n+m'-\xi)xZ=(m'+n-\zeta)Xz$
7. $mxY+n'yz=(m+n'-\zeta')xZ=(n'+m-\xi')Xz$
8. $m'Xy+n'yz=(m'+n'+\eta-\zeta'-\xi)xZ=(m'+n'-\eta')Xz$
9. $mXY+nYz=(m+n-\eta)Xz=(m+n+\eta'-\xi-\zeta')xZ$
10. $m'xy+nYz=(n+m'-\xi')Xz=(m'+n-\zeta')xZ$
11. $mXY+n'yZ=(m+n'-\zeta)Xz=(n'+m-\xi)xZ$
12. $m'xy+n'yZ=(m'+n'+\eta-\zeta-\xi')Xz=(m'+n'-\eta')xZ$
13. $mxY+nYz=(m+n-\eta)xz=(m+n+\eta'-\xi'-\zeta')XZ$
14. $m'Xy+nYz=(n+m'-\xi)xz=(m'+n-\zeta')XZ$
15. $mxY+n'yZ=(m+n'-\zeta)xz=(n'+m-\xi')XZ$
16. $m'Xy+n'yZ=(m'+n'+\eta-\zeta-\xi)xz=(m'+n'-\eta')XZ$

The syllogisms of chapter V are all particular cases of the above list, obtained as follows:—

1.	$m=\eta$	$A'I_{,}I_{,}$	9.	$m=\eta$	$A'O_{,}O_{,}$
	$m=\eta,\ n=\zeta$	$A'A'A'$		$m=\eta,\ n=\zeta'$	$A'E'E'$
	$n=\eta$	$I_{,}A_{,}I_{,}$		$n=\eta$	$I_{,}E_{,}O_{,}$
	$n=\eta,\ m=\xi$	$A_{,}A_{,}A_{,}$		$n=\eta,\ m=\xi$	$A_{,}E_{,}E_{,}$
	$m=\eta,\ n=\eta$	$A'A_{,}I_{,}$		$m=\eta,\ n=\eta$	$A'E_{,}O_{,}$
	$m=\xi,\ n=\zeta$	$A_{,}A'I'$		$m=\xi\ \ n=\zeta'$	$A_{,}E'O'$
2.	$m'=\xi'$	$A'I_{,}I_{,}$	10.	$m'=\xi'$	$A'O_{,}O_{,}$
	$n=\zeta$	$I'A'I'$		$n=\zeta'$	$I'E'O'$
	$m'=\xi',\ n=\zeta$	$A'A'A'$		$m'=\xi',\ n=\zeta'$	$A'E'E'$
3.	$m=\xi$	$A_{,}I'I'$	11.	$m=\xi$	$A_{,}O'O'$
	$n'=\zeta'$	$I_{,}A_{,}I_{,}$		$n'=\zeta$	$I_{,}E_{,}O_{,}$
	$m=\xi,\ n'=\zeta'$	$A_{,}A_{,}A_{,}$		$m=\xi,\ n'=\zeta$	$A_{,}E_{,}E_{,}$
4.	$m'=\eta'$	$A_{,}I'I'$	12.	$m'=\eta'$	$A_{,}O'O'$
	$m'=\eta',\ n'=\zeta'$	$A_{,}A_{,}A_{,}$		$m'=\eta',\ n'=\zeta$	$A_{,}E_{,}E_{,}$
	$n'=\eta'$	$I'A'I'$		$n'=\eta'$	$I'E'O'$
	$n'=\eta',\ m'=\xi'$	$A'A'A'$		$n'=\eta',\ m'=\xi'$	$A'E'E'$
	$m'=\eta',\ n'=\eta'$	$A_{,}A'I'$		$m'=\eta',\ n'=\eta'$	$A_{,}E'O'$
	$m'=\xi'\ \ n'=\zeta'$	$A'A_{,}I_{,}$		$m'=\xi'\ \ n'=\zeta$	$A'E_{,}O_{,}$

5.	$m = \eta$	E,I,O'	13.	$m = \eta$	E,O,I'
	$m = \eta,\ n = \zeta$	E,A'E,		$m = \eta,\ n = \zeta'$	E,E'A,
	$n = \eta$	O'A,O'		$n = \eta$	O'E,I'
	$n = \eta,\ m = \xi'$	E'A,E'		$n = \eta \quad m = \xi'$	E'E,A'
	$m = \eta,\ n = \eta$	E,A,O'		$m = \eta \quad n = \eta$	E,E,I'
	$m = \xi',\ n = \zeta$	E'A'O,		$m = \xi' \quad n = \zeta'$	E'E'I,
6.	$m' = \xi$	E,I,O'	14.	$m' = \xi$	E,O,I'
	$n = \zeta$	O,A'O,		$n = \zeta'$	O,E'I,
	$m' = \xi,\ n = \zeta$	E,A'E,		$m' = \xi,\ n = \zeta'$	E,E'A,
7.	$m = \xi'$	E'I'O,	15.	$m = \xi'$	E'O'I,
	$n' = \zeta'$	O'A,O'		$n' = \zeta$	O'E,I'
	$m = \xi',\ n' = \zeta'$	E'A,E'		$m = \xi',\ n = \zeta$	E'E,A'
8.	$m' = \eta'$	E'I'O,	16.	$m' = \eta'$	E'O'I,
	$m' = \eta',\ n' = \zeta'$	E'A,E'		$m' = \eta',\ n' = \zeta$	E'E,A'
	$n' = \eta'$	O,A'O,		$n' = \eta'$	O,E'I,
	$n' = \eta',\ m' = \xi$	E,A'E,		$n' = \eta' \quad m' = \xi$	E,E'A,
	$m' = \eta',\ n' = \eta'$	E'A'O,		$m' = \eta',\ n' = \eta'$	E'E'I,
	$m' = \xi \quad n' = \zeta'$	E,A,O'		$m' = \xi \quad n' = \zeta$	E,E,I'

We have thus another mode of eſtabliſhing the completeneſs of the ſyſtem of ſyllogiſm, laid down in the laſt chapter : that is, of the ſyſtem in which there is only the common univerſal and particular quantity. Theſe ſyllogiſms of numerical quantity, in which conditions of inference belonging to every imaginable caſe are repreſented by the general forms which numerical ſymbols take in algebra, muſt of neceſſity be the moſt general of their kind. And examination makes it clear that, except the preceding, there can be no ſyllogiſm exiſting between X, Y, Z, and their contraries. Many ſubordinate laws of connexion might be noticed between the general forms and their particular caſes. Thus, each univerſal occurs three times, each fundamental particular twice, and each ſtrengthened particular twice. The firſt form in pages 158, 159, gives only affirmative, the fourth only negative, premiſes : the ſecond and third one of each kind, commencing with a negative in the ſecond, and with an affirmative in the third.

There are two remarkable ſpecies of ſyllogiſm (or rather, which ought to have been remarkable): which I ſhall now proceed to notice.

The diſtinction of larger and ſmaller part, when diviſion into two parts is made, is as much received into the common idiom of language as the diſtinction of whole and part itſelf. 'Moſt of the Xs are Ys,' is nearly as common as 'All the Xs are Ys:' though 'feweſt of the Xs are Ys,' is only ſeen as 'moſt of the Xs are not Ys.' The ſyllogiſms which can be made legitimate by the uſe of this language will do equally well for any fraction, provided we couple with it the fraction complemental to unity (which in the caſe of one half is one half itſelf). Let α and β ſtand for two fractions which have unity for their ſum, as $\frac{3}{7}$ and $\frac{4}{7}$. Let ${}_{\alpha}XY$ and ${}_{\alpha}X:Y$ indicate that leſs than the fraction α of the Xs are or are not Ys. Let ${}^{\alpha}XY$ and ${}^{\alpha}X:Y$ indicate that more than the fraction α of the Xs are or are not Ys.

Then the following ſyllogiſms ariſe from the caſes with the numbers prefixed.

1.	${}^{\alpha}YX + {}^{\beta}YX = XZ$	9.	${}^{\alpha}YX + {}^{\beta}Y:Z = X:Z$
4.	${}^{\alpha}y:X + {}^{\beta}y:Z = xz$	12.	${}^{\alpha}y:X + {}^{\beta}yZ = Z:X$
5.	${}^{\alpha}Y:X + {}^{\beta}YZ = Z:X$	13.	${}^{\alpha}Y:X + {}^{\beta}Y:Z = xz$
8.	${}^{\alpha}yX + {}^{\beta}y:Z = X:Z$	16.	${}^{\alpha}yX + {}^{\beta}yZ = XZ$

It will be ſeen that here are but three really diſtinct forms; of which the ſimpleſt examples are as follows,

Moſt Ys are Xs; Moſt Ys are Zs; therefore ſome Xs are Zs.
Moſt Ys are Xs; Moſt Ys are not Zs; therefore ſome Xs are not Zs.
Moſt Ys are not Xs; Moſt Ys are not Zs; therefore ſome things are neither Zs nor Xs.

It is hardly neceſſary to obſerve that in one of the premiſes 'more than' may be reduced to 'as much as:' but not in both. Thus, if two-ſevenths exactly of the Ys be Xs, and more than five-ſevenths of the Ys be Zs, it follows that ſome Xs are Zs.

The above ſyllogiſms admit a change of premiſe, as follows: If we ſay that more than $\frac{3}{7}$ths of the Ys are Xs, we thereby ſay that leſs than $\frac{4}{7}$ths of the Ys are xs: or ${}^{\alpha}YX$ and ${}_{\beta}Y:X$ are the ſame propoſitions. Thus, 'moſt are' is equivalent to 'a minority (none included) are not.' Hence we have

$${}_{\beta}Y:X + {}_{\alpha}Y:Z = XZ$$

and ſo on. Or we may combine the two forms, as in

$${}^{\alpha}YX + {}_{\alpha}Y:Z = XZ$$

The above are the only ſyllogiſms in which indefinite particulars give concluſions, by reaſon of that approach to definiteneſs which conſiſts in deſcribing what fractions of *the middle term* are ſpoken of, at leaſt, or at moſt. But they are not the only ſyllogiſms of the ſame general ſpecies. In every caſe inference follows when there is a certain preponderance; and the largeneſs of the inference depends upon the extent of that preponderance. Thus in (12) there is an Xz inference when $m'+n'+\eta$ exceeds $\zeta+\xi'$: ſo many units as there are in this exceſs, ſo many Xs (at leaſt) are zs. Now in every caſe, a pair of univerſal premiſes give inference: and in every caſe there muſt be a degree of approach to univerſality at which inference begins. The ordinary ſyllogiſms, I ſuſpect, are, and are meant to be, not ſuch as '*Every* X is Y, *every* Y is Z, therefore *every* X is Z,' but 'generally ſpeaking X is Y, and generally ſpeaking Y is Z, therefore generally ſpeaking X is Z.' And by 'generally ſpeaking' is meant the aſſertion that an enormous majority of inſtances make the aſſertion true. A ſyllogiſm of this ſort is the oppoſite of the *à fortiori* ſyllogiſm; and might be ſaid to be true *ab infirmiori*. If we have X)Y with p exceptions, and Y)Z with q exceptions; then, in form (1.) we have $m=\xi-p$, $n=\eta-q$, and $m+n-\eta=\xi-p-q$. As long, then, as the number of exceptions altogether fall ſhort of the number of Xs, there is inference: if the total number of exceptions be very ſmall, compared with the number of Xs, there is the 'generally ſpeaking' kind of inference. Examine all the univerſal caſes, and it will be found that the ſame law prevails; namely, that there is inference when the numbers of exceptional inſtances in both premiſes together do not amount to the number of inſtances in the univerſal term of the concluſion; and that there is *exceptional* univerſality (as we may call it) in the concluſion, whenever the whole amount of exception is very ſmall, compared with that number of inſtances.

This leads us to what I will call the theory of *exceptional* particular ſyllogiſms. We have ſeen that the eight complex affirmatory ſyllogiſms, which are all *à fortiori* in their concluſions, afford each two particular ſyllogiſms. We have denoted coexiſtence by +; and the coexiſtence of two propoſitions gives more than either. Let us denote exceptive coexiſtence by —: thus, P—Q means that the propoſition P is true except in the instances contained in Q. Thus, X)Y—X:Y means that every X (with some exceptions) is Y. This is, of course, A_1—O_1, and only differs from I_1 in the mode of expression not being 'some more than none at all' but 'some less than all.' In the expression

$$(A_1—O_1)(A_1—O_1)(A_1—O_1)$$

we have the symbol of the *ab infirmiori* syllogism stated above, subject to the possibility of nonexistence if the number of exceptions in the two premises should exceed the number of instances in the universal term of the conclusion. If we look at A_1O_1, as a symbol descriptive of premises, we see one of the inconclusive forms; that is, a form from which we cannot draw an inference. But this is only because our inferences are all positive, and imply assertion of *sufficiency* in the premises. There is no use (except to show the manner in which the parts of a system hang together) in declarations of *insufficiency*: for we know that all collections of premises, whatever they may be sufficient for, will be insufficient for an infinite number of different things. And it is important to remember that while sufficiency is accompanied by *must be*, insufficiency only allows *may be*. From A_1A_1 the conclusion A_1 *must* be true: from A_1O_1 (and as far as these are concerned) it may be false. Accordingly $A_1O_1O_1$ and $O_1A_1O_1$ may serve to express the two defects of $(A_1—O_1)(A_1—O_1)(A_1—O_1)$ from $A_1A_1A_1$, existing in the *ab infirmiori* syllogism, and possibly preventing conclusion altogether: just as $A_1O'O'$ and $O'A_1O'$ show the additional conditions by the fulfilment of which $A_1A_1A_1$ is elevated into the *à fortiori* syllogism $D_1D_1D_1$. It is worth while to dwell upon the varieties of this case. The *ab infirmiori* syllogisms of the strengthened particulars were previously considered.

In all the cases yet treated, we have had, *more or less*, the power of giving instances in common language, without recourse to numerical relation expressed in unusual terms. This of course, is always the case in the syllogisms of chapter V.; and we have given one *common instance* (though never met with in books of logic) from each set of *ab infirmiori* syllogisms. But there are still cases of the same sort to be considered. Though in our definite relation (page 56) of *all*, we usually (in books of logic at leaſt) make the relation exiſt, for each propoſition, between the terms of the propoſition itſelf, yet it may be aſked whether we cannot ſometimes infer ſuch a ſpecies of univerſal as this, 'for every Z there is an X which is Y;' Z being one of the names of the ſecond premiſe. If we examine the firſt two caſes, which will be guide enough, we ſhall find the following reſults from the new ſuppoſitions now made.

1. $m=\zeta$, $n=\xi$, gives $\zeta XY+\xi YZ=\eta' xz$: or if for every Z there be an X which is Y, and for every X a Z which is Y, then, ſo many ys as there are, ſo many things which are neither X nor Z. This ſyllogiſm has little new meaning, and no new application: it requires $\xi=\zeta$, and therefore X)Y and Z)Y.

2. $m'=\zeta$, gives $\zeta\, xy+n\, YZ=nxz$, or if for every Z there be that which is neither X nor Y, and if ſome Ys be Zs, there are as many inſtances which are neither X nor Z. This is a new and effective form.

2. $n=\xi'$, gives $m'xy+\xi' YZ=m'XZ$, a new form.

Theſe two caſes will be preſently further conſidered. Now, obſerve that if $m+n$ in the firſt form, or $m'+n$ in the ſecond, be υ, that is, if the pair m and n be ξ and ξ', or η and η', or ζ and ζ', we have inference of the kind required. The firſt form gives no new ſyllogiſm: ſince υ is more than η, Ys which are Xs, and Ys which are Zs, to the number of υ, give the form (1.) by the main law of inference (page 154). In the ſecond form, if $m'+n=\upsilon$, we diſtribute among the Ys and ys, Zs and xs to the full number of both, ſo that wherever there are not xs (that is, wherever there are Xs) there are Zs: or X)Z as obtained from the form.

But every way of conſtructing $m'xy+nYZ=(m'+n-\xi')XZ$ which gives $m'+n=\upsilon$, is only a caſe of $A_1A_1A_1$. For m' cannot exceed η', and n cannot exceed η: and $m'+n$ being υ or $\eta+\eta'$, we muſt have $m'=\eta'$ and $u\doteq\eta$; whence the aſſertion made. The forms we are now in ſearch of, ſo far as quite new, are all contained in the two new ones above noted; and of theſe, the ſecond is but a transformation of the firſt. The eight varieties derived from uſe of contraries, or from the forms in page 161, beginning with the ſimpleſt, are

$\zeta XY\ +nZ:\eta Y=nX:\zeta Z$	$\zeta' XY\ +nyz\ =nXZ$
$\zeta X:\eta Y+nZY\ =nX:\zeta Z$	$\zeta' X:\eta Y+nY:\zeta Z=nXZ$

$\zeta Y:\xi X + nZ:\nu Y = nxz$	$\zeta' Y:\xi X + nyz = nZ:\xi X$
$\zeta xy + nZY = nxz$	$\zeta' xy + nY:\zeta Z = nZ:\xi X$

These are syllogisms, which exhibit a curious kind of antagonism to the particular syllogisms. Take the syllogism $A_{\prime}O'O'$, the terms being M,Y,Z; we have then $M)Y + Z:Y = Z:M$. Of course the conclusion $M:Z$ is not legitimate from these premises alone: but if M have as many instances as Z, then $M:Z$ *is* legitimate. For if Ms, as many as there are Zs, be among the Ys, and some of the Zs be not among the Ys, though all the rest were, there would not be enough to match all the Ms, or some Ms are not Zs. Now, let M be a name given to an X which is Y, and let such Xs have as many instances as Z, and the above becomes the first of the syllogisms in the last list. Thus, $I_{\prime}O'O_{\prime}$ is legitimate, if the quantity of the subject mentioned in $I_{\prime}$ be taken from the Zs. The second syllogism is $E_{\prime}I_{\prime}O'$, altered into $O_{\prime}I_{\prime}O_{\prime}$ in the same manner.

The reader may find all the results of the above case in the following rule, in which it is understood that all the super-propositions are to be written either way: thus, A' is written x)y, or Y)X, and O' is $nx:\nu y$, or $nY:\xi X$ (page 62). Write down any pair of particulars, followed by I if the pair be of the same sign, and O if the pair be of different signs: as in OOI or IOO. Accent the pair in contradiction to either the direct rule (page 62) as far as the words affirmative and negative are concerned: that is, let a *negative* beginning isolate nothing, and an *affirmative* beginning isolate the middle proposition: or else, accent the pair according to the inverse rule. Thus, $O_{\prime}O_{\prime}I_{\prime}$ and $O'O'I'$ contradict the direct rule, and $O'O'I_{\prime}$ and $O_{\prime}O_{\prime}I'$ preserve the inverse rule. To make these syllogisms good (in the particular way in question) proceed thus:—When the *direct* rule is contradicted, take the quantity of the *first* concluding term from the total of the *second*, if the second premise be affirmative, and from its contrary, if negative. When the inverse rule is preserved, take the quantity of the *second* from the total of the *first*. Thus, in $O'O'I'$ the direct rule is contradicted: and it stands $m'x:\nu'y + n'y:\zeta'z = p'xz$. The second premise is negative, the total of its predicate ζ' instances, that of the contrary ζ. Accordingly, $\zeta x:\nu'y + n'y:\zeta'z = n'xz$, or $\zeta Y:\xi X +$

$n'Z:\eta Y = n'xz$, which is one of the forms already obtained. Again, O'O'I, preferves the inverfe rule, and is $m'x:\eta'y + nZ:\eta Y = pXZ$. The total of the firft term is ξ' inftances. Hence, $m'x:\eta'y + \xi'Z:\eta Y = m'XZ$, or $m'Y:\xi X + \xi'Z:\eta Y = m'XZ$, which is derived from one of the forms given, by interchanging X and Z.

This clafs of fyllogifms *with tranfpofed quantity* naturally leads to the queftion, Is it ufed? Do fuch fyllogifms occur in ordinary or in literary life? If not, there is no reafon for felecting them from the infinite number of cafes which the numerically definite fyftem affords. To try this, fuppofe a perfon, on reviewing his purchafes for the day, finds, by his countercheques, that he has certainly drawn as many cheques on his banker (and may be more) as he has made purchafes. But he knows that he paid fome of his purchafes in money, or otherwife than by cheques. He infers then that he has drawn cheques for fomething elfe except that day's purchafes. He infers rightly enough; but his inference cannot be reduced to a common fyllogifm, with the names in queftion for terms. It is really a fyllogifm of tranfpofed quantity, as follows:—

For every 'memorandum of a purchafe' a 'countercheque' is a 'tranfaction involving the drawing of a cheque.'

Some 'purchafes' are not 'tranfactions involving, &c.'

Therefore fome 'countercheques' are not 'memoranda of purchafes.'

It may be worth while to give one inftance of the verification of the contradictory form. By page 152 it appears that the contradiction of mXY is $(\xi - m + 1)Xy$, or $(\eta - m + 1)xY$, and that of $m'Xy$ is $(\xi - m' + 1)XY$, or $(\eta' - m' + 1)xy$.

To mXY join the contrary of $(m + n - \eta)XZ$, or $(\xi + \eta - m - n + 1)Xz$: we have then

$$mYX + (\xi + \eta - m - n + 1)zX;$$

the inference of which is $(m + \xi + \eta - m - n + 1 - \xi)Yz$, that is, $(\eta - n + 1)Yz$, the contrary of nYZ.

Returning to the forms in page 161, it will be obferved that we have no double inferences. In every cafe we have made ufe of one form of inference: if υ be known, the other is a real equivalent; or elfe it is impoffible, and as we have feen, then the firft is fpurious. If υ be not known, then the fecond is either perfectly indefinite, or elfe identical with the one chofen. Examination will fhow that in every one of the cafes cited in page 161, the neglected form of inference is only faved from perfect indefinitenefs when we are able to apply the word *all* to one or other of the terms: the number being as indefinite as before; the *relation* thus obtained being definite. Take the firft form, and make $n=\eta$; by the firft inference we get then the fyllogifm $I_1A_1I_1$: by the fecond, we get $(m+\upsilon-\xi-\zeta)$xz, indefinite both in number and relation. We do not know what υ, ξ, and ζ are. If we knew as much as that $m+\upsilon$ is lefs than $\xi+\zeta$, we fhould know our inference to be fpurious,* it being not the lefs an inference. Now, add the condition $m=\xi$: the firft inference gives the fyllogifm $A_1A_1A_1$, the fecond inference now becomes $(\upsilon-\zeta)$xz: definite relation enters, and we have z)x, or X)Z, or A_1, as before. And the fame of the other forms.

The reader may perhaps fuppofe that I ought to have commenced this chapter with the complex numerical fyllogifm, in imitation of the method which I followed in treating the ordinary fyllogifm. But in truth there is no fyftem of complex fyllogifm of perfect numerical definitenefs both in premifes and conclufion. To fhow this, let m,XY with the comma, mean that there are exactly m Xs which are Ys, neither more nor fewer. Accordingly m,XY is a fynonyme for mXY$+(\eta-m)$xY. Now combine m,XY and n,ZY, or

$$(m\mathrm{XY}+\overline{\eta-m}\,\mathrm{xY})(n\mathrm{ZY}+\overline{\eta-n}\,\mathrm{zY})$$

We then have
$$\begin{aligned} m\mathrm{XY}+n\mathrm{ZY} &= (m+n-\eta)\mathrm{XZ}\\ (\eta-m)\mathrm{xY}+(\eta-n)\mathrm{zY} &= (\eta-m-n)\mathrm{xz}\\ m\mathrm{XY}+(\eta-n)\mathrm{zY} &= (m-n)\mathrm{Xz}\\ (\eta-m)\mathrm{xY}+n\mathrm{ZY} &= (n-m)\mathrm{xZ}\end{aligned}$$

* I muft again remind the reader, of the diftinction between *fpurious* and *illegitimate*, which exifts in my language. The fpurious inference follows from the premifes, and is perfectly good and true: but from the conftitution of the univerfe, it will always be true, whatever premifes in that univerfe are taken. The illegitimate inference is that which does not follow from the premifes. A conclufion not *known* to be fpurious, that is, there not being the means of knowledge, *is not* fpurious: but an illegitimate conclufion cannot be made legitimate, that is, following from the premifes, by any further knowledge.

Two only of theſe have meaning: let them be the two upper ones. We can aſſign then Z or z to $(m+n-\eta)+(m-n)$, or to $2m-\eta$ of the Xs. But there are not all of the Xs here: for m is leſs than η, and than ξ, whence $2m$ is leſs than $\eta+\xi$, or $2m-\eta$ leſs than ξ. The reſt of the Xs, $\xi+\eta-2m$ in number, may, for aught theſe premiſes declare, be either Zs or zs.

CHAPTER IX.

On Probability.

THE moſt difficult inquiry which any one can propoſe to himſelf is to find out what any thing *is*: in all probability we do not know what we are talking about when we aſk ſuch a queſtion. The philoſophers of the middle ages were much concerned with the *is*, or *eſſence*, of things: they argued to their own minds, with great juſtice, that if they could only find out what a thing is, they ſhould find out all about it: they tried, and failed. Their ſucceſſors, taking warning by their example, have inverted the propoſition; and have ſatisfied themſelves that the only way of finding what a thing is, lies in finding what we can about it; that modes of relation and connexion are all we can know of the eſſence of any thing; in ſhort, that the proverb 'tell me who you are with, and I will tell you what you are,' applies as much to the nature of things as to the characters of men. We are apt to think that we know more of the eſſence of objects than of ideas; or rather, of ideas which have an objective ſource, than of thoſe which are the conſequence of the mind's action upon them. I doubt whether the reverſe be not the caſe: at any rate, when we content ourſelves with inquiry into properties and relations, we have certain knowledge upon our moſt abſtract ideas. The object of this chapter is the conſideration of the degrees of knowledge itſelf. That which we know, of which we are certain, of which we are well aſſured nothing could perſuade us to the contrary, is the exiſtence of our own minds, thoughts, and perceptions, the two ran when actually preſent. This higheſt knowledge, this abſolute certainty, admits of no imagination of the poſſibility of falſehood. We cannot, by ſtopping to conſider, make ourſelves more ſure than we are already, that we exiſt, think, ſee, &c. Next to this, come the things of which we cannot but ſay *at laſt* we are as certain of them as of our own exiſtence; but of which, nevertheleſs, we are obliged to ſay that we arrive at them by proceſs, by reflection. Theſe we call *neceſſary truths* (page 33). The *neceſſity* of admitting theſe things cauſes ſome to imagine that they are merely identities, that they amount to ſaying that when a thing is, it is: but this is not correct. To ſay that two and two make four (which muſt be), and that a certain man wears a black coat (when he does ſo) both involve the pure identity that whatever is, is; and not one more than the other. Nor is two and two identically four, though neceſſarily ſo. Our definitions of number ariſe in the proceſs of ſimple counting. Throw a pebble into a baſket, and we ſay *one*: throw in another, and we ſay *two*; yet one more, and we ſay *three*, and ſo on. The full definitions of the ſucceſſive numbers are ſeen in

$$1 \quad (1+1) \quad \{(1+1)+1\} \quad [\{(1+1)+1\}+1], \text{\&c.}$$

That three and one are four is *definition*: it is our pleaſure to give the name *four* to $3+1$. But that $3+1$ is $2+2$ is neither definition nor pure identity. It is not even true that 'two and two' *is* four; that

$$[\{(1+1)+1\}+1] \textit{ is } (1+1)+(1+1)$$

It is true, no doubt, that 'two and two' is four, in amount, value, &c. but not in form, conſtruction, definition, &c.

There is no further uſe in drawing diſtinction between the knowledge which we have of our own exiſtence, and that of two and two amounting to four. This abſolute and inaſſailable feeling we ſhall call *certainty*. We have lower grades of knowledge, which we uſually call *degrees of belief*, but they are really *degrees of knowledge*. A man knows at this moment that two and two make four: did he know it yeſterday? He feels perfectly certain that he knew it yeſterday. But he may have been ſeized with a fit yeſterday, which kept him in unconſciouſneſs all day: and thoſe about him may have been warned by the medical man not to give him the leaſt hint of what has taken place. He could ſwear, as oaths are uſually underſtood, that it was not ſo: if he could not ſwear to this, no man could ſwear to anything except neceſſary truths. But he could not regard the aſſertion that it was not ſo, as incapable of contradiction: he knows it well, but, as long as it may poſſibly be contradicted, he cannot but ſay that he might know it better.

It may ſeem a ſtrange thing to treat *knowledge* as a magnitude, in the ſame manner as length, or weight, or ſurface. This is what all writers do who treat of probability, and what all their readers have done, long before they ever ſaw a book on the ſubject. But it is not cuſtomary to make the ſtatement ſo openly as I now do: and I conſider that ſome juſtification of it is neceſſary.

By degree of probability we really mean, or ought to mean, degree of belief. It is true that we may, if we like, divide probability into ideal and objective, and that we muſt do ſo, in order to repreſent common language. It is perfectly correct to ſay 'It is much more likely than not, *whether you know it or not*, that rain will ſoon follow the fall of the barometer.' We mean that rain does ſoon follow much more often than not, and that there do exiſt the means of arriving at this knowledge. The thing is ſo, every one will ſay, and can be known. It is not remembered, perhaps, that there is an *ideal probability*, a pure ſtate of the mind, involved in this aſſertion: namely, that the things which have been are correct repreſentatives of the things which are to be. That up to this 21ſt of June, 1847, the above ſtatement has been true, ever ſince the barometer was uſed as a weather-glaſs, is not denied by any who have examined it: that the connexion of natural phenomena will, for ſome time to come, be what it has been, cannot be ſettled by examination: we all have ſtrong reaſon to believe it, but our knowledge is *ideal*, as diſtinguiſhed from *objective*. And it will be found that, frame what circumſtances we may, we cannot invent a caſe of purely objective probability. I put ten white balls and ten black ones into an urn, and lock the door of the room. I may feel well aſſured that, when I unlock the room again, and draw a ball, I am juſtified in ſaying it is an even chance that it will be a white one. If all the metaphyſicians who ever wrote on probability were to witneſs the trial, they would, each in his own ſenſe and manner, hold me right in my aſſertion. But how many things there are to be taken for granted! Do my eyes ſtill diſtinguiſh colours as before? Some perſons never do, and eyes alter with age. Has the black paint melted, and blackened the white balls? Has any one elſe poſſeſſed a key of the room, or got in at the window, and changed the balls? We may be *very ſure*, as thoſe words are commonly uſed, that none of theſe things have happened, and it may turn out (and I have no doubt will do ſo, if the reader try the circumſtances) that the ten white and ten black balls will be found, as diſtinguiſhable as ever, and unchanged. But for all that, there is much to be aſſumed in reckoning upon ſuch a reſult, which is not ſo objective (in the ſenſe in which I have uſed the word) as the knowledge of what the balls were when they were put into the urn. We have to aſſume all that is requiſite to make our experience of the paſt the means of judging the future.

Having made this illuſtration to draw a diſtinction, I now premiſe that I throw away objective *probability* altogether, and conſider the word as meaning the ſtate of the mind with reſpect to an aſſertion, a coming event, or any other matter on which abſolute knowledge does not exiſt. 'It is more probable than improbable' means in this chapter 'I believe that it will happen more than I believe that it will not happen.' Or rather 'I *ought* to believe, &c.:' for it may happen that the ſtate of mind which *is*, is not the ſtate of mind which ſhould be. D'Alembert believed that it was *two* to *one* that the firſt head which the throw of a halfpenny was to give would occur before the third throw: a juſter view of the mode of applying the theory would have taught him it was *three* to *one*. But he *believed* it, and thought he could ſhow reaſon for his belief: to him the probability *was* two to one. But I ſhall ſay, for all that, that the probability *is* three to one: meaning, that in the univerſal opinion of thoſe who examine the ſubject, the ſtate of mind to which a perſon *ought* to be able to bring himſelf is to look three times as confidently upon the arrival as upon the non-arrival.

Probability then, refers to and implies belief, more or leſs, and belief is but another name for imperfect knowledge, or it may be, expreſſes the mind in a ſtate of imperfect knowledge. There is accurate meaning in the phraſe 'to the beſt of his *knowledge* and *belief*,' the firſt word applying to the ſtate of his circumſtances with reſpect to external objects, the ſecond to the ſtate of his mind with respect to the circumstances. But we cannot make any use of the distinction here: what we know is to regulate what we believe; nor can we make any effective use of what we know, except in obtaining and describing what we believe, or ought to believe. According to common idiom, belief is often a lower degree of knowledge: but it is imperative upon us to drop all the *quantitative* distinctions of common life, or rather to remodel them, when we come to the construction of a science of quantity.

I have said that we treat knowledge and belief as magnitudes: I will now put a broad illustration of what I mean. We know, (suppose it *known*) that an urn contains nothing but two balls, one white and one black, undistinguishable by feeling: and we know (suppose this also) that a ball is to be drawn. Disjunctively then we know 'white will be drawn: black will be drawn,' one or the other must be. How do we stand as to 'white will be drawn,' and 'black will be drawn,' separately? Clearly in no preponderance with respect to either. May we then properly and reasonably say that we divide our knowledge and belief of the event 'one or the other' into two halves, and give one half to each. I can conceive much objection to this supposition: but, whether they formally make it or not, I am sure writers on probability act upon it, and are accepted by their readers.

Let us consider what magnitude is, that is to say, how we know we are talking about a magnitude. We know that whenever we can attach a distinct conception of more and less to different instances, so as to say this has more than that, we are talking of comparable magnitudes. We speak of a quantity of talent, or of prudence: we say one man has more talent than another, and one man more prudence than another: but we never say that one man has more talent than another has prudence. If we occasionally say he (the same one man) has more talent than prudence, it is only as an abbreviation: we mean that he has not prudence enough to guide his talent. Just as we might say (though we do not) that there is more cart than horse, when the horse cannot draw the cart: just as, speaking very loosely, we *do* say, the *pressure* of the atmosphere is not fifty *inches*; meaning that it is not enough to balance the pressure of fifty inches of mercury in the barometer. And thus, both up to, and beyond our means of meaſurement, we form to ourſelves diſtinct notions of comparable magnitudes, and incomparable magnitudes, as well as of the meaning of the ſomewhat incorrect, but eaſily amended, figures of ſpeech by which we ſometimes talk of comparing the latter.

But the object of all quantitative ſcience is not merely magnitude, but the *meaſurement* of magnitude. And when are we entitled to ſay that we can meaſure magnitude? As ſoon as we know how, from the greater, to take off a part equal to the leſs: a proceſs which neceſſarily involves the teſt of which is the greater, and which is the leſs, and, in certain caſes, as it may happen, of neither being the greater nor the leſs. As to ſome magnitudes, the clear idea of meaſurement comes ſoon: in the caſe of length, for example. But let us take a more difficult one, and trace the ſteps by which we acquire and fix the idea: ſay *weight*. What weight is, we need not know: the Newtonian, who makes it depend on the earth's attraction, and the Ariſtotelian, who referred it to an impulſe which all bodies poſſeſs to ſeek their *natural places*, are quite at one on their notions of the meaſurable magnitude which their ſeveral philoſophies diſcuſs. We know it as a magnitude before we give it a name: any child can diſcover the *more* that there is in a bullet, and the *leſs* that there is in a cork of twice its ſize. Had it not been for the ſimple contrivance of the balance, which we are well aſſured (how, it matters not here) enables us to poiſe equal weights againſt one another, that is, to detect equality and inequality, and thence to aſcertain how many times the greater contains the leſs, we might not to this day have had much clearer ideas on the ſubject of weight, as a magnitude, than we have on thoſe of talent, prudence, or ſelf-denial, looked at in the ſame light. All who are ever ſo little of geometers will remember the time when their notions of an angle, as a magnitude, were as vague as, perhaps more ſo than, thoſe of a moral quality: and they will alſo remember the ſteps by which this vagueneſs became clearneſs and preciſion.

Now a very little conſideration will ſhow us that, the moment we begin to talk of our belief (the mind's meaſure of our knowledge) of propoſitions ſet before us, we recognize the relations called more and leſs. Does the child feel that the bullet has more ſomething than the cork one bit better than an educated man feels that his belief in the ſtory of the death of Cæſar is more than his belief in that of the death of Remus. Let any one try whether he have not in his mind the means of arranging the following ſet in order of magnitude of belief, including within that term all the range which comes between certain knowledge of the falſehood, and certain knowledge of the truth, of an aſſertion. Let them be 1. Cæſar invaded Britain with the ſole view of benefiting the natives. 2. Two and two make five. 3. Two and two make four. 4. Cæſar invaded Britain. 5. Romulus founded Rome. He will probably diſcover the gradations of neceſſary truth, moral certainty, reaſonable preſumption, utter incredibility, and neceſſary falſehood. Theſe are but names given to different ſtates of the mind with reſpect to knowledge of propoſitions aſſerted; and I ſay they expreſs different ſtates of quantity.

The only difficulty, and a ſerious one it can be made, may be ſtated in the following queſtion;—Are we to conſider the ſort of belief which we have of a neceſſary propoſition (as two and two make four), that is, abſolute knowledge, to which contradiction is glaring abſurdity—as only a ſtrengthened or augmented ſpecimen of the ſort of knowledge which we have of any contingent propoſition (ſuch as Cæſar invaded Britain) which may have been, or might have been, falſe, and can be contradicted without abſurdity? I anſwer, we can eaſily ſhow that the difference of the two caſes is connected with the difference between finite and infinite, not between two magnitudes of different kinds. The mathematician will eaſily apprehend this, and will look upon the various difficulties which ſurround even the explanation as upon things to which he is well accuſtomed, and which he underſtands by many parallel inſtances. We can invent circumſtances under which a contingent propoſition ſhall make any degree of approach to neceſſity which we pleaſe, but ſo that no actual attainment ſhall be arrived at. If an urn contain balls, and if one ball muſt be drawn, then, the balls being all white, it is neceſſary that a white ball muſt be drawn, as neceſſary as that two and two being in any place, there are four in that place: for there are no degrees of neceſſity. But let it be that there are black balls alſo, at the rate of one to a thouſand white ones: the drawing of a white ball is no longer neceſſary; but there is ſtill a ſtrong degree of aſſurance that a white ball will be drawn. We do not readily ſee how much: becauſe the urn has no viſible relation to our uſual caſes of judgment. But let it be made to repreſent the life of a youth of twenty: and let the drawing of a white ball repreſent his living to come of age, and of a black one his death in the interval. There ought to be *ſeven* black balls to the thouſand white ones to make the caſes parallel. And yet we know that our aſſurance of his ſurvival is generally very ſtrong: be it wiſe aſſurance or not, it exiſts, and we act upon it. Now ſuppoſe the rate to be one black to a million of white: the aſſurance is much increaſed, but ſtill there is no neceſſity; the black ball may be drawn. Take one black to a million of million of white, or a million of million of million, &c.: long before we have arrived at ſuch a point, we have loſt all conception of the quantitative difference between our belief in drawing a white ball, and our belief that two and two are four. We ſay it is *almoſt impoſſible* that one trial ſhould give a black ball: and this very phraſe is a recognition of the ſameneſs for which I am contending. Except on the ſuppoſition of ſuch ſameneſs, there is no *almoſt impoſſible*, nor *nearly certain*. Between the impoſſible and the poſſible, the certain and the not certain, there muſt be every imaginable difference, if we do not admit unlimited approach. For it will clearly not be contended that, repreſenting certainty, ſay by 100, we can make an approach to it by an uncertainty counting as, ſay 90, but nothing higher. Repreſenting the ſtate of abſolute knowledge by 100, any one, with a little conſideration, will ſay that the laws of thought fix no numerical limit to our approach towards this ſtate: but that things ſhort of certainty are capable of being brought within any degree of nearneſs to certainty. On ſuch conſiderations, I ſhall aſſume that neceſſity on the one hand, a certainty for, and impoſſibility on the other, a certainty againſt, are extreme limits, which being repreſented by quantities, may allow our knowledge of all contingent propoſitions to be repreſented by intermediate quantities.

It muſt be fully allowed, nay, imperatively inſiſted on, that nothing in the numerical view, tending to connect neceſſary and contingent propoſitions, can at all leſſen the diſtinction between them: nor give the latter any reſemblance to the former, except only in the quantities by which they are indicated. Though there be only one black ball to as many white ones as would fill the viſible univerſe, yet between that caſe and the one of no black balls muſt always exiſt the eſſential difference, that in the former a black ball *may* be drawn, and in the latter it *cannot*. But this very great diſtinction between the neceſſarily certain and the contingent, is it compatible with their being repreſented by numerical quantities as near to one another as we pleaſe? I anſwer that all who are acquainted with the relations of quantity are aware that nearneſs of value is no bar to any amount of difference of properties. A common fraction, for inſtance, may be made as near as we pleaſe in value to an integer: but there do not exiſt, even among propoſitions, more eſſential, or more ſtriking, differences, than thoſe which exiſt between the properties of integers and of fractions. There are crowds of theorems (I ſhould rather ſay unlimited crowds of claſſes of theorems) which are always true when integers are uſed, and never true when fractions are uſed. Let any quantities be named, integer or fractional, and it is eaſy to make claſſes of theorems which are true for thoſe quantities, and not for any others, however near to them. The reader who is not a mathematician muſt rely upon the knowledge of the one who is, that the difference between two quantities, no matter how nearly equal, may be connected with other differences as complete, and by practice as eaſily recognized, as the difference between neceſſary and contingent truth.

I will take it then that all the grades of knowledge, from knowledge of impoſſibility to knowledge of neceſſity, are capable of being quantitatively conceived. The next queſtion is, are theſe quantities capable, *in any caſe*, of meaſurement, or of compariſon with one another. At preſent, we ſtand as the child ſtands with reſpect to the bullet and the cork: perceptive of more and leſs, but without a balance by which to make compariſons. To ſhow the poſtulate on which our balance depends, let us ſuppoſe an urn, which, to our knowledge, contains white, black, red, green, and blue balls, one of each colour. It is within our knowledge that a ball muſt be drawn: accordingly we have full knowledge (and of courſe *entire belief*) that the reſult 'no ball' is impoſſible, and that 'white, or black, or red, or green, or blue' is neceſſary. To the reſult 'white' we accord a certain probability, that is, a certain amount of belief. If a man tell us that white will be drawn, we may hold him raſh, but we do not pronounce his communication incredible : let another tell us that 'black, or red, or green, or blue' will be drawn, and we hold him not ſo raſh, and his communication more credible. We may hold with either, if he will deſcribe his knowledge and belief as partial, and give them their proper amounts. Now, whether we ſhall proceed, or ſtop ſhort at this point, depends upon our acceptance or non-acceptance of the following POSTULATE :—

When any number of events are disjunctively poſſible, ſo that one of them may happen, but not more than one, the meaſure of our belief that one out of any ſome of them will happen, ought to be the amount of the meaſures of our ſeparate beliefs in each one of thoſe ſome.

I mean that any one ſhould ſay, A, B, C, being things of which not more than one can happen, 'my belief that one of the three will happen is the ſum of my ſeparate beliefs in A, and in B, and in C.' This is the poſtulate on which the balance depends ; and there is a ſimilar poſtulate before we can uſe the phyſical balance. The only difference (and that but apparent) is that we are to ſpeak of weights collectively, and of events disjunctively. The weight of the (conjunctive) maſs is the ſum of the weights of its parts : the credibility of the (disjunctive) event is the ſum of the credibilities of its components. There are ſeveral *may-bes*, any one of which may become a *has-been :* when we ſpeak *disjunctively*, it is of the *will-be*, which cannot be ſaid of more than one : the *may-be* of an event deſcribed as contained in 'A, B, C,' is to be repreſented as in quantity the ſum of thoſe in 'A,' in 'B,' and in 'C.'

Is it matter of mere neceſſity that, talking of phyſical weight, the weight of the whole is equal to the ſum of the weights of the parts ? We have learnt to admit this poſtulate, of which no man ever doubted : but no one can ſay that it was neceſſary. The laws of matter and mind being both what they are, the connexion between phyſical collection and mental ſummation is, I grant, neceſſary : the ſimpleſt of manual, and the ſimpleſt of mental, operations, are and, with us, muſt be, concomitants.

But, in the firſt place, it is *not true* that the weight of the whole is equal to the ſum of the weights of the parts, in the manner in which the reader probably imagines it to be true. Let the firſt part we hang on the balance be the weight which is correctly meaſured by W. Then if we hang under it another weight, as correctly repreſented by V, we think we are quite ſure when we ſay that the collective maſs muſt have a weight W + V becauſe its parts have the weights W and V. But its parts have *not* the weights W and V. The weight of V is diminiſhed by the upward attraction of W, and is, ſay, V − M : the weight of W is as much increaſed by the downward attraction of V, and is W + M. And though V − M and W + M added together do give V + W, yet it was not in this way that the reader made out his neceſſary truth. The univerſal equality of action and reaction did not exiſt in the thoughts of the firſt perſon who formed a diſtinct conception of the weight of the whole as compoſed of the ſum of that of the parts: and he was only right by the (ſo far as he was concerned) accidental circumſtance, that two things of which he knew nothing, counterbalanced each other's effects. Nor do we know at this moment, as of neceſſity, that the propoſition is correct. We have much reaſon to think that the law of equality of action and reaction is mathematically true: but, let it fail to the amount of only one grain in a thouſand million of tons, and the propoſition is not true, but only nearly true.

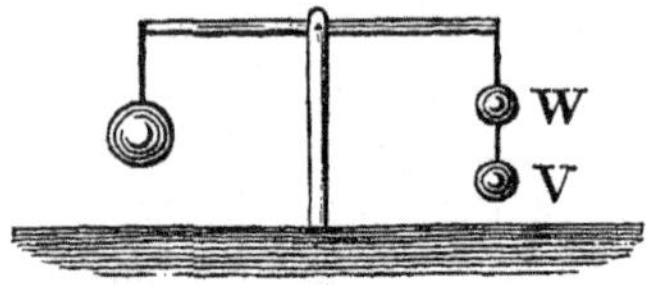

Again, the co-exiſtence of thoſe laws of mind and matter which beſt, ſo to ſpeak, fit each other, and which make the phenomena of the external world, after due conſideration, appear to be almoſt what they *muſt* have been, is not, to our apprehenſions, a neceſſary coexiſtence. We can imagine the following reſult, though we cannot trace what the full conſequences of it would be on the expreſſion of the laws of thought. Conceive ſentient beings, to whom the ſimpleſt mode of arithmetical ſucceſſion is not 0, 1, 2, 3, &c. but 1, 10, 100, 1000, &c. *their* powers of numeration being ſo conſtructed that the ſecond of theſe ſucceſſions has that character of fundamental ſimplicity which we attach to the firſt. Of courſe, their primary ſymbols would be ſignificative of 1, 10, 100, &c. It would be impoſſible for us to conceive any mode by which *ten* or any other number could be thus fundamentally attached to unity, in a manner ſhared by no third number: but, I am not ſaying, 'Imagine how this could be,' but, 'Imagine that it is.' There is no contradiction in the ſuppoſition, either to itſelf, or, till we know much more of the mind than we now do, to anything elſe. Beings ſo conſtituted would have *logarithmic* brains; and if, thus conſtituted, they were placed among our material laws of exiſtence, the manner in which the weight of the whole is to be inferred from thoſe of the parts, would be a profound myſtery for ages, only to be ſolved in an advanced ſtage of mathematical ſcience. A recent mode of conſtructing mathematical tables, which generally carries with it the name of its eminent inventor, Gauſs, would conſtitute one of their principal neceſſities: they would have to uſe it as their only mode (except actual experiment) of finding out that what we repreſent by 156 and 200, together make (and this making would be a complicated proceſs) 356.

Inſtead, then, of trying to eſtabliſh it as perfectly natural and neceſſary to ſay that our belief of 'one of the two A or B, when both cannot happen,' is, quantitatively ſpeaking, the ſum of our belief in A, and our belief in B, I have rather endeavoured to ſhow that the analogous caſes with which we firſt think of comparing this propoſition, other kinds of compoſition, are not ſo natural and neceſſary as is ſuppoſed. There are two ways of levelling; by bringing up the lower, or bringing down the higher. And I particularly wiſh in this chapter to prevent the reader from accepting the arithmetical doctrine of probability quite ſo rapidly as is uſually done. In furtherance of this object, I proceed to the following poſſible objection.

It may be ſaid, you have, by thus formally identifying probability with belief, and ſtating a poſtulate which, in expreſs terms, has not the moſt axiomatic degree of evidence, rendered ſomewhat difficult that which in the ordinary view of ſimple chances, is very eaſy. This charge, I hope, is true: ſuch was my intention, at leaſt. And my reaſon is, that in the ordinary view of the ſubject, one of two things occurs: either probability is ſeparated by definition from ſtate of belief, though it be known that the two words will afterwards be confounded without any permiſſion ; or elſe the poſtulate is tacitly aſſumed, and the difficulty which I ſuppoſe myſelf charged with introducing, is ſlurred over. Take a common queſtion ;—An urn has two white balls and five black ones : there are ſeven equally likely drawings, two white ; therefore the chance or probability of drawing a white ball is called two-ſevenths. But the chance of either particular white ball is one-ſeventh. Now firſt, if any one ſhould ſay that this is mere definition, I can, of courſe, allow it : but it then remains to ſhow what connexion this defined probability has with any ordinary acceptation of the word. But if, probability meaning belief, or ſentiment of probability actually exiſting in the mind, or index of the proper degree of belief, &c. &c.—the above ſtatement be made as fundamentally evident, I ſhould then aſk how it is known that the probability of 'one or the other white ball being drawn' is properly ſet down as the *ſum* of the probabilities of the ſeparate white balls. And I cannot conceive any anſwer except that it is by an aſſumption of the poſtulate. That ſuch aſſumption will finally be knowingly made, on the fulleſt conviction, by every one who ſtudies the theory, I have no doubt whatever : nor that it has been made, no matter in what words, nor with what clearneſs of avowal, by every one who has ſtudied that theory. And therefore I hold it deſirable that the beginner ſhould know what I have here told him.

It is indifferent, as far as the theory is concerned, what numerical ſcale of belief we take. We might, if we pleaſed, copy Fahrenheit's thermometer, ſet down knowledge of impoſſibility as 32°, perfect certainty as 212°, and other ſtates of mind accordingly. Thus, 122° would repreſent perfect indeciſion, belief inclining neither way, an even chance. But this would complicate our formulæ : the uſual and preferable plan is to aſſume 0 as the index of knowledge of impoſſibility, 1 as that of certainty, and intermediate fractions for the intermediate ſtates. This mode of eſtimation makes formulæ and proceſſes ſo much more eaſy than any other, that it muſt be adopted ; but there is a ſtrong objection to it in one point of view : as follows.

When we ſpeak of belief in common life, we always mean that we conſider the object of belief more likely than not : the ſtate of mind in which we rather reject than admit, we call *un*belief. When the mind is quite unbalanced either way, we have no word to exprefs it, becaufe the ftate is not a popular* one. The quantitative theory calls by the name of belief every admiffion of poffibility. When there is only one black ball to a million of white ones, there is fome belief that a black ball will be drawn; a much larger belief in a white one. It would be advantageous in fome refpects that 0 fhould reprefent the ftate of indifference, $+1$, that of knowledge of certainty, and -1, that of knowledge of impoffibility. But this would complicate formulæ too much. I confider it therefore defirable to ufe the common meafures and formulæ, but to affociate them with the one juft propofed, in the following manner.

When a perfon tells us that his belief in an affertion is, fay $\frac{3}{10}$, meaning that he confiders it 3 for and 7 againft, or 7 to 3 againft, we fhould fay in common talk that he difbelieves, but not very ftrongly. In the language of this theory, we fay that he both believes and difbelieves, the latter more ftrongly than the former. Let us add that his *authority* is againft the conclufion. If he fay that it is in his mind an even chance, or that he has no opinion one way or the other, let us fay that he gives no authority either way. If we adapt this definition to the fuppofition that $+1$ and -1 reprefent the extremes of authority for and againft, we have the following rules. The meafure of authority is twice the meafure of belief diminifhed by unity, for, when pofitive, againft, when negative: the meafure of belief is half of unity increafed (algebraically) by the meafure of authority. If a reprefent the meafure of belief, and A that of authority, then

$$A=2a-1 \quad , \quad a=\tfrac{1}{2}(1+A)$$

It is alfo advifable to have a term to reprefent what are ufually called the *odds*. Some might think it defirable to rid the fubject as much as poffible of words derived from gambling: aftronomers have done the fame thing with the phrafes of aftrology, and chemifts with thofe of alchemy. When it is 7 for and 3 againft,

* Many minds, and almoft all uneducated ones, can hardly retain an intermediate ftate. Put it to the firft comer, what he thinks on the queftion whether there be volcanoes on the unfeen fide of the moon larger than thofe on our fide. The odds are, that though he has never thought of the queftion, he has a pretty ftiff opinion in three feconds.

it might be ſaid that the *relative teſtimony* for, is $\frac{7}{3}$, and that againſt, $\frac{3}{7}$. But the brevity of the firſt phraſe will inſure its continuance, let who will try to change it.

The ordinary rule is a conſequence of the notions hereinbefore laid down, and of the particular mode of meaſurement adopted. It is as follows;—When all the things that can happen can be reſolved into a number of equally probable (or credible) caſes, ſome favourable and ſome unfavourable to the event under conſideration, then the fraction which the favourable caſes are of all the caſes, meaſures the probability (or credibility) of the arrival of the event: and the fraction which the unfavourable caſes are of all the caſes, meaſures the probability (or credibility) of the non-arrival. There are, for inſtance, in an urn, 5 white, 4 black, and 3 red balls, 12 in all. It is aſſumed that we know them to be equally likely to be drawn; which here means no more than that we know nothing to the contrary. That one ball muſt be drawn, is ſuppoſed certainly known. Accordingly, our belief in 'one or another' is repreſented by 1: which is, by the poſtulate, the ſum of the ſeveral credibilities of the balls; which laſt are all equal. Therefore each ball has $\frac{1}{12}$: and by the ſame poſtulate, the event 'one or other of the white balls' or the drawing of a white ball, has $\frac{5}{12}$; of a black ball $\frac{4}{12}$; of a red ball, $\frac{3}{12}$.

Inſtances like the above, in which we invent all the caſes and have arbitrary power over their number, are the only ones on which we can employ *à priori* numerical reaſoning. They are alſo the only ones on which we can try experiments. It is important to know whether, as a matter of fact, our belief, numerically formed, will be approximately juſtified by the reſults of trial. And this juſtification is found to exiſt, in the following way. It is a remote, but certain, concluſion from the theory, requiring mathematical reaſoning too complicated to introduce here, that events will, in the long run, happen in numbers proportional to the objective probabilities under which the trials are made. For inſtance;—if a die be correctly formed, ſo that no one face has more tendency than another to fall upwards, the probability of throwing an ace is $\frac{1}{6}$; that of not throwing an ace is $\frac{5}{6}$. The theory tells us its own worthleſſneſs, if in the long run, *not-ace* do not occur five times as often as *ace*. If 60,000 trials were made, the theory would tell us to expect *about* 10,000 aces and about 50,000 *not-aces*. Practice confirms the theory: not, that I know of, in the actual case just cited, but in similar ones. I will state an instance.

Throw a half-penny up, and if it give *tail*, repeat the throw, and so on, till *head* arrives: and let this succession be called a *set*. The probability that a set shall consist of one throw, is shewn by the theory to be $\frac{1}{2}$; that it shall have two throws, $\frac{1}{4}$; three throws, $\frac{1}{8}$; and so on. If a very large number of sets be tried, we are to expect that about half will be of one throw, about a quarter of two throws, about an eighth of three throws; and so on, as long as the number is large enough to give any prospect of something like an average. This experiment has been tried twice: once by the celebrated Buffon, and once by a young pupil of mine, for his own satisfaction; both in 2,048 sets. The results were as follows; the third column showing the number of each kind which the theory asserts to be most probable.

	B	H	
Head at the first throw	1061	1048	1024
No head till the 2nd throw	494	507	512
3rd —	232	248	256
4th —	137	99	128
5th —	56	71	64
6th —	29	38	32
7th —	25	17	16
8th —	8	9	8
9th —	6	5	4
10th —	0	3	2
11th —	0	1	1
12th —	0	0	} 1 (12th to &c)
13th —	0	0	
14th —	0	1	
15th —	0	0	
16th —	0	1	
&c —	0	0	
	2048	2048	2048

In Buffon's trials, there were altogether 1992 tails to 2048 heads, and in Mr. H's there were 2044 tails to 2048 heads.

Instances in which we can command all the cases are to the

mind, in this theory, what acceſſible lengths are to the eye. We can meaſure the latter by a rule, and ſo train the organ to judge of lengths which cannot be approached, or caſes in which the rule is not at hand.

I ſhall now refer the reader to other works on the ſubject, for further details on the operative part, and proceed to juſt as much as is neceſſary for the particular purpoſe of the next chapter, namely, the application of the hypotheſis of meaſure of belief to queſtions of argument and teſtimony. Two theorems will be enough: the firſt relating to independent events, the ſecond to the probability of events which are neither wholly independent, nor wholly conſequent, either upon the other. The word *event* is uſed in the wideſt poſſible ſenſe: it does not even neceſſarily mean future event. Unleſs our knowledge, either of the circumſtances, or of the event itſelf, thereby undergo ſome alteration, it is nothing to us now whether it has happened, or is to happen.

Let there be two events, P and Q, of which the probabilities are the fractions a and b; and let them be wholly independent of one another, the arrival or non-arrival of either being perfectly independent of that of the other. The probability that both ſhall happen is the product of a and b: and ſimilarly for more events than two. Suppoſe, to take an inſtance, that a is $\frac{4}{7}$ and b is $\frac{3}{5}$. We muſt then conſider P as an event which has 3 ways of failing to 4 of happening: if we would have an urn from which the credibility of drawing a white ball ſhould be that of the happening of P, we muſt put in 4 white balls and 3 not white (ſay black) balls. Similarly to repreſent Q, we muſt have an urn of 3 white and 2 black balls. Now to aſcertain the proſpect of drawing white from both urns, we muſt count all the caſes. A ball from the urn of 7 may be combined with one from the urn of 5, in 7×5 or 35 ways. But a *white* ball from the firſt urn may be combined with a *white* ball from the ſecond, in 4×3 or 12 different ways. There are then 35 caſes in all, 12 of which are favourable: hence the probability in favour of white from both (which is that of the two events both happening) is

$$\frac{12}{35} \quad \text{or} \quad \frac{4\times 3}{7\times 5} \quad \text{or} \quad \frac{4}{7}\times\frac{3}{5} \quad \text{or} \quad ab.$$

Similar reaſoning may be applied to more events than two. This theorem has a large. number of conſequences, ſome of which we may notice.

When a is the probability for, $1-a$ is the probability againſt. This I ſhall always denote by a': ſimilarly b' will ſtand for $1-b$; and ſo on.

Required the probability that of a number of independent events, P,Q,R, &c one or more ſhall happen. Let a,b,c, &c. be the ſeveral probabilities, then that of their all failing is the product $a'b'c'$ .... and that of their *not* all failing (or of one or more happening) is $1-a'b'c'$ .... Accordingly, if there be only two events, for 'one or both' we have $1-(1-a)(1-b)$ which is $a+b-ab$. If the number of events be n, and all equally probable (ſo that $a=b=c$, &c.) for 'one or more' we have $1-a'^n$ or $1-(1-a)^n$.

It is a conſequence of this laſt that, however unlikely an event may be, it is ſure (in the common ſenſe of the word) to happen, if the trial can be repeated as often as we pleaſe. However ſmall a may be, or however near to unity $1-a$, n may be taken ſo great that $(1-a)^n$ ſhall be as ſmall as we pleaſe, or $1-(1-a)^n$ as near to unity as we pleaſe, or the probability that the unlikely event will happen once or more in n times, as great as we pleaſe. Let $a=1:(k+1)$, which means that the odds are k to 1 againſt the event on any one trial: the following rough deductions will ſhow what kind of reſults the formula gives, true within an inſtance or two when k is conſiderable. In $\frac{7}{10}k$ inſtances it is an even chance that the event happens once or more; in $2{\cdot}3k$, it is 9 to 1; in $4{\cdot}6k$, 99 to 1; $6{\cdot}9k$, 999 to 1; $9{\cdot}2k$, 9999 to 1: and in $23k$, it is ten thouſand millions to 1. Thus, ſuppoſe at each trial it is a hundred to one againſt ſucceſs. Then of thoſe who try 70 efforts, as many will ſucceed once or more as will altogether fail, in the long run. Of thoſe who try 6900 times, only one of a thouſand will always fail. A perſon who will not examine an aſſertion that comes to him with ten to one againſt it, muſt count it an even chance that he throws away one or more truths, if he follow his plan ſeven times.

Let us now ſuppoſe that there are reaſons why the ſeveral inſtances which can arrive are not equally credible. Suppoſe the urn to contain a white, a black, and a red ball, and ourſelves to have reaſons to think the balls not equally probable or credible, but that 6, 5, and 2 are the proportions of the degrees of belief we ſhould accord to them ſeverally. If then $6x$ repreſent the probability of a black ball, $5x$ and $2x$ will repreſent thoſe of the other two ſeverally. By the poſtulate, $13x$ repreſents that of one or the other. But this is certainty; whence x muſt be $\frac{1}{13}$, and $\frac{6}{13}$, $\frac{5}{13}$, and $\frac{2}{13}$ are the probabilities of the white, black, and red balls. That is to ſay, when the ſeveral inſtances are unequally probable, we muſt count each inſtance as though it occurred a number of times proportioned to its probability, and then proceed as in the caſe of equally probable inſtances. Thus, in the above, inſtead of ſaying (as we ſhould do if the balls were equally probable) that the probability of the white ball is

$$\frac{1}{1+1+1}, \quad \text{we ſay it is} \quad \frac{6}{6+5+2}; \quad \text{or} \quad \frac{6m}{6m+5m+2m}$$

would do, m being any number or fraction whatſoever.

Now ſuppoſe two urns, one of all white balls, and the other of all black ones. If we actually draw a ball, and find it white, we know that the urn choſen to draw from muſt have been the firſt: the ſecond *could not* have given that drawing. But ſuppoſe the firſt urn to have 99 white balls to one black, and the ſecond one white to 1000 black. If we now draw again, and draw a white one, not knowing from which we drew, we feel almoſt certain, from the drawing, that we have choſen the firſt urn. We ſtill feel *almoſt* certain that the ſecond urn would have given a black ball. This inverſion of circumſtances, this concluſion that the circumſtances under which the event did happen, are moſt probably thoſe which would have been moſt likely to bring about the event, is of the utmoſt evidence to our minds: but the queſtion now before us is, are we to call it a ſecond poſtulate, or is it deducible from the other one? It is ſo deducible, and is not a ſecond poſtulate; but it has not been uſual to give a very diſtinct account of the deduction.* If it could not be made,

* So well eſtabliſhed is this ſpecies of inverſion in the mind, that both Laplace and Poiſſon, the two moſt eminent mathematical writers on the ſubject, of the preſent century, have in a certain caſe aſſumed that an equation which gives the moſt probable value of x in terms of y, is therefore the one which gives the moſt probable value of y in terms of x. This is carrying the principle too far.

the following proceſs would, no doubt, be ſufficient: it has often been held ſo. Let the urns have 6 white balls to 1 black, and 2 white balls to 9 black. Then the probabilities of drawing a white ball from the two are $\frac{6}{7}$ and $\frac{2}{11}$, which are in the proportion of 33 to 7. If, becauſe when we chooſe the firſt urn, we have nearly five times as much chance of a white ball as the ſecond one would give, we conclude that a known white ball from an unknown urn is in that proportion more likely to have come from the firſt urn; we ſhall have $\frac{33}{40}$ and $\frac{7}{40}$ for the proper degrees of belief in the two urns. For if $33x$ be that for the firſt urn, then $7x$ muſt, by the aſſumption, be that for the ſecond: and for one or the other, we have $40x$. But this is certainty; whence x muſt be $\frac{1}{40}$.

To reduce this reſult to dependence upon the firſt poſtulate, proceed as follows. The probability that two events are *connected*, our belief, that is, in the *connexion*, muſt be the ſame whether the two events, or either of them, have happened, or whether they be yet to happen: unleſs there be ſomething in the happening which alters our knowledge, and puts us in a different ſtate for forming a judgment. Suppoſe I make up my mind, rightly or wrongly, as to how far I will believe that a white ball, *if* drawn, *will have been* drawn from the firſt urn. An inſtant after, I am told that the trial I anticipated has been made, and the contingency which I ſuppoſed has occurred; a white ball has been drawn. I know no more than I took myſelf to know in my hypotheſis; and cannot therefore have any means of altering my opinion. Now, without altering the proportions in the urns, change the numbers of the balls, ſo that there may be the ſame total number in each: let them be

{66 white, 11 black}   {14 white, 63 black}

Now put each ball in an urn by itſelf, 154 urns in all. This gives $\frac{1}{154}$ to any one ball, if I chooſe an urn at hazard. But it was ſo before: as to the firſt of the two urns for inſtance, $\frac{1}{2}$ was the probability of chooſing that urn, and $\frac{1}{77}$ that of chooſing one particular ball from it: and $\frac{1}{2} \times \frac{1}{77}$ is $\frac{1}{154}$. If we then remove all the urns with black balls, ſo that a white ball muſt be drawn, the chance of its being one of the 66 is $\frac{66}{80}$ or $\frac{33}{40}$. If without removing the black balls, we think of the probability of a white ball, if drawn, being of the 66, or of the 14, the credibilities of thofe fuppofitions are as 66 to 14. If, having chofen an urn, we find it contains a white ball, the fame probabilities are ftill in that proportion.

The rules derived from fimilar reafoning, whether for judging of the probabilities of precedents from an obferved confequent, or for judging of the probabilities of events which reftrict each other, are precifely the fame, as follows. If the probability of the obferved event, fuppofed ftill future, from the feveral poffible precedents, feverally fuppofed actually to exift, be a,b,c, &c : then, when the event is known to have happened, the probabilities that it happened from the feveral precedents are

$$\frac{a}{a+b+c+\ldots} \text{ for the firft, } \frac{b}{a+b+c} \ldots \text{ for the fecond, \&c.}$$

Again, if there be feveral events, which are not all that could have happened; and if, by a new arrangement (or by additional knowledge of old ones) we find that thefe feveral events are now made all that can happen, without alteration of their relative credibilities: their probabilities are found by the fame rule. If a, b, c, &c. be the probabilities of the feveral events, when not reftricted to be the only ones: then, after the reftriction, the probability of the firft is $a \div (a+b+\ldots)$, of the fecond, $b \div (a+b+\ldots)$ and fo on.

We may obtain a very diftinct notion of this laft theorem, as follows. Suppofe two events, which are *among* thofe that can happen, and let one, fay, be twice as probable as the other. This means, that among all the independent, and equally likely, cafes, there are twice as many favourable to the firft as to the fecond. Now, fuppofe by fome alteration of fuppofitions, the introduction of new knowledge, for inftance, it is found, all the cafes remaining as before, that all are prevented from happening except thefe two events. This new ftate of things does not alter the cafes in number: accordingly, the *proportion* of the probabilities of the two events is as before, two to one. But now one of them muft happen: or the fum of thefe probabilities muft be unity. It follows then that one of them is $\frac{2}{3}$, and the other $\frac{1}{3}$. The fame reafoning may be applied to more complicated cafes.

It frequently happens, when different problems are folved by the ſame formula, that they may be conſidered as the ſame problem in two different points of view: and alſo that one and the ſame problem may be conſidered as belonging to either claſs. For inſtance;—Let there be two witneſſes, whoſe credibilities (or the probabilities that in any given inſtance they are correct) are *a* and *b*. As long as we do not know that they are talking about the ſame thing, the probability that both will tell truth is *ab*. But the moment we know that they both aſſert the ſame thing, the problem is changed: they muſt now be either both right or both wrong; before, one might have been right and the other wrong. To take the firſt view of the problem, we have now an obſerved event, both ſtate that the circumſtance did happen. There are two precedents; the event did, or did not, happen. If it did, the probability of the obſerved event (which is then that both are right) would be *ab*; if it did not, it would then be $(1-a)(1-b)$. Accordingly, the probability that the obſerved event did happen, will be, by the rule above, *ab* divided by $ab+(1-a)(1-b)$.

If we take the ſecond view, we have, before the reſtriction, four poſſible caſes, the probabilities of which are ab, $a(1-b)$, $b(1-a)$ and $(1-a)(1-b)$. After the reſtriction, only the firſt and fourth are poſſible: whence the concluſion is as juſt given. Full exemplifications of theſe methods will be found in the next chapter.

CHAPTER X.

On probable Inference.

THERE are two ſources of conviction, *argument* and *teſtimony*, reaſon why the thing ſhould be, ſtatement that the thing is. When the argument is neceſſarily good, we call it *demonſtration*: when the ſtatement can be abſolutely relied on, we call it *authority*. Both words are uſed in lower than their abſolute ſenſes; thus, very cogent arguments are often called demonſtration, and very good evidence, authority.

I ſhall ſuppoſe all the arguments I ſpeak of to be logically valid; that is, having conclusions which certainly follow from the premises. If then the premises be all true, the conclusion is certainly true. If *a*, *b*, *c*, &c. be the probabilities of the independent premises, or the independent propositions from which premises are deduced, then the product *abc*. . . is the probability that the argument is every way good.

Argument being offer of proof, its failure is only failure of proof: and the conclusion may yet be true. But testimony is an assertion of the truth of the conclusion; and its failure can only be failure of truth. If a proposition of Euclid turn out to be badly demonstrated, the enunciation need not therefore be false. An argument may prove, disprove, or neither prove nor disprove: a testimony cannot be true, false, or neither true nor false. This distinction generally gains no more than a one-sided admission: persons begin to see it when some over-zealous brother writes weakly on their own side of a question; but they are very apt to think, with respect to the other side, that answering the arguments is disproving the conclusion.

Testimony is, for the above reason, more easily understood than argument. It is the most effective mode of conveying knowledge to the uneducated. But it must not be supposed that, in any stage of reason, argument can be the only vehicle of information, even on subjects called argumentative. This point is one of great importance.

When argument is demonstration, it establishes its conclusion against all testimony. The idea of an infallible witness bearing evidence against a demonstrated conclusion, is a contradiction. That *n* consecutive numbers have a sum which is divisible by *n*, whenever *n* is odd, is demonstrated. If a thousand of the best qualified witnesses that ever lived, both for honesty and arithmetic, were to swear that they had discovered 101 very high consecutive numbers, the sum of which is not divisible by 101, any mere beginner in mathematics would be more sure that a thousand good witnesses had lost their wits or their characters, than any one else can be of anything not admitting of demonstration.

But when argument does not amount to demonstration, not only is the truth or falsehood of the conclusion matter of credibility, but the issue of the argument is not that mere truth or falſehood. It does not ſtand thus: 'According as this argument is good or bad, ſo is the concluſion true or falſe,' but 'According as this argument is good or bad, ſo is the concluſion *true in this way*, or *not true in this way*, (that is, either falſe, or true in ſome other way).' If we were to ſay 'men are trees, and trees have reaſon, therefore men have reaſon,' we have a perfectly logical argument, falſe in the matter of both premiſes: but we cannot deny the concluſion.

Suppoſe now that an argument is preſented to us of which we are ſatisfied that the like will prove their concluſions to be true in the particular modes aſſerted, in nine caſes out of ten. What are we to ſay of the truth or falſehood of the concluſion? We have $\frac{9}{10}$ of belief to its being true in one particular way: how much ſhall we add for other poſſible ways? Are we to reſt in the concluſion as having 9 to 1 for it, or are we to allow more? We cannot ſay, let us confine ourſelves to the grounds we have got, and believe or diſbelieve, not in the concluſion, but in the concluſion as obtained in that one way.

I take it for granted that the mind muſt have a ſtate with reſpect to every aſſertion preſented to it, with reaſon, or without reaſon. Every propoſition, the terms of which convey any meaning, at once, when brought forward, puts the hearer into ſome degree of belief, or, if we uſe the common phraſe, of belief or unbelief: including, of courſe, the intermediate ſtate, which is as clearly marked upon our ſcale as any other. Men who are accuſtomed to ſuſpend their opinion, as it is called, that is, to throw themſelves into the intermediate ſtate when they have no definite reaſon to think either way, are intereſted in this queſtion as much as any others. If there be ſome ſtate, though not numerically appreciable, in which their belief muſt be, there is ſome ſtate, which they would rather know numerically than not, in which it ought to be. In the preceding caſe, ſuppoſe it known that 9 to 1, or $\frac{9}{10}$, is granted to the concluſion from the argument alone, and any one wiſhes to ſuſpend his opinion as to the remaining $\frac{1}{10}$. Is he to grant half of that $\frac{1}{10}$, and ſay that $\frac{9}{10}+\frac{1}{20}$ or $\frac{19}{20}$ is what he would wiſh to make the meaſure of his belief, if he knew how? The conſideration of this queſtion will enter among others.

The manner in which he deals with the reſult of the argument muſt depend upon *teſtimony*, uſing the word in its wideſt ſenſe. Firſt, every man has, as juſt noticed, a teſtimony in his own mind as to every propoſition. He may ſet out with the intermediate ſtate: he may have no reaſon to lean either way, *and may know it;* that is to ſay, he may have to apply an argument of $\frac{9}{10}$ to an exiſting probability of $\frac{1}{2}$. Or he may have previous good reaſon, or bad reaſon, which makes him lean to the aſſertion or denial; and the meaſure of this leaning muſt then be combined with $\frac{9}{10}$. Or he may have other teſtimony to combine with that of his own previous ſtate. Any way, he cannot have a definite opinion on the bare truth or falſehood of the concluſion of the argument, without appeal to the previous ſtate of his own mind at leaſt, if not to that of others.

It is generally ſaid that we are to throw away authority, and judge by argument alone; that our reaſon is to be convinced, and not biaſſed by the opinion of others; that no concluſions are worth anything, except thoſe which a man forms for himſelf. All the forms in which this frequent caution is expreſſed, I take to be diſtortions of the very needful warning not to allow authority more weight than is properly due to it: a warning, by the way, which is juſt as much wanted with reſpect to argument as to authority. For every miſtake which has been made by taking authorities *on truſt* (that is, taking bad witneſſes to prove the goodneſs of aſſerted good ones), one miſtake at leaſt has been made by taking arguments *on preponderance:* that is, treating them as proving their concluſion, as ſoon as they ſhow it to be more likely than its contradiction.

To form the habit of allowing authority *no more* weight than is due to it, and the ſame of argument, is undoubtedly one great object of mental cultivation: but it ought not to be forgotten that it is another and juſt as great an object to form the habit of allowing them *no leſs*. Suppoſe an argument of value $\frac{9}{10}$ is preſented, and that at the ſame time we have the teſtimony of a witneſs againſt the concluſion, of whom we *know* that he leads us right 1000 times for each once that he miſleads us. Is there any ſenſe in reducing this witneſs to one of no authority, or of an even chance, upon the principle of depending on argument only? Except the argument be demonſtration, we muſt be prepared to admit that a witneſs may be as good as an argument, or better.

I ſhall now proceed to the ſeveral problems which this ſubject requires, conſidering firſt teſtimony alone, next argument alone, and then the two in combination.

Problem I. There are independent teſtimonies to the truth of an aſſertion, of the value μ, ν, ρ, &c. (one of them being the initial teſtimony of the mind itſelf which is to form the judgment): required the value of the united teſtimony.

Let μ' be $1-\mu$, &c. as in page 187. Here is a problem of the ſame claſs as in page 190; the reſtrictions are, that all the teſtimonies are right, or all wrong, the independent chances of which are $\mu\nu\rho\ldots$ and $\mu'\nu'\rho'\ldots$ Hence the probabilities are

$$\frac{\mu\nu\rho\ldots}{\mu\nu\rho\ldots+\mu'\nu'\rho'\ldots}\text{ for; }\qquad\frac{\mu'\nu'\rho'\ldots}{\mu\nu\rho\ldots+\mu'\nu'\rho'\ldots}\text{ againſt.}$$

Obſerve, firſt, that any numbers proportional to μ, μ' &c. will do as well: and if the products have a common denominator, (as generally they have) the numerators only need be uſed. Secondly, the eaſieſt way of expreſſing the reſult is by ſaying that it is $\mu\nu\rho\ldots$ to $\mu'\nu'\rho'\ldots$ for, or $\mu'\nu'\rho'\ldots$ to $\mu\nu\rho\ldots$ againſt.

For inſtance, let it be in my mind 99 to one againſt an aſſertion, that is, I bear only the teſtimony $\frac{1}{100}$ *in favour* of it. Let four witneſſes, for whoſe accuracy it is 2 to 1, 3 to 1, 4 to 1, 5 to 1, depoſe in favour of it: I want to know how it ought to ſtand in my mind. The teſtimonies for and againſt, are

$$\frac{1}{100},\ \frac{2}{3},\ \frac{3}{4},\ \frac{4}{5},\ \frac{5}{6};\ \text{and}\ \frac{99}{100},\ \frac{1}{3},\ \frac{1}{4},\ \frac{1}{5},\ \frac{1}{6};$$

Hence, neglecting the common denominator, it ought to be $1\times2\times3\times4\times5$ to $99\times1\times1\times1\times1$, or 120 to 99, or 40 to 33, for the aſſertion.

Obſerve that in ſaying the witneſs gives teſtimony, ſay $\frac{2}{3}$, it is of no conſequence whether it be a queſtion of judgment, or of veracity, or of both together. I mean that, come how it may, I am ſatisfied that when he ſays anything, it is 2 to 1 he ſays what is correct.

An eaſy rule for the more common modes of expreſſion preſents itſelf thus. The combined relative teſtimony is the product of the ſeparate relative teſtimonies. Thus, two witneſſes of 6 truths to one error, and of 7 truths to one error, are equivalent to one witnefs of 42 (or 6×7) truths to one error. Three witneffes of 8, 6, 5 truths to 7, 3, 11 errors are equivalent to one witnefs of 80 truths to 77 errors.

A jury of twelve equally truftworthy perfons, after conferring together, agree to an affertion on which previoufly I had no leaning. Suppofing me fully fatisfied that fuch agreement gives 100 to 1 for their refult, what am I to think of the *deliberate* opinion of any one among them, that is, of his opinion after he has had the advantage of difcuffion with others.

Let μ be the value of fuch teftimony from any one; then by the queftion

$$\mu^{12} : (1-\mu)^{12} :: 100 : 1, \quad \text{or } \mu : 1-\mu :: \sqrt[12]{100} : 1$$

fay as 1·468 to 1. That is, I think inconfiftently if I rely on the united verdict as upon 100 to 1, unlefs I am prepared to think it 1468 to 1000, or about 3 to 2, for each juror alone.

Of $m+n$ equally truftworthy jurors, a majority m are for, and n againft, a conclufion. If μ be the value of the teftimony of each, then the odds are to be taken as being $\mu^m(1-\mu)^n$ for, and $\mu^n(1-\mu)^m$ againft. But

$$\mu^m(1-\mu)^n : \mu^n(1-\mu)^m :: \mu^{m-n} : (1-\mu)^{m-n}$$

which are exactly as if the majority $m-n$ had been all, and unanimous. From the original formula it will appear that two equally good teftimonies on oppofite fides produce no effect on the refult.

If then, the unanimity of the jury box in this country could be confidered as that of deliberate conviction, we might fay that a larger jury, with the condition that the majority fhould exceed the minority by 12 at leaft, would be always as good, and often better. But there are various confiderations which prevent the above refult from being applicable. The neceffity of being unanimous, as our law ftands, may lower the value of the verdict. On the other hand, a jury of 30, required to find by a majority of 12, would generally proceed to a vote before they had put the matter to each other with the real defire to gain opinion which the prefent practice produces: confequently, the value of their verdict would perhaps be lower than that of the majority only, required to be unanimous.

The theory thus appears to confirm the notion on which we often act, that a given excess of majority over minority, is of the same value whatever the numbers in the two may be. And this might be the case, if the thing called deliberation in a large body, were as well adapted to the discovery of truth as the same thing in a smaller one. The reader must remember that this test does not compare the one witness on his own judgment with a number after common deliberation; but the first, after common deliberation with others, is compared with the whole.

But in this, and all the problems of this chapter, the distinction must be carefully drawn between the credibility of a circumstance at one time and at another. For example, a witness enters with 10 to 1 in his favour, and owing to combination with others, the result comes out that it is 100 to 1 he is in error in the particular matter on which he gives evidence. We cannot believe both that it is 10 to 1 he is right, and 100 to 1 that he is wrong. What we believe is the latter, for the case in question.

As another instance, suppose m independent witnesses of equal goodness (μ) unite in affirming that a certain ball was drawn from a lottery of n balls: collusion being supposed impossible. My knowledge of the circumstances of the affirmation here alters the problem. If n be considerable, it is almost impossible that the witnesses, by independent falsehood or error, should all pitch on the same wrong ball. To find the bias this ought to give me to the conclusion that they have told the truth, I must observe that there being $n-1$ balls not drawn, whichever of these any one chooses, by error, the chance of any one of the rest choosing the same is $1 \div (n-1)$, the probability that all the $m-1$ shall choose the same is $1 \div (n-1)^{m-1}$. Hence, the odds are as μ^m to $(1-\mu)^m$ multiplied by the last-named expression, or as $(n-1)^{m-1}\mu^m$ to $(1-\mu)^m$. If n be very great, the odds may be enormous for the assertion, even though μ, the credibility of each witness, may be small. In cases of ordinary evidence, the thing asserted is usually one out of almost an infinite number of equally possible assertions, and the agreement of even two witnesses (for when m is *two* or upwards, n appears in the formula) is certain conviction, if, as assumed, we know the two witnesses to be not in collusion. If $\mu = 1 \div n$, which is as much as to say that the evidence of each witness makes a ball no more likely to have been the one drawn, becaufe he fays it, that it was on our mere knowledge that a ball *had* been drawn, it turns out 1 to $n-1$ for the truth of the affertion, juft as it was before the evidence. But let $\mu=(1+\alpha)\div n$, α being any fraction, however fmall, that is, let each witnefs make the affertion more probable than at firft, however little: then the odds for its truth become

$$(1+\alpha)^m \quad \text{to} \left(1-\frac{\alpha}{n-1}\right)^{m-1}(n-1-\alpha)$$

which odds may be made as great as we pleafe, by fufficiently increafing m. That is to fay, however little each witnefs may be good for, in real fupport of the affertion, or in making it more probable than it is of itfelf, a fufficient number of witneffes, certainly independent, will give it any degree of credibility whatever.

The ftudent of this fubject is always ftruck by the frequency of the problems in which the fcience confirms an ordinary notion of common life, or is confirmed by it, according to his ftate of mind with refpect to the whole doctrine. It is impoffible to fay that we have a theory *made to explain* common phenomena, and hence affording no reafon for furprife that it does explain them. The firft principles are too few and two fimple, the train of deduction ends in conclufions far too remote. I believe hundreds of cafes might be cited in which the refults of this theory are found already eftablifhed by the common fenfe of mankind: in many of them, the mathematical fciences were not powerful enough to give the modes of calculation, when the principles of the theory were firft digefted.

There are problems, however, in which we cannot eafily come into poffeffion of data on which many will agree. The fimple queftion of independent witneffes is not one of them: but the queftion of collufion is. One of the difficulties is as follows. We cannot inftitute independent hypothefes upon the goodnefs of the witneffes and the probability of their having conferred upon their evidence. They declare, expreffly or by implication, that they have not done fo: if they have, there is falfehood in one part of their evidence; or, which makes the difficulty ftill greater, there may have been general, but (as they affert or imply) not particular conference: they may have been biaffed by each other, without knowing how or to what extent. The firſt ſtep in one view of the problem is eaſily made, as follows.

Let μ be the value of the evidence of each witneſs, m their number, n the number of aſſertions they have power to chooſe from, all as before. Let λ be the probability that there has been particular conference between them. There are then four caſes to which the problem is reſtricted:—(1) they have conferred and agreed to ſpeak truth; (2) they have not conferred and all ſpeak truth; (3) they have conferred and agreed on a falſehood; (4) they have not conferred and have all lighted upon the ſame falſehood. The à priori probabilities of theſe four caſes are

$$\lambda\mu^m, \quad (1-\lambda)\mu^m, \quad \lambda(1-\mu)^m, \quad \frac{(1-\lambda)(1-\mu)^m}{(n-1)^{m-1}}:$$

and the odds that they ſpeak the truth (ſuppoſing n ſo great that we may reject the fourth caſe) are μ^m to $\lambda(1-\mu)^m$. Now comes the practical difficulty of this queſtion;—How are λ and μ to be connected? Every caſe which is worth examining ſuppoſes that the greater the chance of there having been particular conference, the leſs is the witneſs worth from that very circumſtance. For it is to be remembered that we are not generally able to give the witneſs a character wholly independent of his evidence in the caſe before us; in hiſtorical queſtions, for inſtance, it frequently happens that we have nothing but the witneſſes to try* the caſe by, and nothing but the caſe to try the witneſſes by. A very common occurrence is this;—that a caſe is one in which no one would throw any doubt upon the witneſſes, except for ſuſpicion of conference, and juſt as much doubt as there is ſuſpicion of conference. This makes $\mu=1-\lambda$, and gives $(1-\lambda) : \lambda^{m+1}$ for the odds in favour of the aſſertion. On this ſuppoſition, it follows that whenever the chances are againſt all the witneſſes having conferred particularly, their number, if great enough, ought to give any degree of credibility to the aſſertion.

* This gives riſe to two great tendencies, which very nearly divide the world among them. Some ſettle the caſe in their own minds, and then try the witneſſes: ſome ſettle the witneſſes and then try the caſe: not a few bring their ſecond reſult back again to juſtify their firſt aſſumption. When there are two unknown quantities with only one equation, it is eaſy for thoſe who will aſſume either to find the other. But the difficulty is to find the moſt probable value of both.

Problem 2. Let there be any number of different aſſertions, of which one muſt be true, and only one: or of which one may be true, and not more than one: or of which any given number may be true, but not more: required the probability of any one poſſible caſe.

The ſolution of all theſe varieties depends on one principle, explained in page 190; requiring the previous probabilities of all the conſiſtent caſes to be compared. As an inſtance, ſuppoſe four aſſertions, A,B,C,D, and ſuppoſe μ,ν,ρ,σ, to be the probabilities from teſtimony, for each of them. If either of them have ſeveral teſtimonies, their united force muſt be aſcertained by the laſt problem. Firſt, let it be that one of them muſt be true, and one only. The probabilities in favour of A,B,C,D, are in the proportion of $\mu\nu'\rho'\sigma'$, $\nu\mu'\rho'\sigma'$, $\rho\mu'\nu'\sigma'$, and $\sigma\mu'\nu'\rho'$. Either of theſe, divided by the ſum of all, repreſents the probability of its caſe. Secondly, let it be that one of them only can be true, and all may be falſe. Put on the fifth quantity $\mu'\nu'\rho'\sigma'$, for the caſe in which all are falſe. For example, there are four diſtinct aſſertions, not more than one of which can be true. The ſeparate evidences for theſe four aſſertions give them the probabilities $\frac{2}{7}$, $\frac{3}{11}$, $\frac{1}{8}$ and $\frac{4}{5}$. There is a certain aſſertion which is true if either of the firſt three be true: required the probability of that aſſertion. Here, neglecting the common denominator, which is $7 \times 11 \times 8 \times 5$ in every caſe, the probabilities of the ſeveral aſſertions, and that of all being falſe, are as 2.8.7.1, 3.5.7.1, 1.5.8.1, 4.5.8.7, and 5.8.7.1, or as 112, 105, 40, 1120, and 280. The odds for one of the firſt three caſes againſt one of the other two are $112+105+40$ to $1120+280$ or as 257 to 1400; or it is 1400 to 257 againſt the truth of the aſſertion.

Suppoſe the condition were that two of the aſſertions, but not more, *may* be true, and that one *muſt* be true. Then the poſſible caſes, meaning by an accent that the aſſertion is not true, are AB'C'D', BA'C'D', CA'B'D', DA'B'C', ABC'D', ACB'D', ADB'C', BCA'D', BDA'C', CDA'B'. Conſequently, the probabilities of theſe caſes are in the proportion of $\mu\nu'\rho'\sigma'$, $\nu\mu'\rho'\sigma'$, $\rho\mu'\nu'\sigma'$, &c. And the odds in favour of, ſay A, being true, are as the ſum of all the terms which contain μ, to the ſum of thoſe which contain μ'.

When we wiſh to ſignify that no evidence is offered either for or againſt one of the aſſertions, we muſt put it down as having the teſtimony $\frac{1}{2}$. To put down 0 in the place of $\frac{1}{2}$ would be to make an infallible witneſs declare that it is not true. Suppoſe there are four aſſertions, one of which muſt be true and one only: evidence of goodneſs $\frac{4}{7}$ is offered for the firſt, and none either way for the others. Required the probability of the firſt. The probabilities of the four aſſertions are in the proportion of 4.1.1.1, 1.3.1.1, 1.3.1.1, and 1.3.1.1, and it is 4 to 9 for the firſt, or 9 to 4 againſt it.

Problem 3. Arguments being ſuppoſed logically good, and the probabilities of their proving their concluſions (that is, of all their premiſes being *true*) being called their validities, let there be a concluſion for which a number of arguments are preſented, of validities a, b, c, &c. Required the probability that the concluſion is proved.

This problem differs from thoſe which precede in a material point. Teſtimonies are all true together or all falſe together: but one of the arguments may be perfectly ſound, though all the reſt be prepoſterous. The queſtion then is, what is the chance that one or more of the arguments proves its concluſion. That all ſhall fail, the probability is $a'b'c'$.... that all ſhall not fail, the probability is $1-a'b'c'$.... Accordingly, if we ſuppoſe n equal arguments, each of validity a, the probability that the concluſion is proved is $1-(1-a)^n$. And, as in page 187, if the odds againſt each argument be k to 1, then, the number of ſuch arguments being as much as ·7 k, the concluſion is rendered as likely as not.

But are we really to believe, having arguments againſt the validity of each of which it is 10 to 1, that ſeven ſuch arguments make the concluſion about as likely to be true as not. If ſuch be the caſe, the theory, uſually ſo accordant with common notions, is ſtrangely at variance with them. This point will require ſome further conſideration.

In this problem I conſider only argument, and not teſtimony, which, neverthelеſs, cannot be finally excluded (ſee page 194). If the concluſion be one on which our minds are wholly unbiaſſed to begin with, it may ſeem that we have no eſcape from the preceding reſult. And to it we muſt oppoſe, for conſideration at leaſt, the common opinion of mankind that ſtrong arguments are the preſumption of truth, weak arguments of falſehood.

If a controverſialiſt were to bring forward a hundred arguments, and if his opponent were ſo far to anſwer them as to make it ten to one againſt each, there can be no doubt that the latter would be conſidered as having fairly contradicted the former.

We muſt not forget that argument, in a great many caſes, involves and produces the effect of teſtimony, and this in an eaſily explicable and perfectly juſtifiable manner. If I were to pick up a bit of paper in the ſtreets, on which an argument is written, for a concluſion on which I have no previous opinion, and by an unknown writer, and if I could ſay that that argument left on my mind the impreſſion of ten to one againſt its validity, I might be prepared to allow it to ſtand as giving $\frac{1}{11}$ of probability, and upon that ſuppoſition to combine it with my previous opinion, $\frac{1}{2}$, as in the next problem. But ſuppoſe it is on a queſtion of phyſics, and Newton is the propoſer of it, and that it is his only argument, and therefore, I conclude, his beſt. The caſe is now entirely altered: poſſibly the concluſion is one on which the following argument would have great probability: 'If this concluſion were true, it could be proved; if it could be proved, Newton could have proved it; therefore if it were true, Newton could have proved it: but Newton cannot prove it; therefore it is not true.' If the caſe be ſuch that the two premiſes of this laſt argument have each 9 to 1 for it, or $\frac{9}{10}$; then, though the original argument give $\frac{1}{11}$ for the concluſion, the mere circumſtance of Newton bringing this argument as his beſt is $\frac{81}{100}$ againſt it. If Newton at the ſame time declare his belief in the concluſion, we have on one ſide his argument and his authority, on the other ſide the argument ariſing from his being reduced to ſuch an argument.

That ſuch conſiderations have weight, we know: and that they ought to have weight, we may eaſily ſee. It is of courſe, dependent upon the particular concluſion what weight ſhall be attached to the aſſertion, 'if this concluſion were true it could be proved.' The courts of law conſtantly act upon this principle. They conſider (very juſtly I think) that evidence, however good it may be, is much lowered by not being the beſt evidence that could be brought forward. If a man be alive, and capable of being produced with ſufficient eaſe, they will not take any number of good witneſſes to the fact of his having been very recently alive. In enumerating the arguments, then, for or againſt a propoſition, thoſe muſt be included, if any, which ariſe out of the nature, mode of production, or producers, of any among them. And until this has been properly done, we are not in a condition to apply the methods of the preſent chapter.

Problem 4. A concluſion and its contradiction being produced, one or the other of which muſt be true, and arguments being produced on both ſides, required the probability that the concluſion is proved, diſproved (*i. e.* the contradiction proved), or left neither proved nor diſproved.

Collect all the arguments for the concluſion, as in the laſt problem, and let a be the probability that one or more of them prove the concluſion. Similarly, let b be the probability that one or more of the oppoſite arguments prove the contradiction. Both theſe caſes cannot be true, though both may be falſe. The probabilities of the different caſes are thus derived. Either the concluſion is proved, and the contradiction not proved, or the concluſion not proved and the contradiction proved, or both are left unproved. The probabilities for theſe caſes are as $a(1-b)$, $b(1-a)$ and $(1-a)(1-b)$, and the probability that the concluſion is *proved* is $a(1-b)$ divided by the ſum of the three, and ſo on. The fraction $(1-a)(1-b)$ divided by this ſum may be called the *inconcluſiveneſs* of the combined arguments. The manner in which this inconcluſiveneſs is to be diſtributed between the hypotheſis of the truth and falſehood of the concluſion *muſt* depend upon teſtimony, in the complete ſenſe of the word.

The predominance of one ſide or the other, as far as arguments only are concerned, depends on which is the greateſt, $a(1-b)$ or $b(1-a)$, or ſimply on which is the greateſt, a or b. If the arguments on both ſides be very ſtrong, or a and b both very near to unity, then, though $a(1-b)$ and $b(1-a)$ are both ſmall, yet $(1-a)(1-b)$ is very ſmall compared with either. The ratio of $a(1-b)$ to $b(1-a)$ on which the degree of predominance depends, may, conſiſtently with this ſuppoſition, be anything whatever. But we cannot pretend that, when oppoſite ſides are thus both nearly demonſtrated, the mind can take cognizance of the predominance which depends upon the ratio of the ſmall and imperceptible defects from abſolute certainty. The neceſſary conſequence is, that the arguments are evenly balanced, and are as if they were equal: there is no ſenſible notion of predominance. This is the ſtate to which moſt well conducted oppoſitions of argument bring a good many of their followers. They are fairly outwitted by both ſides, and unable to anſwer either, and the concluſion to which they come is determined by their own previous impreſſions, and by the authorities to which they attach moſt weight; and theſe are, of courſe, thoſe which favour their own previouſly adopted ſide of the queſtion.

When no argument is produced on one ſide of the queſtion, the caſe is very different from the caſe of the preceding problems, in which no teſtimony is produced. Here the queſtion is, 'Has the concluſion been proved or not proved;' and when no argument is produced, we are certain it has not been proved. Accordingly, if no argument were urged for the contradiction, we ſhould have $1-b=1$, or $b=0$.

If, in the preceding problem, the two ſides of the queſtion be not contradictions, but ſubcontradictions, of which neither need be true, but both cannot be, the problem is ſolved in the ſame way, for the caſes are juſt the ſame. But we may introduce a diſtinction which the former caſe would not admit. When one muſt be true, every argument againſt one is of equal force for the other; which is not the caſe when neither need be true. Let there, then, be arguments for the firſt concluſion and againſt it, and let a and p be the probabilities that one or more of the arguments for, prove it, or againſt, diſprove it. Let b and q be the ſimilar probabilities for the ſecond concluſion. Then, there are theſe caſes:—1. The arguments (or ſome of them) for the firſt are valid, againſt it invalid, and thoſe *for* the ſecond are invalid (it matters nothing whether thoſe *againſt* the ſecond be valid or invalid). 2. The arguments for the firſt are invalid, thoſe for the ſecond valid, and againſt it invalid. 3. The arguments againſt the firſt are valid, and thoſe for it invalid. 4. The arguments againſt the ſecond are valid, and thoſe for it invalid. 5. All the arguments are invalid. Accordingly, the probabilities that the firſt is proved, that it is diſproved, that the ſecond is proved, that it is diſproved, and that neither of the two is proved nor diſproved, are in the proportion of $a(1-p)(1-b)$, $(1-a)p$, $b(1-q)(1-a)$, $(1-b)q$, and $(1-a)(1-b)(1-p)(1-q)$.

Problem 5. Given both teſtimony and argument to both ſides of a contradiction, one side of which must be true, required the probability of the truth of each side.

This is the most important of our cases, as representing all ordinary controversy. Collect all the testimonies, and let their united force for the first side be μ, and, from the nature of this case, $1-\mu$ for the other side. Let a and b be the probabilities that the first side and the second side are proved by one or more of the arguments in their favour. Now, observe that, for the *truth* of either side, it is not essential that the argument for it should be valid, but only that the argument against it should be invalid. Accordingly, the probabilities of the two sides are in the proportion of $\mu(1-b)$ and $(1-\mu)(1-a)$, and the probabilities of the two sides are represented by

$$\frac{\mu(1-b)}{\mu(1-b)+(1-\mu)(1-a)} \qquad \frac{(1-\mu)(1-a)}{\mu(1-b)+(1-\mu)(1-a)}$$

First, let there be no testimony either way: we must then have $\mu=\frac{1}{2}=1-\mu$; consequently, these probabilities are as $1-b$ to $1-a$. Let no argument have been offered for the second side, or let $b=0$. Then we have 1 to $1-a$, for the odds, or $1\div(2-a)$ for the probability of the first side being true. It has been usual to say that if an argument be presented of which the probability is a, the truth of the conclusion has also the probability a. Probably the above was the case intended as to testimony, &c., and the probability should then have been

$$\frac{1}{2-a} \quad \text{or} \quad a+\frac{(1-a)^2}{2-a}$$

which is always greater than a. Or, as we might expect, the possibility of the conclusion being true, though the argument should be invalid, always adds something to the probability of its being true. Moreover, $1\div(2-a)$ is always greater than $\frac{1}{2}$: or any argument, however weak, adds something to the force of the previous probability. The same thing is true in every case. Suppose a new argument to be produced for the first side, of the force k. The effect upon the formula is to change $1-a$ into $(1-a)(1-k)$, and the odds in favour of the conclusion are increased in the proportion of 1 to $1-k$. But this is to be underſtood ſtrictly in the ſenſe deſcribed in page 202, namely, we are to ſuppoſe that the newly produced argument *is* ſingle, that is, does not by the circumſtances of its production cauſe itſelf to be accompanied by an argument for the ſecond ſide, or againſt the firſt. If this laſt ſhould happen, and the argument thus created for the ſecond ſide have the force l, the odds are altered in the proportion of $1-l$ to $1-k$.

From the above it appears that oppoſite arguments of the force a and b are exactly equivalent to a teſtimony the odds for the truth of which are as $1-b$ to $1-a$. Thus, ſuppoſe we have for a concluſion witneſſes whoſe teſtimonies are worth $\frac{2}{3}$, $\frac{2}{3}$, $\frac{4}{7}$, $\frac{9}{10}$; arguments for of the ſeveral forces, $\frac{1}{5}$, $\frac{11}{12}$, $\frac{5}{8}$; and arguments againſt of the forces $\frac{2}{9}$, $\frac{2}{11}$, $\frac{6}{7}$. Writing numerators only, we put down

For, 2, 2, 4, 9; 7, 9, 1:
Againſt, 1, 1, 3, 1; 4, 1, 3.

Hence it is, 2. 2. 4. 9. 7. 9. 1 to 1. 1. 3. 1. 4. 1. 3, or 252 to 1 for the concluſion.

An argument, we ſhould infer beforehand, is better than a teſtimony of the ſame force; for the failure of the argument is nothing againſt the concluſion, but the failure of the teſtimony is its overthrow. So ſays the formula alſo: the introduction of a teſtimony of the value k, not before received, alters the exiſting odds in the proportion of k to $1-k$: but the introduction of an argument of the ſame force alters them in the greater proportion of 1 to $1-k$. Thus, the introduction of the teſtimony of a perſon who is as often wrong as right ($\frac{1}{2}$) alters the odds in the proportion of 1 to 1, or does not alter them at all: but the introduction of an argument which is as likely as not to prove the concluſion, alters them in the proportion of 1 to $1-\frac{1}{2}$, or of 2 to 1.

Are we not in the habit, unconſciouſly, of recognizing ſome ſuch diſtinction? Do we not give much more weight to argument than to teſtimony? I ſuſpect the anſwer ſhould be in the affirmative: that an argument of 3 to 1 does convince us much more than a teſtimony of 3 to 1. I ſuſpect we ſhow it, not in numerical appreciation, of courſe, but in liſtening to and allowing weight to arguments, when we ſhould refuſe teſtimony of the ſame character.

It may be doubted, however, whether we have much ſcope for experiment on the lower degrees either of teſtimony or argument. Perhaps it is not often we meet a witneſs, whether as bearing teſtimony of veracity to a fact, or of judgment to a concluſion, whoſe evidence is as low as $\frac{1}{2}$; and the ſame perhaps of an argument.

I have ſpoken, in the previous part of this chapter, of the rejection of authority, that is, of teſtimony, authority being only high teſtimony. Let us now examine by the formula and ſee what it amounts to. Let a be the probability that the argument proves its concluſion: and let us therefore perſiſt in ſaying that a is the probability for the truth of the concluſion. In the formula, b being$=0$, let μ be made $a \div (1+a)$, it will be found that the probability for the concluſion, μ divided by $\mu+(1-\mu)(1-a)$, comes out a, as required. Conſequently, in the caſe of a ſingle argument, the total rejection, as it would be thought, of all teſtimony, is really equivalent to accompanying every argument by a teſtimony leſs than $\frac{1}{2}$, depending upon its own force. It is to declare that, by the laws of thought, an argument of $\frac{7}{10}$ is of its own nature accompanied by a witneſs of $\frac{7}{17}$, one of $\frac{3}{4}$ by a witneſs of $\frac{3}{7}$, and ſo on; this is clearly not what was meant. Nor, I ſuppoſe, can it be meant that we are arbitrarily to ſtart with the teſtimony $\frac{1}{2}$, and to reduce our own evidence, and that of all others, to the ſame. If there be any ſenſe in which the rejection of authority is defenſible, it muſt be when we are required to proceed as if we were in perfect ignorance what the value of the authority is. We cannot ſuppoſe it to be as likely to have one value as another. Suppoſe, for inſtance, that the arguments have unknown propoſers: we cannot treat their authorities as if they were juſt as likely to be exceſſively high or low as to be very near to none at all. The more rational ſuppoſition is that the authority ſhould be more likely to be ſmall than great, as likely to be againſt as for, and very unlikely to be exceſſively great either for or againſt. I cannot here enter into the mode in which ſuch an hypotheſis can be expreſſed or uſed: but the reſult of the ſimpleſt formula which ſatisfies the above conditions, is as follows:— Let $r=(1-b)\div(1-a)$, b and a meaning as above; then the probability that the conclusion is true, which has a for the validity of its argument, &c. is

$$r(r^3-6r^2+3r+6r\log r+2)\div(r-1)^4$$

where $\log r$ means the *Naperian* logarithm (99-43rds of the common logarithm will be near enough for the present purpose). If, for instance, $r=2$, which, on the supposition of no previous balance of testimony, would give 2 to 1 for the conclusion, the formula just written gives ·636, or 636 to 364, something less than 2 to 1.

In the case first discussed in page 202, it may be thought that the weakness of a proposed argument, from one who should have brought a better, if there had been one, may be considered as a *testimony* against the conclusion rather than an *argument*. Suppose his argument, for instance, to have only the probability $\frac{1}{10}$. He tells us then, that after he has done his best, it is 9 to 1 against the proposition being proved. If we are very confident that it could be proved, if true, and that he could do it, if any one, he comes before us as a testimony of 9 to 1 against the *truth* of the conclusion, or very nearly so. If we take, then, all that his argument wants of demonstration, as so much evidence from him against the conclusion, this amounts to supposing that, a being the validity of his argument, a is also his testimony for the conclusion (and $1-a$ that against it). If there be only argument for, and none against, and if our minds be previously unbiassed, we represent this case by putting a for μ in the formula, and the odds for the conclusion are then as a to $(1-a)^2$. On this supposition, which I incline to think well worthy of attention, we should not consider an unopposed argument from an acute reasoner as giving the conclusion to be as likely as not, unless $a=(1-a)^2$ or $a=\cdot 382$, a little more than $\frac{1}{3}$. Were it not for our peculiar introduction of testimony, then, the conclusion being as likely as not to begin with, an argument which has any probability of proving it, would have made it more likely than not, as before seen.

But that the introduced testimony should be exactly as above, is a mere supposition. If it were a mathematical proposition, for instance, and Euler were to declare himself unable to give more than a probability of proof, I, for one, should consider him as giving a much higher rate of teſtimony againſt the truth of the aſſertion than is ſuppoſed in the preceding. But all this has reference to the queſtion how to meaſure teſtimonies and validities in particular caſes, which is quite a diſtinct thing from the inveſtigation of the way to uſe them when meaſured.

In caſes in which the number of arguments is multiplied, it generally happens that they ſtand or fall together, in parcels: namely, that the ſame failure which makes one invalid, neceſſarily makes others invalid. In this caſe, independent arguments muſt be ſelected, and the probabilities for them alone employed.

We ſee in this problem an illuſtration of the commonly obſerved reſult, that the ſame argument produces very different final concluſions in two different minds; and this when, ſo far as can be judged, both are diſpoſed to give the ſame probabilities to the ſeveral premiſes of the argument. The initial odds, come how they may, or μ to $1-\mu$, ſhould be altered by the arguments in the proportion of $1-b$ to $1-a$. Accordingly, b and a being the ſame to both parties, their belief in the concluſion may have any kind of difference, if μ be not the ſame thing to both.

Problem 6. Given an aſſertion, A, which has the probability a; what does that probability become, when it is made known that there is the probability m that B is a neceſſary conſequence of A, B having the probability b? And what does the probability of B then become?

Firſt, let A and B not be inconſiſtent. The caſes are now as follows, with reſpect to A. Either A is true, and it is not true that both the connexion exiſts and B is falſe: or A is falſe. This is much too conciſe a ſtatement for the beginner, except when it is ſuppoſed left to him to verify it by collecting all the caſes. The odds for the truth of A, either as above or by the collection, are $a\{1-m(1-b)\}$ to $1-a$. As to B, either B is true, or B is falſe and it is not true that A and the connexion are both true. Accordingly, the odds for B are as b to $(1-b)(1-ma)$.

The reader muſt remember that when B neceſſarily follows from A, B muſt be true when A is true, but may be true when A is falſe; while A muſt be falſe when B is falſe. And now we ſee that a propoſition is not neceſſarily unlikely, becauſe it is very likely to lead to an incredibility, or even to an abſolute impoſſibility. Let $b=0$, or let B be impoſſible: then the odds for A are as $a(1-m)$ to $1-a$. Say that it is 9 to 1 that the connection exiſts; then theſe odds are as a to $10(1-a)$. If a be greater than $\frac{10}{11}$, ſtill A remains more likely than not, even when it is 9 to 1 that it leads to the abſurdity B.

Secondly, let A and B be inconſiſtent, ſo that both cannot be true. Either then A is true, B falſe, and the connexion does not exiſt; or A is falſe. The odds for A are then as $a(1-b)(1-m)$ to $1-a$. With reſpect to B, either B is true and A is falſe, or B is falſe, and A and the connexion are not both true. The odds for B are then as $b(1-a)$ to $(1-b)(1-ma)$.

Among the early ſophiſms with which the Greeks tried the power of logic, as a formal mode of detecting fallacies, was the conſtruction of what we may call *ſuicidal* propoſitions, aſſertions the truth of which would be their own falſehood. If a man ſhould ſay 'I lie,' he ſpeaks neither truth nor falſehood; for if he ſay true, he lies, and if he lie, he ſpeaks truth. Such a ſpeech cannot be interpreted. Again, the Cretan, Epimenides, ſaid that all the Cretans were incredible liars; is he to be believed or not? If we believe him, we muſt, he being a Cretan, diſbelieve him. Some ſtated it thus;—'If we believe him, then the Cretans are liars, and we ſhould not believe him; then there is no evidence againſt the Cretans, or we may believe him, ſo that the evidence againſt the Cretans revives, &c. &c. &c. Refer ſuch a propoſition to the theory of probabilities, and the difficulty immediately diſappears. Whatever the credit of Epimenides as a witneſs may be, that is, whatever, upon his word, the odds may be for his propoſition, the ſame odds are there againſt him from the propoſition itſelf. Theſe equal conflicting teſtimonies balance one another (problem 1) and leave the effect of other teſtimonies to the ſame point unaltered. The ſophiſm of Epimenides, as ſtated, is but an extreme caſe of the ſecond of the problems before us. The propoſition B is inconſiſtent with A, and the connexion is certain $(m=1)$: the odds for B muſt then be as $b(1-a)$ to $(1-b)(1-a)$, or as b to $1-b$, exactly what they are independently of the previous aſſertion.

CHAPTER XI.

On Induction.

THE theory of what is now called *induction* muft occupy a large fpace in every work which profeffes to treat of the matter of arguments; but there is not much to fay upon the genuine meaning of the word, in any fyftem of formal logic. And that little would be lefs, if it were not for the miftaken oppofition which it has long been cuftomary to confider as exifting between the inductive procefs and the reft of our fubject.

By induction (ἐπαγωγή) is meant the inference of a univerfal propofition by the feparate inference of all the particulars of which it is compofed: whether thefe particulars defcend fo low as fingle inftances or not. Thus if X be a name which includes P,Q,R, fo that every thing which is X muft be one of the three: then if it be fhown feparately that every P is Y, and that every Q is Y, and that every R is Y; it follows that every X is Y. And this laft is faid to be proved by induction. Thus (Chapter VI).

$$X)P,Q,R + P)Y + Q)Y + R)Y = X)Y$$

is an inductive procefs. In form, it may be reduced as in page 123, to one ordinary fyllogifm.

Complete induction is demonftration, and ftrictly fyllogiftic in its character. In the preceding procefs we have y)p, y)q, y)r, which give y)pqr: and X)P,Q,R is pqr)x; whence y)x, or X)Y. It is a queftion of names, that is, it depends upon the exiftence or nonexiftence of names, whether a complete induction fhall preferve that form, or lofe it in the appearance of a *Barbara* fyllogifm, formed by help of the conjunctive poftulate of Chapter VI.

But when the number of fpecies or inftances contained under a name X is above enumeration, and it is therefore practically impoffible to collect and examine all the cafes, the final induction, that is, the ftatement of a univerfal from its particulars, becomes impoffible, except as a *probable* ftatement: unlefs it fhould happen that we can detect fome law connecting the fpecies or inftances, by which the refult, when obtained as to a certain number, may be inferred as to the reft.

This laſt named kind of *induction by connexion*, is common enough in mathematics, but can hardly occur in any other kind of knowledge. In an innumerable ſeries of propoſitions, repreſented by P_1, P_2, P_3, P_4, &c, it may and does happen that means will exiſt of ſhowing that when any conſecutive number, ſuppoſe three, of them are true, the next muſt be true. When this happens, a formal induction may be made, as ſoon as the three firſt are eſtabliſhed. For by the law of connexion, P_1, P_2, and P_3, eſtabliſh P_4; but P_2, P_3, and P_4, eſtabliſh P_5; and then P_3, P_4, and P_5, eſtabliſh P_6; and ſo on *ad infinitum*. It is to be obſerved that this is really *induction*: there is no way, in this proceſs, of compelling an opponent to admit the truth of P_{100} without forcing him, if he decline to admit it otherwiſe, through all the previous caſes.

As an eaſy inſtance, obſerve the proof that the *ſquare* of any number is equal to the ſum of as many conſecutive odd numbers, beginning with unity, as there are units in that number: as ſeen in

$$6 \times 6 = 1 + 3 + 5 + 7 + 9 + 11$$

Take any number, n; and write n ns (repreſenting a unit by a dot) in rank and file. To enlarge this figure into $(n+1)(n+1)$s, we muſt place n more dots at each of two adjacent ſides, and one more at the corner. So that the ſquare of n is turned into the ſquare of $n+1$ by adding $2n+1$, which is the $(n+1)$th odd number. Thus 100×100 is turned into 101×101 by adding the 101ſt odd number, or 201. If then the theorem alleged be true of $n \times n$, it is therefore true of $(n+1) \times (n+1)$. But it is true of the firſt number, 1×1 being 1; therefore it is true of the ſecond, or $2 \times 2 = 1 + 3$; therefore it is true of the third, or $3 \times 3 = 1 + 3 + 5$; and ſo on.

But when we can neither examine every caſe, nor frame a method of connecting one caſe with another, no abſolutely demonſtrative induction can exiſt. That which is uſually called by the name is the declaration of a univerſal truth from the enumeration of ſome particulars, being the aſſumption that the unexamined particulars will agree with thoſe which have been examined, in every point in which thoſe which have been examined agree with one another. The reſult thus obtained is one of probability; and though a moral certainty, or an unimpeachably high degree of probability, can easily be obtained, and actually is obtained, and though most of our conclusions with respect to the external world are really thus obtained, yet it is an error to put the result of such an induction in the same class with that of a demonstration. There is no objection whatever to any one saying that the former results are *to his mind* more certain than those of the latter: the fact may be that they are so. The difference between necessary and contingent propositions lies in the qualities from which they receive those adjectives, more than in difference of credibility. I know that a stone *will* fall to the ground, when let go: and I know that a square number *must* be equal to the sum of the odd numbers, as above: and though, when I stop to think, I do become sensible of more assurance for the second than for the first, yet it is only on reflection that I can distinguish the certainty from that which is so near to it.

The rule of probability of a *pure induction* is easily given. Supposing the simple question to be whether X *is* or *is not* Y, there being no previous circumstances whatsoever to make us think that any one X is more likely than not to be Y, or less likely than not. These are the circumstances of what I call a *pure* induction. To begin with, it is 1 to 1 that the first X examined shall be a Y: if this be done, and X_1 be a Y, then it is 2 to 1 that X_2 shall be a Y; should it so happen, then it is 3 to 1 that X_3 shall be a Y. Generally, when the first m Xs have all been examined, and all turn out to be Ys, it is $m+1$ to 1 that the $(m+1)$th X shall be a Y.

The simplicity of this rule must not lead the student to suppose he can find a simple reason for it. Let 10 Xs have been examined and found to be Ys: what do we assert when we say it is 11 to 1 that the 11th X shall be a Y? We assert that if an *infinite* number of urns were collected, each having white balls and black balls in *infinite number* but in a definite ratio, and so that every possible ratio of white balls to black ones occurs once; and if every possible way of drawing eleven balls, the first ten of which are white, were selected and put aside: then, of those put aside, there are eleven in which the eleventh ball is white, for one in which the eleventh ball is black. The reader will find some difficulty in forming a distinct conception of this, and of courſe will find it impoſſible to have any axiomatic perception of the truth or falſehood of the reſult.

It may be worth while to ſhow that a ſuppoſition making ſome degree of approach to the preceding circumſtances will give ſome approach to the reſult. Firſt, in lieu of an infinite number of balls in each box, which is ſuppoſed only that withdrawal of a definite number may not alter the ratio, let each ball drawn be put back again, which will anſwer the ſame purpoſe. Let there be only ten urns with ten balls in each, of which let the firſt have one white, the ſecond two white, &c. and the laſt all white. The number of ways of drawing eleven white balls ſucceſſively out of any one urn is the eleventh power of the number of white balls in the urn: that of drawing ten white balls followed by one black one is the tenth power of the number of white balls multiplied by the number of black ones. If we were to put together all the firſt, and then all the ſecond, we ſhould find about 21 times as many ways of arriving at the firſt reſult (ten white, followed by a white) as the ſecond (ten white followed by a black). But if we now increaſed the number of urns, and took a hundred, having one, two, &c. white balls, we ſhould find inſtead of 21, a number much nearer to 11; and ſo on.

Accordingly, when without any previouſly formed bias, we find that *m* Xs, ſucceſſively examined, are each of them a Y, we ought then to believe it to be $m+1$ to 1 that the next, or $(m+1)$th X, will be a Y. And further, *a* being a fraction leſs than unity, we have a right to ſay there is the probability $1-a^{m+1}$ that the Xs make up the fraction *a* or more, of the Ys. Or thus;—if the fraction *a* be, ſay $\frac{6}{7}$, and if *m* be 10: then if the 10 firſt Xs be all Ys, the probability that $\frac{6}{7}$ or more of the Xs are Ys is juſt that of drawing one or more black balls in 11 drawings, from an urn in which $\frac{6}{7}$ of the balls are always white.

If, for example, the firſt 100 Xs were all Ys, it would be found to be 1000 to 1 that $93\frac{4}{10}$ per cent, at leaſt, of all the Xs are Ys.

If as before, the firſt *m* Xs obſerved have all been Ys, and we aſk what probability thence, and thence only, ariſes that the next *n* Xs examined ſhall all be Ys, the anſwer is that the odds in favour of it are $m+1$ to *n*, and againſt it *n* to $m+1$. No induction then, however extenſive, can by itſelf, afford much probability to a universal conclusion, if the number of instances to be examined be very great compared with those which have been examined. If 100 instances have been examined, and 1000 remain, it is 1000 to 101 against all the thousand being as the hundred.

This result is at variance with all our notions; and yet it is demonstrably as rational as any other result of the theory. The truth is, that our notions are not wholly formed on what I have called the *pure induction*. In this it is supposed that we know no reason to judge, except the mere mode of occurrence of the induced instances. Accordingly, the probabilities shown by the above rules are merely *minima*, which may be augmented by other sources of knowledge. For instance, the strong belief, founded upon the most extensive previous induction, that phenomena are regulated by uniform laws, makes the first instance *of a new case*, by itself, furnish as strong a presumption as many instances would do, independently of such belief and reason for it.

With this however I have nothing farther to do, except to observe that, in the language of many, induction is used in a sense very different from its original and logical one. It is made to mean, not the collection of a universal from particulars, but the mode of arrival at a common cause for varied, but similar, phenomena. A great part of what is thus called induction consists in discovery of *differences*, not *resemblances*. Under this confused use of language, the usual theory is introduced, namely, that Aristotle was opposed to all induction, that Bacon was opposed to every thing else, that the whole world up to the time of Bacon followed Aristotle, that the former was the first who showed the way to oppose the latter, that each had a logic of his own, &c. &c. The whole of this account abounds with misstatements. The admitted and sufficiently striking difference between the philosophy of modern and ancient times, in all natural and material branches of inquiry, is not so easily explained as by choosing two men, one to bear all the blame, the other all the credit: nor are Copernicus, Gilbert, Tycho Brahé, Galileo, and the other predecessors of the *Novum Organum*, destined to be always deprived of their proper rank.

What is now called induction, meaning the discovery of laws from instances, and higher laws from lower ones, is beyond the province of formal logic. Its instruments are induction properly ſo called, ſeparation of apparently related, but really diſtinct particulars (the neglect of which was far more hurtful to the old philoſophy than a neglect of induction proper would have been, even had it exiſted) mathematical deduction, ordinary logic, &c. &c. &c. It is the uſe of the whole box of tools : and it would be as abſurd to attempt it here, as to append a chapter on carpentry to a deſcription of the mode of cutting the teeth of a ſaw.

The proceſſes of Ariſtotle and of Bacon are equally thoſe which we are in the habit of performing every day of our lives. But ſome perform them well, and ſome ill. It is extraordinary that there ſhould be ſuch diviſion of opinion on the queſtion whether a careful analyſis of them, and ſtudy of the parts into which they decompoſe, is of any uſe towards performing them well. On this point, and on the character of Bacon's office in philoſophy, a living writer, to whom I ſhould think it likely that many yet unborn would owe their firſt notions of Bacon's writings, expreſſes himſelf in a manner which I quote, and comment on at length, as the beſt expoſition I can find, of a claſs of opinions which is very prevalent, and, I fully believe, to the prejudice of ſober thought and accurate knowledge.

The vulgar notion about Bacon we take to be this, that he invented a new method of arriving at truth, which method is called Induction, and that he detected ſome fallacy in the ſyllogiſtic reaſoning which had been in vogue before his time. This notion is about as well founded as that of the people who, in the middle ages, imagined that Virgil was a great conjuror. Many who are far too well informed to talk ſuch extravagant nonſenſe, entertain what we think incorrect notions as to what Bacon really effected in this matter.

The inductive method has been practiſed ever ſince the beginning of the world, by every human being. It is conſtantly practiſed by the moſt ignorant clown, by the moſt thoughtleſs ſchoolboy, by the very child at the breaſt. That method leads the clown to the concluſion that if he ſows barley, he ſhall not reap wheat. By that method a ſchoolboy learns that a cloudy day is the beſt for catching trout. The very infant, we imagine, is led by induction to expect milk from his mother or nurſe, and none from his father.

Not only is it not true that Bacon invented the inductive method ; but it is not true that he was the firſt perſon who correctly analyſed that method and explained its uſes. Ariſtotle had long before pointed out the abſurdity of ſuppoſing that ſyllogiſtic reaſoning could ever conduct men to the diſcovery of any new principle, had ſhown that ſuch diſcoveries muſt be made by induction, and by induction alone, and had given the history of the inductive process, concisely indeed, but with great perspicuity and precision.

Again, we are not inclined to ascribe much practical value to that analysis of the inductive method which Bacon has given in the second book of the *Novum Organum*. It is indeed an elaborate and correct analysis. But it is an analysis of that which we are all doing from morning to night, and which we continue to do even in our dreams. A plain man finds his stomach out of order. He never heard Lord Bacon's name. But he proceeds in the strictest conformity with the rules laid down in the second book of the *Novum Organum*, and satisfies himself that minced pies have done the mischief. "I eat minced pies on Monday and Wednesday, and I was kept awake by indigestion all night." This is the *comparentia ad intellectum instantiarum convenientium*. "I did not eat any on Tuesday and Friday, and I was quite well." This is the *comparentia instantiarum in proximo quæ natura data privantur*. "I ate very sparingly of them on Sunday, and was very slightly indisposed in the evening. But on Christmas-day I almost dined on them, and was so ill that I was in great danger." This is the *comparentia instantiarum secundum magis et minus*. "It cannot have been the brandy which I took with them; for I have drunk brandy daily for years without being the worse for it." This is the *rejectio naturarum*. Our invalid then proceeds to what is termed by Bacon the *Vindemiatio*, and pronounces that minced pies do not agree with him.

We repeat that we dispute neither the ingenuity nor the accuracy of the theory contained in the second book of the *Novum Organum*; but we think that Bacon greatly overrated its utility. We conceive that the inductive process, like many other processes, is not likely to be better performed merely because men know how they perform it. William Tell would not have been one whit more likely to cleave the apple if he had known that his arrow would describe a parabola under the influence of the attraction of the earth. Captain Barclay would not have been more likely to walk a thousand miles in a thousand hours, if he had known the place and name of every muscle in his legs. Monsieur Jourdain probably did not pronounce D and F more correctly after he had been apprised that D is pronounced by touching the teeth with the end of the tongue, and F by putting the upper teeth on the lower lip. We cannot perceive that the study of grammar makes the smallest difference in the speech of people who have always lived in good society. Not one Londoner in ten thousand can lay down the proper rules for the use of *will* and *shall*. Yet not one Londoner in a million ever misplaces his *will* and *shall*. Dr. Robertson could, undoubtedly, have written a luminous dissertation on the use of these words. Yet, even in his latest work, he sometimes misplaced them ludicrously. No man uses figures of speech with more propriety because he knows that one figure of speech is called a metonymy, and another a synecdoche. A drayman in a passion calls out 'You are a pretty fellow,' without suspecting that he is uttering irony, and that irony is one of the four primary tropes. The old systems of rhetoric were never regarded by the most experienced and discerning judges as of any use for the purpose of forming an orator. "Ego hanc vim intelligo" ſaid Cicero "eſſe in præceptis omnibus, non ut ea ſecuti oratores eloquentiæ laudem ſint adepti, ſed quæ ſuâ ſponte homines eloquentes facerent, ea quoſdam obſervaſſe, atque id egiſſe; ſic eſſe non eloquentiam ex artificio, ſed artificium ex eloquentia natum." We muſt own that we entertain the ſame opinion concerning the ſtudy of Logic, which Cicero entertained concerning the ſtudy of Rhetoric. A man of ſenſe ſyllogizes in *celarent* and *ceſare* all day long without ſuſpecting it: and though he may not know what an *ignoratio elenchi* is, has no difficulty in expoſing it whenever he falls in with it.—('*Lord Bacon*,' in *Critical and Hiſtorical Eſſays contributed to the Edinburgh Review*. By Thomas Babington Macaulay.)

This brilliant paſſage has, I have no doubt, appeared to many completely deciſive of the queſtion which it affirms: and, as ſo often happens in like caſes, there is a certain exaggeration againſt which it is of truth. It is good againſt thoſe who confound analyſis and recombination of exiſting materials with introduction of them: and who might profeſs to ſee in agriculture ſomething which would have benefited mankind, though plants and animals had not been natural products of the ſoil. But I now proceed to examine it, againſt thoſe who affirm that Ariſtotle and Bacon are of *no* uſe, and who very frequently fall into the common logical fallacy of ſuppoſing that their caſe is proved, as ſoon as it is made out that they are not of *all the* uſe: which Mr. Macaulay himſelf has done, except as againſt the exaggerators aforeſaid.

We reaſon inductively from morning till night, and even in our dreams. True: and how badly we often do it, particularly in ſleep. A plain man is then produced, to reaſon on Bacon's principles: and Mr. Macaulay has imitated a plain man better than he intended, by making him do it wrongly. Look over the induction, and it will appear that the caſe is not made out; an excluſion is wanting: it may have been the *mixture* of minced pies and brandy which did the miſchief. The plain man ſhould have tried minced pies without brandy; but he had drunk the latter daily for years, and it never ſtruck him. This is preciſely one of the points in which we are moſt apt to deceive ourſelves, and for which we moſt need to have recourſe to the completeneſs of a ſyſtem of rules; ſomething is left taken for granted. The things of courſe, our daily habits, are neglected in the conſideration of anything of a leſs uſual character: the plain man left off the minced pies upon trial; but not the brandy: Chriſtmas miſchief muſt be referred, he thinks, entirely to Chriſtmas fare, if at all.

But even if this omiſſion had been ſupplied, and the reſult found to confirm the concluſion, yet the plain man has ſtopped where the plain man frequently does ſtop, at what Bacon calls the *Vindemiatio prima*, the rudiments of interpretation. Completeneſs is ſeldom anything but ſtudy and ſyſtem. Philoſophy ought to bring him to the reſult that daily brandy has made that ſpirit ceaſe to give the ſtimulus which, were its uſe only occaſional, would enable his ſtomach to bear an unuſually rich diet for a ſhort time. Our plain friend is preciſely in the poſition of a bankrupt who curſes the times, on reaſoning ſtrictly Baconian as far as it goes, and forgets that a caſual tightneſs in the money market would never have upſet him, if it had not been for the previous years of extravagant living and raſh ſpeculation.

But there are many proceſſes which are not better performed becauſe men know "how they perform them." Mr. Macaulay here means "becauſe men know the laws of that part of the proceſs which nature does for them." That men ſhould not know better how to perform for knowing how *they* perform is almoſt a contradiction in terms. William Tell knew how to ſhoot all the better for knowing which end of the arrow *he* was accuſtomed to fit to the ſtring: had he wanted this knowledge, his chance of cleaving the apple would have been much diminiſhed. But he would not have been improved by knowing that his arrow deſcribed a parabola. True, becauſe it did not do ſo. The *centre of gravity* of the arrow would deſcribe a parabola, if it were not for the reſiſtance of the air; or ſomething ſo near it as to be undiſtinguiſhable. But, taking the deſcription as roughly correct, William Tell did know, inductively, that the arrow deſcribes a curve, concave to the earth: and had made thouſands of experiments in connexion of the *two ends* of that curve, which were all that he was concerned with. It is no argument againſt the ſtudy, as a ſtudy, of *induction*, that the amount of uſeful reſult which it had recorded in the mind of William Tell in the ſhape of habit, would not have been augmented by *deductive* knowledge of an intermediate *ſtatus* with which *he* had nothing to do. But let knowledge advance, under both modes of progreſs, and Tell becomes an artillery officer, the rude arrow a truly shaped and balanced ball, means of meaſurement are applied, the true curve is more correctly repreſented than by the parabola, and thirty pounds of iron are thrown to four times the diſtance which an arrow ever reached, and with a certainty almoſt equal to that of the legend.

But if Captain Barclay had known the places and names of the muſcles, he would not have been more likely to walk a thouſand miles in a thouſand hours. The inſtance is far fetched: becauſe the feat conſiſted in the exhibition of power of endurance acquired by practice. If my denial ſeem as far fetched, it is the fault of the propoſer. Captain Barclay muſt, by habit, by induction, have acquired facility in varying his pace and geſture ſo as to eaſe the muſcles. Had he been well acquainted with the *diſpoſition and uſes* of theſe organs to begin with (towards which knowledge of their *places and names* would have contributed) he would have learnt this art more eaſily. Though not altogether *ad elenchum*, yet I may ſay that in this caſe the effect of ſuch knowledge would have been that he would have been *leſs* likely to have performed the feat. Had he directed his attention to ſome ſcience of obſervation, he would not have needed to have ſought fame, or exhauſtion of remarkable energy, in ſuch a trifling purſuit. And further, in a very common caſe, mechanics has taught what few ever learn by induction, though they have conſtant opportunities of doing it: namely, that in walking, the ordinary practice of ſwinging the arms is injurious and tiring; that a very trifling amount of it tells ſeriouſly in a long journey. Here is one uſeful reſult, which natural induction does not commonly teach, and there may be many more of the ſame kind: the queſtion between it and regular ſtudy requires the conſideration, not only of what is done, and whether it might be done better, but of what is not done.

Next, M. Jourdain did not pronounce D and F more correctly after his attention had been called to the details of the act of pronunciation. None but Molière ever knew whether he did or not: but all who have watched the progreſs of inſtruction know that the bad habits or natural imperfections of children are removed or alleviated by making them practice mechanical pronunciation, with perceptive adoption of rules. In every one of a few detached inſtances in which I have ſeen children at their reading lessons in France, I have noticed that a return upon the habits of pronunciation is always a part of the exercise: and that the letters are pronounced with that distinct effort which makes the pupil sensible of the action required. I have always attributed to this practice the more uniform standard of pronunciation which prevails among the educated French, as compared with ourselves.

But the study of grammar makes no difference in the speech of people who have always lived in good society. If Mr. Macaulay mean merely as to the use of *shall* and *will*, and the like, it may certainly be said that the perpetual use of speech (which is not reasoning) does enable every one to form the habits of those about him. But that grammar, as a whole, produces no effect upon the speech of good society, is one side of a balanced matter of opinion. Many contend that it has produced, in our generation and the one above it, a very unfortunate effect: they aver that the purity and character of our English has been deteriorated by Lindley Murray and his school, and that we much want better grammar teaching. On the subject of *shall* and *will*, it is remarkable that Mr. Macaulay, whom a vigorous faculty of illustration, combined with immense reading, enables to strew his path with instances, has to invent his case, and to refer to a treatise which Robertson could have written. But it is not enough: if we grant that such a treatise would have been *luminous*, we may be safe; but would it have been *correct?* And further, knowledge must abdicate at once, if we pronounce useless all that has been clearly explained by those who have not rightly practised. Bacon himself might have taken *exsors ipsa secandi* for his motto.

Next, it is said that no man uses figures of speech more correctly because he knows that one is *metonymy* and another *synecdoche*. True; and in like manner no man consults his books more easily because he has a bookcase. But, having the bookcase, he arranges his books in it, and then he knows where to find them. Mr. Macaulay dwells throughout upon nomenclature. I might insist upon its superstructure: but even mere naming is useful, when the meaning of the name is clearly understood. A mind well stocked with understood names cannot keep itself from being constantly in the act of classification, which contains induction. The mere involuntary reference of instance number two to instance number one, which is made when we remember that the second must have the same name as the first, is comparison and induction, leads to reflection, cultivates taste, and gives power. The drayman, who calls out in a passion, "You are a pretty fellow!" without knowing that he is uttering irony, is an incomplete picture: there is omitted a wish relative to the eyes of his opponent, and an adjective which is (in such quarrels) sometimes prophetically, but seldom descriptively, true. The value of the difference between this savage irony and the more elegant form of it which is so pleasing in the description of the plain man's induction quoted above, is not within the comprehension of the drayman: the foundation of a better mode of expression than undisciplined rhetoric furnishes, so far as its adoption is matter of taste, was laid by those who placed irony among the primary tropes. Good taste is a result of comparisons, which could not have been made without nomenclature.

Did Cicero declare that systems of rhetoric are not of any use? The very quotation appears to mean that these systems, *præcepta*, have their power; that men get them by observation, and put them into practice. The *ea secuti oratores* refers to what was done in the first instance, by the first eloquent men, *suâ sponte*. Most truly does he say that the art of rhetoric is derived from eloquence, and not *vice versâ*: most falsely, as far as can be judged, does he seem to insinuate that it was all done at one step; first, some one or more consummate orators, secondly, a finished system, drawn from observation of their methods. Perhaps he intended a particular reference to a certain orator then nameless: the sentence, thus construed, contains nothing but matter which Tully is likely enough to have whispered to Cicero.

A system is a tool, and it must be employed upon materials which different men furnish from their different means. But the coat must be cut according to the cloth, both in size and quality: no reproach to the scissors, nor prejudice to their superiority over the sharpened wood of the savage, even though practice will enable him to use the latter better than any civilized man who is not a tailor can use the former. The formation of tools, mental or material, is a cyclical process. The first iron was obtained by help of wood; one of the first uses of it was to make better tools, to get more iron, with which better tools still were made, and so on. And in this way we may trace back any art to natural tools, and to materials which are to be had for the gathering. The assertion made by Mr. Macaulay, and many others, that in logic only, of all the abstract sciences, our natural means are as good as those which result from diligent analysis, is one which terminates in an issue of fact. The instances given are contained in the assertion that a *man of sense* syllogizes in *cesare* and *celarent* all day long without *suspecting it*, and though he does not know what an *ignoratio elenchi* is, can always detect it when he meets with it.

Mr. Macaulay begins with an indefinite term, a man *of sense*: and the clause is deficient in logical perspicuity. First, what is a man *of sense?* I grant that I should doubt the sense of a man who could not make the inferences described by *cesare* and *celarent*. But do men become men of sense by nature, without education? if yes, I deny the assertion that men of sense reason (correctly) in *cesare*, &c. The man of sense who is not educated is as likely to assert that *cesaro* is all that can be obtained, or to invent the form *fesape*, as the plain man to forget to try the mince pies without brandy before he concludes. If no, then the assertion is itself *ignoratio elenchi*: for the very question is how to make men of sense; can they not be, *ceteris paribus*, formed better and faster with study of logic than without: it being agreed on all hands that *this* man of sense is always a practical logician.

Next, a man of sense reasons, &c. without suspecting it. Suspecting what? that he is reasoning, or that he is reasoning in *cesare?* I suppose the latter: that is to say, I take it to be meant that a man of sense *may* (not *must*, for some Aristotelians are men of sense) not know that the logicians call the form of reasoning he uses *cesare*. This is easily granted: but what is it but the celebrated *ignoratio elenchi* of Locke, who fancied that he raised an objection against the pretensions of the logicians, when he declared he never could believe that God had made men only two-legged, and left it to Aristotle to make them rational. No one ever denied that men reasoned before Aristotle, and would have reasoned still if he had never lived.

Mr. Macaulay, probably without so much as a new application

to the inkſtand, after falling into the *ignoratio elenchi*, ſingles out this very fallacy as the one which a man of ſenſe is ſure to detect. But if there be a fallacy which is the ſtaple of paralogiſm, it is this one. *Delectat domi*, for ordinary diſcuſſion (eſpecially after dinner) is little elſe; *impedit foris*, for three fourths of public debate, from the Houſes of Parliament downwards, is made up of it. A man who expoſes it in converſation is conſidered a tireſome, and if he do it often, an uncourteous perſon: he "has no converſation," he "harps upon one ſubject," he "won't let you ſpeak."

I have made the above comments upon a very marked paſſage of an eminent writer, in preference to introducing their ſubſtance as a diſſertation of my own, that I might have the advantage of the reader ſeeing that I meet real arguments, inſtead of my own verſion or ſelection. It would probably be difficult to find a better concentration of the ſubſtance of the antagoniſt views, with reſpect to the formal ſtudy of reaſoning, than is contained in my quotation from Mr. Macaulay: and I may ſafely take his adoption of them as proof that theſe views yet require the notice of a writer on logic.

There is one reſult of the theory of probabilities, cloſely connected with induction proper, which it will be adviſable to notice here.

When the ſyllogiſm is declared illegitimate, on account of both premiſes being particular, a probable concluſion of great ſtrength may be admitted in many caſes. This muſt be the more inſiſted on, becauſe it is too common to attend to nothing but the demonſtrative ſyllogiſm, leaving all of which the concluſions are only probable, however probable, entirely out of view.

I take as the inſtance the ſyllogiſm, or imperfect ſyllogiſm, 'Some Xs are Ys, ſome Zs are Ys, therefore there is ſome probability that ſome Xs are Zs.' If the number of Xs and Zs together exceed the number of Ys (as in Chapter VIII) there is a certainty that ſome Xs are Zs. Let us then ſuppoſe this is not the caſe.

Let the whole number of Ys in exiſtence be n, and let m and

n be the numbers of Xs and Zs which are among them. I ſhall conſider two diſtinct caſes:—Firſt, when the diſtribution of the Xs and Zs among the Ys is utterly unknown; ſecondly, when their diſtribution is that of *contiguity*, that is, when the Ys being for ſome reaſon arranged in a particular order, the Xs which are Ys are ſucceſſive Ys, and the ſame of the Zs which are Ys.

For the firſt caſe a very rough notion will do, confined to the ſuppoſition that few Xs and Zs are mentioned, compared with the whole number of Ys. When the Xs and Zs together make a large proportion of the Ys in number, then, if we have no reaſon for making them contiguous, or otherwiſe limiting the equally probable arrangements, it may be ſaid to be a moral certainty that ſome Xs are Zs.

In the firſt caſe, if we divide 43 times the product of m and n by 100 times ν, it gives us a ſufficient notion (not large enough) of the common logarithm of k, the odds in favour of ſome Xs being Zs being k to 1. Say there are 1000 Ys, and that 100 Xs are Ys and 100 Zs are Ys. Then $43 \times 100 \times 100$ divided by 100×1000 is 4·3, which is the logarithm of 20,000. It is then more than 20,000 to 1 that, in this caſe, one or more Xs are Zs. A more exact rule is as follows. To $43mn$ divided by 100ν add its hundredth part, and to the reſult add ſuch a fraction of itſelf as $m+n$ is of 2ν. Thus $43mn \div 100\nu$ being 4·3, which, with its hundredth part is 4·343, and $m+n$ (200) being the tenth part of 2ν (or 2000), we add to 4·343 its tenth part, giving 4·777, which is about the logarithm of 60,000, ſtill under the mark. It is more than 60,000 to 1 that ſome Xs are Zs. When the fractions are very ſmall, this rule is accurate enough, if ν be conſiderable. Its reſult is, that if ν be very conſiderable, and if a perceptible fraction of the Ys be Xs, and a perceptible fraction Zs, *and if we really have no reaſon to make the limitation of contiguity or the like*, then we are juſtified in treating it as a moral certainty that ſome Xs are Zs. But I ſuſpect the relation of contiguity, to which I now proceed, better repreſents the actual ſtate of the caſe in ordinary argument.

When the Xs which are Ys are contiguous, and alſo the Zs which are Ys, the probability that no Xs are Zs is the fraction having the product of $\nu-m-n+1$ and $\nu-m-n+2$ for numerator, and the product of $\eta - m + 1$ and $\eta - n + 1$ for denominator. Thus in the example above proposed, 1000 Ys containing among them 100 Xs and 100 Zs (each set contiguous) we have 801×802 for numerator and 901×901 for denominator. This fraction is about 8-tenths; so that it is now 8 to 2, or 4 to 1, *against* any Xs being Zs.

In order to find the probability against the number of Xs which are Zs exceeding k, add k to both the multipliers in the numerator, which then become $\eta - m - n + k + 1$ and $\eta - m - n + k + 2$. For example, there are 100 Ys, containing 30 Xs and 60 Zs (each set contiguously): what is the chance against the number of Xs which are Zs exceeding 10? The numerator is 21×22: the denominator is 71×41. This fraction is 462 by 2911; whence it is 462 to 2449 *against*, or 2449 to 462 (more than 5 to 1) *for*, the number of Xs which are Zs exceeding 10.

The chances, it is to be remembered, are all *minima:* except when we mean that m Xs, *and not more*, are Ys, &c. These questions may serve to give some notion of the manner in which arguments not logically conclusive, may be morally so.

What is called *circumstantial* evidence is a species of induction by probability. The thing required to be found has the marks P,Q,R,S, &c.: this Y has the marks P,Q,R,S, &c.: there is then a certain amount of circumstantial evidence that this Y is the thing we want to find. If it can be shown that there is but one thing which has all these marks, then the circumstantial evidence is demonstrative. But if there were, say 100 Ys, of which 5 have the mark P, 5 the mark Q, &c., then having ascertained one Y which has all the marks, the question is, what chance is there against another Y having them all: the same chance, at least, is there that the Y found is the one sought. Instead however, of attempting the problem in this way, which is never resorted to for want of data (I mean that the resemblance which the rough processes of our minds bear to those of the theory of probabilities does not here exist) I take it as follows. If the possession of the mark P give a certain probability to the Y found being that sought, it is as a witness whose testimony has a certain credibility. Similarly for Q,R,S, &c. Compound these testimonies, when known, by the rule in page 195, and the result is the value of the circumstantial evidence.

CHAPTER XII.

On old Logical Terms.

IN this chapter I propose to say something on a few terms of the old Logic, which though they keep their places in works on the subject, and have some of them passed into common language, are very little used. They relate generally to the simple notion, and the name by which it is expressed: and have little of special reference, either to the proposition or syllogism. They are mostly derived from Aristotle, whose incidental expressions became or give rise to technical terms, and whose single sentences were amplified into chapters. And here, as in other places, I have nothing to do with the degree of correctness with which Aristotle's meaning was apprehended, nor even with how much was drawn from Aristotle and how much added to him, but only with the actual phrases and their usual meaning.

The words *logic* and *dialectics** are now usually taken as meaning the same thing: the old distinction is that dialectics is the part of logic in which common and probable, but not necessary, principles, are used. But the distinction is neither clearly laid down, nor faithfully adhered to, even by Aristotle himself.

The *term* (in this work always called *name*) was divided into *simple* and *complex:* the simple term was the mere name, the complex term was what all moderns call the *affirmative proposition.* Thus *man* and *run* were simple terms: *man runs*, a complex term. Later writers rejected this confusion: and divided the acts of the mind considered in logic into *apprehension*, *judgment*, and *discourse*, taking cognizance of notions, propositions, and arguments. The common meaning of the word *discourse*,

* Our language is capricious with regard to the use of singular and plural of words in *ic:* thus we have logic and dialectics, arithmetic and mathematics, physic and physics for medicine and natural philosophy. Some modern writers are beginning to adhere uniformly to the singular, in which I cannot follow them, for I am afraid an English ear would not bear with *mathematic* as a substantive. Would it not better consist with the genius of our language if the plurals were to be always used, and the singulars made adjectives without the termination *al*?

(which now generally applies to ſomething ſpoken) is derived from its place in this diviſion. The word *argument*, which is now equivalent to *reaſoning againſt oppoſition expreſſed or implied*, was originally nothing but the middle term of a ſyllogiſm.

The ſimple term was *univerſal* or *ſingular :* univerſal, when of more inſtances than one, as man, horſe, ſtar ; ſingular, when of one inſtance only, as the ſun, the firſt man, the pole-ſtar, this book. Singular names were called *individuals*, from the etymology of the word, as belonging to objects not diviſible into inſtances to each of which the name could be applied. I have not dwelt upon the diſtinction between ſingular and univerſal, becauſe it is ineffective in inference. And moreover, a ſingular propoſition is only objectively ſingular, but ideally plural. 'Julius Cæſar was a Roman': in point of fact, there was but one Cæſar. But take any imaginary repetition of the circumſtances of Cæſar's life; ſuch, for inſtance as occurs to thoſe who have thought of the poſſibility of the ſame courſe of events returning into exiſtence after a certain cycle: and then the term Cæſar becomes plural. Or, even without ſo forced a ſuppoſition, we may ſay that, if we deſcribe Cæſar, we muſt deſcribe a Roman: that our definition of Cæſar is ſo cloſe as to fit only one man that ever lived, makes no eſſential difference in the character of the propoſition.

But a further diſtinction which was made divided ſingular terms into ſubjects of univerſal, and ſubjects of particular, propoſitions. A determinate (or definite) individual, as Cæſar, this man, was the former: a vague (or indefinite) individual, as a certain man, the firſt comer, was the latter. The diſtinction is that of 'ſome man' and 'this one man.'

Certain notions of eſſence or relation, accompanying the apprehenſion of a name, were called *categories*, or *predicaments*, meaning 'modes of aſſertion with reſpect to' the object named. Ariſtotle gave ten categories, and might have given ten hundred. In their uſual Latin form they were *ſubſtantia*, *quantitas*, *qualitas*, *relatio*, *actio*, *paſſio*, *ubi*, *quando*, *ſitus*, *habitus*.

The word tranſlated by ſubſtance, ȣσία, means mode of being: and its literal Latin is *eſſentia*, eſſence. It is called *ſubſtance* (that which ſtands under) as ſupporting *accidents*, preſently explained. It is far too metaphyſical a term to come into common life without ſome degradation: and accordingly it there means that of which a thing is compoſed, whether material or not. Accordingly we have the material ſubſtance of a coat, the intellectual ſubſtance of an argument. But, as we uſe the word, its meaning belongs to the other predicaments. In fact, the ſubſtance of the old logicians ſtands, as to exiſtence, in the ſame ſituation as *matter* (page 30) with reſpect to our ſenſible perceptions, or *object* with reſpect to our ideas. The ſubſtance, it was ſaid, is *per ſe ſubſiſtens*, while the accident could not be ſaid *eſſe*, but *ineſſe*. The diſtinction between the ſubſtance (mode of being) and the material ſubſtance (in the modern ſenſe) may be helped by the diſtinction between *ſubſtantia prima* and *ſubſtantia ſecunda*, the firſt referring to the individual, the ſecond to the general term. Thus the ſubſtance of John, as John, was *ſubſtantia prima*; as man, *ſubſtantia ſecunda*. All theſe very metaphyſical notions were the ſtudent's firſt introduction to logic, and were conſidered as of the utmoſt importance.

The predicament of *quantity*, derived from the notion of whole and part, was conceived as either *continuous* or *diſcrete*. In continuous quantity, the unit was diviſible, in diſcrete, indiviſible. Thus ten feet is continuous, ten men diſcrete. The diſtinction is preciſely that of magnitudinal and numerical.

Quality was ſubdivided into 1. *Habit* and *diſpoſition*, the latter term being uſed for the imperfect ſtate of the former 2. *Power* and want of it 3. *Patibilis qualitas* and *paſſio*, applied to the ideas of that which is undergone, the firſt permanently, the ſecond for a time. 4. *Form* and *figure*.

Relation then, as now, referred to the ſuggeſtions derived from compariſon of two things or ideas. It was divided into verbal and real (*ſecundum dici* and *ſecundum eſſe*). Thus the relation of *profit* to *profitable* was verbal: that of *father* to *ſon*, or of *above* to *below*, real. The two things related, or *correlatives*, were called *ſubject* and *term*: ſo that of two correlatives, giving two oppoſite relations, the ſubject of either was the term of the other. The *fundamentum* of the relation was that in which it took its riſe, when it had a beginning.

Action and *paſſion*, the production and reception of an effect, requiring the producing *agent*, and the receiving *patient*, were divided into *immanent*, or enduring in the agent, and *tranſient*, or paſſing out to another. Actions were *univocal*, or *æquivocal*, according as their effects were of the ſame or different ſpecies. A few years before the publication of Newton's Principia, it was taught in a work imported into Cambridge that when mice bred mice, the action was univocal, but when *the sun* bred mice (the writer muſt have been thinking of Ariſtotle and ſome of the ſchoolmen) æquivocal. There was alſo the *terminus à quo* and the *terminus ad quem* to repreſent the ſtate before and the ſtate after the action. Thus, when all this nonſenſe was ſent to Coventry, the *terminus à quo* was an immenſe quantity of univocally bred learning of the preceding kind; the *terminus ad quem* was the rooting up of the wheat of logic with the tares.

The *where* (as to abſolute poſition), the *when*, and the *ſite* (relative poſition) gave no peculiar terms of ſubdiviſion. The *habitus* (ἔχειν) referring to *poſſeſſion* generally in the firſt inſtance, was materialized by ſome of the old logicians till it related to *dreſs* only, or *habit* in the thence acquired meaning.

The word predicament (and category as well) has been introcuced into common language to ſignify a ſet of circumſtances under which any thing takes place. It is then no longer confined to the above predicaments, nor is there any occaſion that it ſhould be.

The *predicables* (κατηγορέμενα) are diſtinguiſhed from *predicaments* (κατηγορίαι) in that the former belong to any ſimple notion or name, and may be predicated of it: the latter belong to the connexion (when affirmative) between two names. They are ſaid to be five in number, *genus*, *ſpecies*, *differentia*, *proprium*, and *accidens*.

The words *genus* and *ſpecies* have preſerved their old meaning. If there be a number of names of which each is ſubidentical of the one which follows, ſay V, W, X, Y, Z: then of any two, ſay W and X, X is a *genus* containing the *ſpecies* W. Here Z is the *ſummum genus*, and V the *infima ſpecies*: X is the *genus proximum* of W, Y the *genus remotum*. In what I have called a *univerſe*, which is a *ſummum genus*, having for its *infima ſpecies* the individual inſtance of any name in it, the ſuperidentical is the genus, the ſubidentical the ſpecies. Subcontraries (and contraries) are *oppoſite ſpecies*; ſupercontraries and complex particulars have no ancient name.

The *differentia* is that by which one claſs (be it ſpecies or genus, the difference being accordingly termed *ſpecific* or *generic)* is diſtinguiſhed from another. Thus the difference (or one difference) ſeparating the ſpecies *man* from the other ſpecies of the genus *animal*, is the epithet *rational*.

The *proprium* (or property) is that which belongs to the ſpecies *only*, whether it be to all or only to ſome: thus to ſtudy, and to ſpeak, are equally *propria* of man. But the old commentators give definitions of the property as follows. There are four kinds. 1. That which belongs to the ſpecies alone, but not to all. 2. To all the ſpecies, but not to that alone. 3. To the ſpecies only, and to all of it, but not at all times. 4. To the ſpecies alone, to all, and always.

The *accidens* (or accident) is that which may ſometimes belong to the individual of a ſpecies, but not neceſſarily, nor to that ſpecies alone. In modern language, the term is limited to what is unuſual and unexpected.

The word *cauſe* was uſed by the ancients in a wider ſenſe than by us: more nearly in the ſenſe of the Latin *cauſa*, or the Italian *coſa*. Cauſes were diſtinguiſhed into *material*, *formal*, *efficient*, and *final*. The *material* cauſe was the very matter of a thing, conſidered as a kind of giver of exiſtence; the *formal* cauſe was its form, in the ſame light; the *efficient* cauſe (our common Engliſh word) the agent or precedent; and the *final* cauſe, the ultimate end or object, conſidered as a reaſon for the exiſtence of the thing. Sometimes writers ſtill talk of final cauſes, and are as unintelligible to moſt readers as if they had talked of final beginnings.

The word form was uſed in a wider ſenſe than that of figure or ſhape, to mean, as it were, law of exiſtence, mode, diſpoſition, arrangement. Mere figure or ſhape was only one of the *accidental forms*, as diſtinguiſhed from *ſubſtantial forms*, belonging to the ſubſtance. And *motion* was as widely uſed as *form*: it meant any alteration. Thus, *corruption* was one of the *motions* of matter. Change from place to place, to which the modern word is confined, was *local* motion.

The original uſe of the terms *ſubject* and *object* is to denote a thing conſidered as that which may have ſomething inherent in it, or attached to it, or ſpoken of it, &c.; and as that which may be *objected* to the mind or reafon, or made to come in its way. Thus it was faid that matter is the *fubject* of thofe properties which are the *objects* of the mind in natural philofophy. The tranfition to the modern fenfe of *object*, namely, end propofed, is natural enough. In modern times, fubject and object are ufed * with refpect to *knowledge:* the fubject being the mind in which it is, the object being the external fource from which it comes. For *fubjective* and *objective* I have in this work ufed *ideal* and *objective* (page 29). *Adjunct* was the technical term for that which is in the fubject.

A *modal* propofition was one in which the affirmation or negation was expreffed as more or lefs probable: including all that is technically under probability (Chapter IX) from neceffity to impoffibility. The theory of probabilities I take to be the unknown God which the fchoolmen ignorantly worfhipped when they fo dealt with this fpecies of enunciation, that it was faid to be beyond human determination whether they moft tortured the modals, or the modals them. Their gradations were *neceffary*, *contingent*, *poffible*, *impoffible;* contingent meaning more likely than not, poffible lefs likely than not. Thefe they connected with the four modes of enunciation, A, I, O, E, and when by *fome* is meant *more than half*, the connexion is good. The controverfy about *modal* forms continues up to this day among logicians who are not mathematicians: I fhould fuppofe that the latter would never give it a thought, except as a branch of the theory of probabilities, and except as to the confideration how the terms by which the non-mathematical logician indicates his degrees of belief are to be placed upon the numerical fcale. In like manner he reads the thermometer by graduation, and though he admits the freezing and boiling point, which have an origin in nature, he leaves temperate, fummer heat, blood heat, &c. to the fancy of thofe who choofe to employ them.

At the fame time it is clear that thefe modal forms were confidered not merely as ufeful in expreffion of the nature and amount of belief, but as fuggeftive of real branches of inquiry, fubfervient to that great *à priori* inquiry into the nature of things to which

* See a full account of thefe words in Sir William Hamilton's notes to Reid, p. 806, &c.

mediæval logic was applied. We are not fit to judge of the inſtrumental part of this philoſophy, unleſs we conſider alſo the materials on which it was founded. In an age in which much more faith was demanded of the ſtudent than now; when he was much more frequently required to decide in one way or the other upon a ſingle teſtimony; when, in addition to the non-mythic wonders recorded in ancient writers, which there was no mode of contradicting, all that was known of immenſe regions and countries reſted upon very few accounts, and thoſe filled with ſtories quite as ſtrange:—the abſence of other means of diſtinguiſhing truth from falſehood obliged thoſe who thought to lay much ſtreſs upon *à priori* conſiderations. It matters little to us whether we infer the *neceſſity* of man being a walking animal from the non-arrival of exceptions, and thence the *univerſality* of the rule, or the univerſality from the ſuppoſed perfect induction of inſtances, and thence the neceſſity. But it was of much more conſequence to the old logician: of more *real* conſequence. He did not know but that any day of the week might bring from Cathay or Tartary an account of men who ran on four wheels of fleſh and blood, or grew planted in the ground like Polydorus in the Æneid, as well evidenced as a great many nearly as marvellous ſtories. As he could not pretend to inductive and demonſtrative univerſality, even upon the queſtion of the form of his own race, he was obliged to combine with his argument the antecedent teſtimony of his own and other minds, in the manner which the real doctrine of modals (page 205) ſhows to be neceſſary in all non-demonſtrated concluſions. It is true that he frequently confounded the prediſpoſition of minds with the conſtitution of objects; the teſtimony with the thing teſtified about.

We ſhall never have true knowledge of the ſchools of the middle ages, until thoſe who have ſtudied both their philoſophy, their phyſics, and their ſtate of tradition, will look at their weapons of controverſy as both offenſive and defenſive, and give a fair account of the amount of protection afforded by the firſt, in the exiſting ſtate of the ſecond and third. It would alſo be adviſable to conſider whether, looking at the power of communication by land and ſea, and all the circumſtances of literary intercourſe, it would have been practicable to place the knowledge of the earth and its details upon any better footing of evidence.

One leading feature of the ſchoolmen, acute as they were, and as to repreſentation of notions, inventive, and which is ſhared by many more modern writers who have not diſciplined themſelves mathematically, is ſeen in their employment of quantity: there are inſtances of the ſtrange uſe, the wrong uſe, and the no-uſe. Moſt of them ariſe from indiſtinct apprehenſion of continuity, which obliges them to accept ſuch ſtages of quantity as are expreſſed by exiſting terms, without any effort to fill up gaps. There is alſo a ſlovenlineſs of definition in what relates to quantity. Thus dozens of inſtances might be given in which the *ſome* of the particular propoſition is ſo defined that we might ſuppoſe it is 'ſome, not all,' inſtead of 'ſome, it may be all,' and the former is the expreſs definition of ſome writers: and it is only when we find in rules that XY does not allow us to infer X:Y, nor to contradict X)Y, that we aſcertain the real intended meaning. "Logicians," ſays Sir William Hamilton, "have referred the quantifying predeſignations *plurimi*, and the like, to the moſt oppoſite heads; ſome making them univerſal, ſome particular, and ſome between both." They muſt have had curious ideas of quantity who made the propoſition 'moſt Xs are Ys' either univerſal, or between the univerſal and particular: I ſhould ſuppoſe that thoſe who did the latter muſt have imagined *ſome* to refer to a *minority*.

There is a ſtrange notion of quantity revived in modern times, which conſiſts in making *plurality of attributes* a part of the quantity of a notion. It is called its *intenſive quantity*, or its *intenſion*, or *comprehenſion*. It is oppoſed to *extenſive quantity*, or *extenſion*, which is the more common notion of quantity, referring to the number of ſpecies or of individuals (it may be either, the individual is the real *infima ſpecies*) contained under the name. Thus *man* is not ſo extenſive as *animal*, but more intenſive; the attribute *rational* gives greater comprehenſion. But 'man reſiding in Europe' is leſs extenſive and more comprehenſive than either. It is ſaid that the greater the intenſive quantity the leſs the extenſive, but this is not true, unleſs no two of the ſigns of intenſion be properties of the ſame ſpecies. Thus, according to ſuch ſtatements as I have ſeen, 'man, reſiding in Europe, drawing breath north of the equator, ſeeing the ſun riſe after thoſe in America,' would be a more intenſively quantified notion than 'man reſiding in Europe';

but certainly not more *extenſive*, for the third and fourth elements of the notion muſt belong to thoſe men to whom the firſt and ſecond belong. Thus, in the Port-Royal Logic, one of the earlieſt modern works (according to Sir W. Hamilton), in which the diſtinction is drawn, it is ſaid that the *comprehenſion* of the idea of a triangle includes ſpace, figure, three ſides, three angles, and the equality of the angles to two right angles. But the idea of *rectilinear three-ſided figure* has juſt as much extenſion.

The relation between comprehenſion and extenſion exiſts, and is uſeful: but not, I think, as that of different kinds of *quantity*. In page 148, where I hold that the propoſition *is contained* in its neceſſary conſequence, the view is one of extenſion: the ordinary view is one of comprehenſion. 'Every caſe in which P is true, is a caſe in which Q is true,' tells us that all the P-caſes are contained, as to extent (number and location of inſtances), among the Q-caſes. But, as to comprehenſion, every P-caſe contains all that diſtinguiſhes a Q-caſe from other things. When, in page 47, it is ſaid that the idea of man is contained in that of animal, I ſpeak of extenſion: all the inſtances to which the firſt idea applies are among thoſe to which the ſecond applies. But, as to comprehenſion, the idea of animal is contained in that of man: all that defines animal goes to the definition of man, and other things beſides. In page 50, the "*is* of poſſeſſion of all eſſential characteriſtics," refers to comprehenſion; the "*is* of identity" to extenſion: both poſſeſſing equally the characters under which the verb may occur in logic. There is no diſtinction which affects inference: for X)Y has exactly the ſame properties whether we interpret it as expreſſing that Y has all the extenſion of X, and may be more; or that X has all that Y has in comprehenſion, and may be more.

In pages 115, &c. we have the mode of repreſenting names of more or leſs comprehenſion. Thus, P, Q, R, &c. being characteriſtics, the obvious propoſition PQ)P, illuſtrates the theorem that where the comprehenſion of one name has all that of a ſecond (as PQ has that of P) the extent of the ſecond is at leaſt as great as that of the firſt. And the ſelf-evident poſtulate in page 115, by which we may diminiſh the extent of a term univerſally uſed, or increaſe that of one particularly uſed, may be expreſſed in language of comprehenſion. That is, we may augment the comprehenſion of a univerſal, or diminiſh that of a particular. Thus, X)Y gives XP)Y, and X.Y gives XP.Y: but X)YP gives X)Y.

It will be eaſily ſeen that comprehenſion has the firſt attribute of quantity (page 174): there is *more* and *leſs* about it. But it is not of the *meaſurable* kind (page 175). As to extent, 200 inſtances bear a definite ratio to 100, which we can uſe, becauſe our inſtances are *homogeneous*. But different qualities or deſcriptions can never be numerically ſummed as attributes, to any purpoſe ariſing out of their number. Does the idea of *rational animal*, two deſcriptive terms, ſuggeſt any uſeful idea of *duplication*, when compared with that of *animal* alone. When we ſay that a chair and a table are more furniture than a chair, which is true, we never can cumulate them to any purpoſe, except by abſtracting ſome homogeneous idea, as of bulk, price, weight, &c. To give equal quantitative weight to attributes, as attributes, ſeems to me abſurd: to uſe them numerically otherwiſe, is at preſent impoſſible.

The reader will have ſeen the origin of ſeveral very common terms, which are uſed in a ſenſe coinciding with, or at leaſt much reſembling, that put upon them by the ſchoolmen. But there is one which has diametrically changed its meaning; it is the word *inſtance*. The word *inſtantia* (and alſo ἔνστασις) implied a caſe *againſt*, not *for*; the latter was *exemplum*: ſo that *inſtance to the contrary* would have been tautology.

I have referred the word *enthymeme* to this chapter, though it is always regularly explained in connexion with the ſyllogiſm. According to Ariſtotle, Ἐνθύμημά ἐστι συλλογισμὸς ἀτελὴς ἐξ εἰκότων καὶ σημείων, an enthymeme is an imperfect ſyllogiſm from probables and ſigns: the modern critics reject the word ἀτελὴς, *imperfect*, as interpolated. The word *ſign* ſeems to mean indication, ſymptom, or effect, which makes the cauſe almoſt neceſſary or highly probable. But the ſchools took the word *enthymeme* to mean a ſyllogiſm with a ſuppreſſed and implied premiſe, ſuch as 'He muſt be mortal, being a man.' I cannot help ſuſpecting that Ariſtotle*

* He ſays all that is communicated (λέγεται) of the predicate, will be aſſerted in words (ῥηθήσεται) of the ſubject. Theſe two different tenſes of two different verbs are often both tranſlated by *dicitur*. Why did they

made no difference between a ſuppreſſed premiſe, clearly intended and diſtinctly received, and one formally given. It ſeems to me that we might as well diſtinguiſh a written from a ſpoken ſyllogiſm, as to the logical character of the two.

CHAPTER XIII.

On Fallacies.

THERE *is* no ſuch thing as a claſſification of the ways in which men may arrive at an error: it is much to be doubted whether there ever *can be*. As to mere inference, the main object of this work, it is reducible to rules: theſe rules being all obeyed, an inference, as an inference, is good; conſequently a bad inference is a breach of one or more of theſe rules. Except, then, by the production of examples to exerciſe a beginner in the detection of breaches of rule, there is nothing to do in a chapter on fallacies, ſo far as thoſe of inference are concerned. Nevertheleſs, there are many points connected with the matter of premiſes, to which it is very deſirable to draw a reader's attention: and above all to queſtions in which it is not at firſt obvious whether the miſtake be in the matter or in the form; or in which it may be the one or the other, according to the ſenſe put upon the words.

If there be anything *ridentem dicere verum quod vetat*, writers on logic have in all ages moſt grievouſly neglected the prohibition in treating this ſubject, and have given the ſtudent a preſcriptive right to ſome amuſement. One reaſon of this was, that the

occur? For various reaſons, I allow myſelf to ſuſpect, though not ſcholar enough to maintain, that λόγος generally meant communication, paſſage from one mind to another by any means, as much at leaſt with reference to the receiving, as to the imparting, mind: and that it is here oppoſed to ῥῆσις, ſpeech, in that ſenſe. Throw the verbs back to their primary meanings, and it will be 'That which *is picked up* of the predicate, ſhall *flow out* about the ſubject.' If my conjecture be correct, the modern enthymeme is here put on the ſame footing as the fully expreſſed ſyllogiſm.

Greeks endeavoured to try the new art by inventing inferences the falſehood of which could not be detected by its rules. Theſe, as may be ſuppoſed, were whimſical efforts of reaſoning: nevertheleſs, they have been handed down from book to book, unſurpaſſed in their way. Another reaſon is, that jeſts, puns, &c. are for the moſt part only fallacies ſo obvious that they excite laughter; and the greater number of them can be ſhown to break one or another of the rules of logic. Accordingly, they furniſh ſtriking examples of theſe rules; the application of which, in ſerious terms, has itſelf a taſte of the ludicrous. Boccacio has, by his inimitable mode of narration, made a good ſtory the jeſt of which could be deſcribed as conſiſting in nothing more than the aſſumption that what can be predicated of ſtorks* in general can be predicated of roaſted ſtorks: which is what logicians would call the *fallacia accidentis*, or arguing *a dicto ſimpliciter, ad dictum ſecundum quid.*

The terms *fallacy*, *ſophiſm*, *paradox*, and *paralogiſm*, are applied to offences againſt logic; but not with equal propriety, *Fallacy* and *ſophiſm* may technically have been firſt applied to arguments in which there is a failure of logic: but it is now very common to apply them alſo to arguments in which there is a falſehood of fact, or error of principle, though logically treated; and if this laſt uſe be not correct, writers on logic have ſanctioned it in their examples. Many perſons go further, and call the erroneous ſtatement itſelf a fallacy: that men are in the habit of walking on their heads, they would ſay is a very obvious fallacy. A *paradox* is properly ſomething which is contrary to general opinion: but it is frequently uſed to ſignify ſomething ſelf-contradictory: thus the newſpaper which recently avowed

* A ſervant who was roaſting a ſtork for his maſter was prevailed upon by his ſweetheart to cut off a leg for her to eat. When the bird came upon table, the maſter deſired to know what was become of the other leg. The man anſwered that ſtorks had never more than one leg. The maſter, very angry, but determined to ſtrike his ſervant dumb before he puniſhed him, took him next day into the fields where they ſaw ſtorks, ſtanding each on one leg, as ſtorks do. The ſervant turned triumphantly to his maſter: on which the latter ſhouted, and the birds put down their other legs and flew away. "Ah, Sir," ſaid the ſervant, "you did not ſhout to the ſtork at dinner yeſterday: if you had done ſo, he would have ſhown his other leg too."

its opinion that the repeal of the corn laws would make food both cheap and dear is ſaid to have maintained a paradox. The modern uſe of the word implies diſreſpect, but it was not ſo formerly. Thus in the ſixteenth century the opinion of the earth's motion was ſtyled the *paradox of Copernicus* by writers who meant neither praiſe nor blame, but only reference to the opinion of Copernicus as *an unuſual one*. The more preciſe writers of our day uſe the word paradox for an opinion ſo very ſingular and improbable, that the holder of it is chargeable with an undue bias in favor of ſingularity or improbability for its own ſake. *Paralogiſm*, by its etymology, is beſt fitted to ſignify an offence againſt the formal rules of inference. It has been frequently abuſed by mathematical writers, who have ſignified by it errors of ſtatement, and undue aſſumptions: but it is not completely ſpoiled for the purpoſe, and I ſhall therefore uſe it to denote a formal error in inference, as a particular claſs of fallacy or ſophiſm, words which it would now be difficult to diſtinguiſh in meaning. Some have defined *paralogiſm* to be that by which a man deceives himſelf, and *ſophiſm* that by which he tries to deceive others: on what grounds I do not know.

The queſtion of a premiſe being right or wrong in fact or principle, unleſs indeed it contradict itſelf, does not belong to logic: nor could it ſo belong unleſs logic were made, in the wideſt ſenſe, that attempt at the attainment of the *cognitio veri* which ſome have defined it to be. All that relates to the collection of true premiſes with reſpect to the vegetable world belongs to botany; with reſpect to the heavenly bodies, to aſtronomy; with reſpect to the relation of man to his Creator, to theology. Even were it within the province of logic, it would be impoſſible, in leſs ſpace than an encyclopædia, to enter upon queſtions connected with the matter of ſyllogiſms. With regard to paralogiſms, or *logical* fallacies, (ſo called, as an error about the meaſure of ſpace is called a *geometrical* error) the claſſification under breach of rules would be good in form, but would afford no baſis for the treatment of the ſubject. Thoſe who bring them forward ſeldom proceed in direct defiance of rule, but in various modes of evaſion. Theſe it would be almoſt impoſſible to arrange in ſatisfactory order.

Ariſtotle made a claſſification of fallacies, which was of courſe adhered to by the writers of the middle ages. In this, as in every other place, when I ſpeak of Ariſtotle and his ſyſtem, I ſpeak of it as underſtood by thoſe writers. How far they diſtinctly comprehended their maſter is a queſtion into which I could not enter here, even if I were competent to write on the ſubject. It is, however, ſufficiently apparent that the logic of Ariſtotle is not of the purely formal character which marked the dialectics of the middle ages: there is a much more decided introduction of the attempt to write on the matter of ſyllogiſm than many perſons think there is. The claſſification of fallacies ſeems to be one proof of this: and the interpretation of that claſſification by the middle writers ſeems to add their teſtimony to the aſſertion: in this part of the ſubject they abandon technicalities almoſt entirely.

It ought to be eſpecially remembered that we are very diffeently ſituated from thoſe writers, not as to what is fallacy, but as to what the ſpecimens of it produced are likely to be. Out of a world of general principles declared by authority, or declared to be ſelf-evident by authority, they had to produce logical deductions; and, of courſe, the pure ſyllogiſm and its rules were to them as familiar as the alphabet. The idea of an abſolute and glaring offence againſt the ſtructure of the ſyllogiſm being ſupported one moment after it was challenged, would no more ſuggeſt itſelf to the mind of a writer on logic than it would now occur to a writer on aſtronomy that the accidental error (which might happen to any one) of affixing four ciphers inſtead of five in multiplying by a hundred thouſand would be maintained after expoſure. Accordingly, their formal chapters on fallacies would naturally relate, if not entirely to fallacies of matter, at leaſt to thoſe in which the fallacy of matter very cloſely hinges upon that of form. And ſo it is in all the old ſyſtems which I have examined. The Ariſtotelian diviſion (or rather ſelection, for it is far from including everything) lends itſelf eaſily to this adaptation.

We, on the contrary, live in an age in which formal logic has long been nearly baniſhed from education: entirely, we may ſay, from the education of the habits. The ſtudents of all our univerſities (Cambridge excepted) may have heard lectures and learnt the forms of ſyllogiſm to this day: but the practice has been ſmall: and out of the univerſities (and too often in them) the very name of logic is a bye-word.

The philosophers who made the discovery (or what has been allowed to pass for one) that Bacon invented a new species of logic which was to supersede that of Aristotle, and their followers, have succeeded by false history and falser theory, in driving out from our system all study of the connexion between thought and language. The growth of inaccurate expression which this has produced, gives us swarms of legislators, preachers, and teachers of all kinds, who can only deal with their own meaning as bad spellers deal with a hard word, put together letters which give a certain resemblance, more or less as the case may be. Hence, what have been aptly called "the slipshod judgments and crippled arguments which every-day talkers are content to use." Offences against the laws of syllogism (which are all laws of common sense) are as common as any species of fallacy: not that they are always offences in the speaker's or writer's mind, but that they frequently originate in his attempt to speak his mind. And the excuse is, that he meant differently from what he said: which is received because no one can throw the first stone at it, but which in the middle ages would have been regarded as a plea of guilty. The current notions about what logic is, are beautiful and wonderful. I have heard a disputant, an educated man, a graduate, escape from allowing himself to be convinced that he was arguing with a middle term particular in both premises by declaring that *facts* were better than *syllogisms*: the form of his argument would have proved that men are plants, because both require air. "I" he said, "produce you *facts*, like Bacon: you quibble about their combination, like Aristotle."

The Aristotelian system of fallacies contains two subdivisions. In the first, which are *in dictione*, or *in voce*, the mistake is said to consist in the use of words: in the second, which are *extra dictionem*, or *in re*, it is said to be in the matter.

Of the first set six kinds were distinguished, as follows:—

1. *Æquivocatio* or *Homonymia*, in which a word is used in two different senses; giving really no middle term (if the middle term be in question) or a term in the conclusion which is not the same name as that used in the premises. For example, 'All criminal actions ought to be punished by law: prosecutions for theft are criminal actions; therefore, prosecutions for theft ought to be punished by law.' Here the middle term is doubly ambiguous,

both *criminal* and *action* having different ſenſes in the two premiſes. But here, as in many other caſes, the choice lies with the ſophiſt to bring the fallacy under the head to which we refer it or not. It may pleaſe him to aſſert that he means the ſame thing by *criminal action* in both premiſes; in which caſe, the inference is logical, but one or the other premiſe muſt be denied as to the matter. Again, 'Finis rei eſt illius perfectio; mors eſt finis vitæ; ergo mors eſt vitæ perfectio.' Here the ambiguity may be thrown either on *finis* or on *perfectio*. The following example can be traced through books for three centuries. 'Every dog runs on four legs; Sirius (the dog-ſtar) is a dog; therefore Sirius runs on four legs.' It has been the defect of many old works on logic that *all* their examples have been of that obvious abſurdity, which is well enough in one or two inſtances. Such as 'Nothing is better than wiſdom and virtue; dry bread is better than nothing; therefore, dry bread is better than wiſdom and virtue.' Some of the old examples are 'A mouſe eats cheeſe; a mouſe is one ſyllable; therefore one ſyllable eats cheeſe.' And again, 'Iſte pannus eſt de Anglia; Anglia eſt terra; ergo, iſte pannus eſt de terra.'

Where the ſyllogiſm is formally put, equivocation of the middle term is generally ſeen with great eaſe. The moſt difficult exception is, I think, the old fallacy, in which giving the name of the genus is confounded with giving the name of the ſpecies, and thereby, of courſe, giving the name of the genus. As in 'To call you an animal is to ſpeak truth; to call you an aſs is to call you an animal; therefore, to call you an aſs is to ſpeak truth.' This equivocation will puzzle a beginner as to its form, and the more ſo from the evident falſehood of the matter. The middle term is "He who ſays that you are *one* among all animals." He ſpeaks truth; and the one who calls you an aſs or a gooſe, certainly ſays that you are *one* among all animals. The equivocation is in the two different uſes of the word one; in the firſt premiſe, it is an entirely indefinite *one;* in the ſecond it is a leſs indefinite *one*. This *one* is not attached to the quantity of the middle term, which is univerſal in the firſt premiſe, and particular in the ſecond: but is part of the middle term itſelf.

The manner in which the ſerious fallacy of equivocation moſt frequently appears, is in the connection of the old aſſociations of a word which has ſhifted its meaning with the altered meaning of the ſame. The word loyal, for inſtance, originally meaning no more (and no leſs) than *lawful*, which, as applied to a man, meant one who reſpected the laws, and had not forfeited any right by miſbehaviour, now means attached to the Crown and to the title of the holder of it. In conteſts for ſucceſſion, the winner would, of courſe, aſſume that *lawful* men were on his ſide. In more recent times, the term was always ſelf-applied, at elections, by thoſe who ſupported the party which had the confidence of the Crown for the time being: but on ſuch occaſions, abſtinence from the fallacy which the French call the *voie du fait* is the utmoſt which can be expected of human nature.

The word *publication* has gradually changed its meaning, except in the courts of law. It ſtood for *communication to others*, without reference to the mode of communication, or the number of recipients. Gradually, as printing became the eaſieſt and moſt uſual mode of publication, and conſequently the one moſt frequently reſorted to, the word acquired its modern meaning: if we ſay a man publiſhes his travels, we mean that he writes and prints a book deſcriptive of them. I ſuſpect that many perſons have come within the danger of the law, by not knowing that to write a letter which contains defamation, and to ſend it to another perſon to read, is *publiſhing a libel;* that is, by imagining that they were ſafe from the conſequences of publiſhing, as long as they did not print. In the ſame manner, the well-eſtabliſhed rule that the firſt publiſher of a diſcovery is to be held the diſcoverer, unleſs the contrary can be proved, is miſunderſtood by many, who put the word printer in the place of publiſher. I could almoſt fancy that ſome perſons think rules ought to travel in meaning, with the words in which they are expreſſed.

A ſimilar change has taken place in the meaning of the word to *utter*, the ſenſe of which is to *give out*, but which now means uſually to give out of the mouth in words. As yet, I am not aware that any perſon charged with the *utterance* of counterfeit coin has pleaded that no one ever uttered coin except the princeſs in the fairy tale: but there is no ſaying to what we may come, with good example, and under high authority.

It may almoſt be a queſtion whether, in the time of Ariſtotle, ſucceſsful equivocation, that is, undetected at the moment, would not have been held binding on the diſputant who had failed to detect it. The genius of uncultivated nations leads them to place undue force in the verbal meaning of engagements and admiſſions, independently of the underſtanding with which they are made. Jacob kept the bleſſing which he obtained by a trick, though it was intended for Eſau: Lycurgus ſeems to have fairly bound the Spartans to follow his laws till he returned, though he only intimated a ſhort abſence, and made it eternal: and the Hindoo god who begged for three ſteps of land in the ſhape of a dwarf, and took earth, ſea and ſky in that of a giant, ſeems to have been held as claiming no more than was granted. The great ſtreſs laid by Ariſtotle on ſo many different forms of verbal deception, may have ariſen from a remaining tendency among diſputants to be very ſerious about what we ſhould now call play upon words.

Governments permit what would otherwiſe be equivocation to take a ſtrong air of truth, by legiſlating in detail againſt the principles of their own meaſures. The window-tax is a ſpecial inſtance. A newſpaper calls it a tax upon the light which God's beneficence has given to all. The anſwer would be plain enough, namely, that it is an income tax levied upon a uſe of that light which (how truly matters not here) is aſſerted to be a fair criterion of income. But this anſwer is deſtroyed by the permiſſion to block up windows, and thereby evade the tax: which is thus made to fall upon the light uſed, and not upon the means of uſing it which the ſize of the houſe affords. According to the principle of this impoſt, the blocked window is as fair a criterion of the income of the occupant as the open one, and ſhould have been ſo conſidered.

Among the forms which the fallacy of equivocation frequently aſſumes, is that of the ſophiſt altering or qualifying the known meaning of a word in his own mind, without giving the other party any notice: ſo that there may be, if not two meanings in one mind, yet different meanings in the two minds concerned. A perſon aſſerts that 'Nobody denies, &c. &c.' Should this go down, the point is gained; what nobody denies muſt be undeniable. But ſhould it be conteſted (and it will generally be found that the things which nobody denies are matters of ſome difference of opinion, while thoſe which nobody *can* deny are quite ſure to be points of conſtant controverſy) the evaſion is ready. It is no ſenſible perſon, or nobody that underſtands the ſubject, nobody that is anybody, in ſhort: while perhaps it cannot be ſettled who does, or who does not, underſtand the ſubject, until, among other things, the very point in diſpute is determined.

There is a wide range of equivocations ariſing out of meanings which are ſometimes implied and ſometimes not. A large claſs of them is made by the uſual, but not univerſal, practice, of giving to the thing the name of that which it is intended to be, whether the attempt be ſucceſsful or not. This is now abbreviation or courteſy; but it was the rule. According to old definitions, bad reaſoning is reaſoning, *ſyllogiſmus ſophiſticus* is a ſyllogiſm, and in an old book now before me, the fruits and effects of demonſtration are ſcience, opinion, and *ignorance*, the latter containing belief of falſehood derived from *bad* demonſtration, which we ſhould now call *no* demonſtration.

One fallacy of our time, and a very favourite one, is the ſettlement of the merit of a perſon, or an opinion, not by arguing the place of that perſon or opinion in its ſpecies, but by arbitrary alteration of the boundary of the ſpecies, with the intent of excluding the individual in queſtion altogether.

It is ſomewhat analogous to the proceeding of the landlord who unroofs the houſe to get rid of a tenant. Thus we have had the controverſy whether Pope was a *poet*, not whether he was a good poet or a bad one, but whether he was a poet at all. The diſputants, or ſome of them, claimed a right to define a poet, and decided that none but verſe-makers of a certain goodneſs (to be ſettled by themſelves) were poets. They might juſt as well have decided, on their own authority, that none but men of a certain amount of reaſoning power were *men*. Had they done this laſt, as long as they fixed the amount at a figure which included themſelves under the name, nobody would have thought they materially altered the extent of the term: it is not eaſy to ſee why they have rights ſo arbitrary, over words the objective definitions of which are nearly as well fixed as that of man.

Another form of the fallacy of equivocation is the aſſuming, without expreſs ſtatement, that the meaning of a phraſe can be determined by joining the meanings of its ſeveral words: which is not always true in any language. When two words come together, it often happens that their dictionary meanings would never enable us to arrive at their known and ufual (and therefore proper) compound meaning: though they might help us in explaining how that laft meaning arofe. A perfon undertakes to crofs a bridge in an incredibly fhort time: and redeems his pledge by croffing the bridge as one would crofs a ftreet, that is, by traverfing the breadth. Now, though it be true that, in general, to crofs is to go over the breadth, or fhorter dimenfion, yet in the cafe before us, the phrafe is elliptical, and fignifies croffing *the river* upon the bridge. Nor can it be faid that this common meaning is incorrect: that which is common and well known is, in language, always correct. No reafonable perfon would fay that a French newfpaper is wrong in reporting an army to be *à cheval fur la rivière*, becaufe a river is not a horfe. This literal (or rather unlettered) mode of interpretation is adopted among gamblers in fettling bets: and is of itfelf enough to raife a ftrong prefumption that their occupation is not that of well-educated men.

It is common enough in controverfy, for one fide or the other to have fixed meanings of words in his own mind, on which he proceeds without any inquiry as to whether thofe meanings will be *conveyed* by the words to the other fide, or to the reader. It is very difficult to avoid this form of the fallacy, without giving the meanings of the moft effential terms, on the firft occafions of their occurrence. It is not uncommon to meet with a writer who appears to believe, at leaft who certainly acts upon, the notion that the right over words refides in him, and that others are wrong fo far as they differ from him. I do not only mean that there are many who have an undue belief in their own judgments, both as to words and things: but I fpeak of thofe who, though fhowing a proper modefty in refpect to their own conclufions, feem to be unable to do the fame with refpect to their definitions of words. If all mankind had fpoken one language, we cannot doubt that there would have been a powerful, perhaps a univerfal, fchool of philofophers who would have believed in the inherent connexion between names and things; who would have taken the found *man* to be the mode of agitating the air which is effentially communicative of the ideas of reafon, cookery, bipedality, &c. The writers of whom I fpeak, are more or leſs of this ſchool; they treat words as abſolute images of things by right of the letters which ſpell them. "The French," ſaid the ſailor, "call a cabbage a *ſhoe*; the fools! why can't they call it a cabbage, when they muſt know it is one?"

Equivocation may be uſed in the form of a propoſition; as for inſtance, in throwing what ought to be an affirmative into the form of a qualified negative, with the view of making the negative form produce an impreſſion. Thus a controverſial writer will aſſert that his opponent has not attempted to touch a certain point, except by the abſurd aſſertion, &c. &c. &c. To which the other party might juſtly reply, "Your own words ſhow that I have made the attempt, though your phraſe has a tendency, perhaps intended, to make your reader think that there is none, or at leaſt to blind him to the difference between *none* and *none that you approve of*."

2. The *fallacia amphiboliæ*, or *amphibologiæ*, differs in nothing from the laſt, except in the equivocation being in the conſtruction of a phraſe, and not in a ſingle term: as in confounding that which is Plato's (property) with that which is Plato's (writing). Or, as in 'Qui ſunt domini ſui ſunt ſui juris; ſervi ſunt domini ſui; ergo ſervi ſunt ſui juris.' The ambiguities of conſtruction in our language, ariſing from want of inflexions and genders are tolerably (and intolerably) numerous. The difficulty of determining the emphatic word often gives a doubt as to the meaning. But very often indeed there is a want of the diſtinction which the algebraiſt makes when he writes three-and-four tens as diſtinguiſhed from three and four-tens: $(3+4).10$ and $3+4.10$. It cannot, for inſtance, be ſaid whether 'I intend to do it and to go there to-morrow' means that it will be done to-morrow or not. It may be either—(I intend to do it and to go there) to-morrow, or—I intend to do it and (to go there to-morrow). The preſumption may be for the firſt conſtruction: but it is only a preſumption, not a rule of the language. In an inſtance cited by Dr. Whateley—"If this day happen to be Sunday, this form of prayer ſhall be uſed and the faſt kept the next day following," the conſtruction is ambiguous, and the intended meaning probably againſt the preſumption. There is a book of the laſt century, written by a "teacher of mathematics, and writing maſter to Eton College." Were mathematics taught at Eton,

or not? Punctuation may be an assistance; but it so often happens that the author leaves that point to the printer, that it is hardly safe to rely upon it. Printers punctuate correctly when the meaning is clear: but when it is ambiguous, they may be as apt to take the wrong meaning as any other readers.

3, 4. The *fallacia compositionis*, and *fallacia divisionis*, consist in joining or separating those things which ought not to be joined or separated. If we may say that A is X and B is Y, so that A and B is X and Y, we have no right to infer that we may form the compound and collective names 'A and B,' and 'X and Y,' and say that 'A and B' is 'X and Y.' Thus two and three are even and odd: but five is not even and odd. Again, two and five are four and three; but neither is two four, nor five three. It must be remembered that the word *all*, in a proposition, is not necessarily significative of a universal proposition: it may be a part of the description of the subject. Thus in 'all the peers are a house of Parliament,' we do not use the words *all the peers* in the same sense as when we say 'all the peers derive their titles from the Crown.' In the second case the subject of the proposition is *peer;* and the term *all* is distributive, synonymous with each and every. In the first case the subject is *all the peers*, and the term *all* is collective, no more distinguishing one peer from another than one of John's fingers is distinguished from another in the phrase, 'John is a man.' The same remarks may be made on the word *some;* as in 'some peers are dukes,' and 'some peers are the committee of privileges.' The *all* and *some* of the quantity of the proposition are distributive terms; the all and some of the subject are collective. Again, all men are a species (of animals) which no number of men are, wanting the rest. *All men* here make the one individual object of thought of a singular proposition. This amounts to an ambiguity of construction, an *amphibologia*, as do most sources of fallacy falling under this head, which can therefore hardly be considered as anything more than a case of the last. We want another idiom or the algebraical distinction, as in 'All (peers) hold of the Crown; (all peers) are a house of Parliament.'

5. The *fallacia prosodiæ* or *accentus* was an ambiguity arising from pronunciation, and its introduction seems to lead to very minute subdivision of the subject, and to ensure the entrance of

none but ludicrous examples. Burgerſdicius does not think it unworthy of himſelf to deſcend to the following, 'Omnis equus eſt beſtia; omnis juſtus eſt æquus, ergo omnis juſtus eſt beſtia.' An older writer has 'Tu es qui es; quies eſt requies; ergo, tu es requies.' Theſe are mere puns; and the makers of them were fairly beaten by the contriver of 'Two men eat oyſters for a wager, one eat ninety-nine, the other eat two more, for he eat a hundred and won.' But more ſerious fallacies may be referred to this head. A very forced emphaſis upon one word may, according to uſual notions, ſuggeſt falſe meanings. Thus, 'thou ſhalt not bear falſe witneſs againſt thy neighbour,' is frequently read from the pulpit either ſo as to convey the oppoſite of a prohibition, or to ſuggeſt that ſubornation is not forbidden, or that anything falſe except evidence is permitted, or that it may be given *for* him, or that it is only againſt *neighbours* that falſe witneſs may not be borne.

A ſtatement of what was ſaid, with the ſuppreſſion of ſuch tone as was meant to accompany it, is the *fallacia accentus*. Geſture and manner often make the difference between irony or ſarcaſm, and ordinary aſſertion. A perſon who quotes another, omitting anything which ſerves to ſhow the *animus* of the meaning; or one who without notice puts any word of the author he cites in italics, ſo as to alter its emphaſis; or one who attempts to heighten his own aſſertions, ſo as to make them imply more than he would openly ſay, by italics, or notes of exclamation, or otherwiſe, is guilty of the *fallacia accentus*.

To this fallacy I ſhould refer one of very common occurrence, the alteration of an opponent's propoſition ſo as to preſent it in a manner which is logically equivalent, but which alters the emphaſis, either as noticed in page 134, or in any other manner. It is generally not reaſoning, but retort, which is the object of the alteration: for inference cannot be altered by changing a propoſition into a logical equivalent, but a ſmart repartee may be very effective againſt 'Some Xs are Ys,' but flat enough againſt 'ſome Ys are Xs.' And even when the proponent miſtakes his own meaning, and miſcalculates his own emphaſis, ſtill, if the miſtake be obvious, there is fallacy in taking advantage of it; for he who communicates in ſuch incorrect terms as ſhow what the correct ones are, does, in fact, communicate in correct terms, to all who ſee the ſhowing. Of courſe, reſpect for logic never ſtood in the way of a ſucceſsful retort from the time of Ariſtotle till now, nor will on this ſide of the millenium. A ſpeculator once wrote to a ſcientific ſociety, to challenge them to an (on his part) anti-Newtonion controverſy, relying on it that he could contend in mechanics, though avowedly ignorant of geometry. He was anſwered by a recommendation to ſtudy mathematics and dynamics. His rejoinder was an angry pamphlet, in which, indignant at the unfairneſs, as he took it to be, of the recommendation, he exclaimed, 'I did not confeſs my ignorance of dynamics.' Had he been worth the anſwering, it would have been impoſſible to reſiſt the reply 'No, but you ſhowed it.' Had he written, as he meant 'It was not dynamics of which I confeſſed ignorance,' and had an opponent written, as many would have done, 'You ſay, ſir, that you did not confeſs your ignorance of dynamics: indeed you did not, you contented yourſelf with an ample diſplay of it,' he would have uſed the *fallacia accentus*. Nor would he, in my opinion, have been clear of it though he had only taken advantage of a wrong, but evidently wrong, placement of emphaſis on the part of the aſſailant. The uſe of ſuch a weapon, as to its legitimacy, depends entirely upon the manner in which the queſtion ſhall be ſettled how far irony is allowable. Where the anſwer is in the affirmative, a very obvious fallacy, as a ſarcaſm, may be permitted. But I may here obſerve, that irony itſelf is generally accompanied by the *fallacia accentus*; perhaps cannot be aſſumed without it. A writer diſclaims attempting a certain taſk as above his powers, or doubts about deciding a propoſition as beyond his knowledge. A ſelf-ſufficient opponent is very effective in aſſuring him that his diffidence is highly commendable, and fully juſtified by the circumſtances.

6. The *fallacia figuræ dictionis*, as explained, means literally a miſtake in grammar and nothing elſe; as that becauſe *fluvius* is *aqua* it is *humidA*, or that becauſe *aqua* is feminine, ſo is *poeta*.

All theſe fallacies *in dictione* come under the head of ambiguous language, and amount to nothing but giving the ſyllogiſm four terms, two of them under the ſame name. The fallacies *extra dictionem* are ſet down as follows.

1. The *fallacia accidentis*; and 2. That *à dicto ſecundum quid ad dictum ſimpliciter*. The firſt of theſe ought to be called that

of *à dicto ſimpliciter ad dictum ſecundum quid*, for the two are correlative in the manner deſcribed in the two phraſes. The firſt conſiſts in inferring of the ſubject with an accident that which was premiſed of the ſubject only: the ſecond in inferring of the ſubject only that which was premiſed of the ſubject with an accident. The firſt example of the ſecond muſt needs be 'What you bought yeſterday, you eat to-day; you bought raw meat yeſterday; therefore, you eat raw meat to-day.' This piece of meat has remained uncooked, as freſh as ever, a prodigious time. It was raw when Reiſch mentioned it in the *Margarita Philoſophica* in 1496: and Dr. Whateley found it in juſt the ſame ſtate in 1826. Of the firſt, we may give the inſtance 'Wine is pernicious; therefore, it ought to be forbidden.' The expreſſed premiſe refers to wine uſed immoderately: the concluſion is meant to refer to wine however uſed. This ſpecies of fallacy occurs whenever more or leſs ſtreſs is laid upon an accident, or upon any view of the ſubject, in the concluſion, than was done in the premiſes. As in the following:—'All that leads to ſuch philoſophy as that of the ſchoolmen, with their logic, muſt be unworthy to be ſtudied, except hiſtorically.' The intent of ſuch a ſentence is not formally to propoſe the falſe ſyllogiſm, 'The ſchoolmen had that which led them to a falſe philoſophy; the ſchoolmen had logic; therefore, logic led them to a falſe philoſophy,' but only to take the chance of the ſtreſs thus laid upon logic producing a diſpoſition to ſuppoſe that the logic was in fault. The premiſes are really:—

The philoſophy of the ſchoolmen (who paid particular attention to logic)	is	a falſe philoſophy.
Every falſe philoſophy	is	that the guides to which ſhould be neglected, except as hiſtory.

whence it is rightly inferred that the guides to ſuch a philoſophy as that of the ſchoolmen (who ſtudied logic) are only of hiſtorical uſe. And the ſame thing might equally be inferred of the ſchoolmen who ate mutton, a practice to which moſt of them were as much addicted, no doubt, as to making ſyllogiſms. The art of the ſophiſt conſiſts in making the accident which is either unfairly introduced, or withdrawn, or ſubſtituted, have an apparently relevant relation to the ſubject itſelf. Undoubtedly, the ſchoolmen's logic has a connexion with their philoſophy which the mutton they ate has not: but as long as it is not *the* connexion which permits the inference, it is abſolutely irrelevant.

All the fallacies which attempt the ſubſtitution of a thing in one form for the *ſame thing* (as it is called) in another, belong to this head: ſuch as that of the man who claimed to have had one knife twenty years, giving it ſometimes a new handle, and ſometimes a new blade. The anſwer given by the calculating boy (page 54, note) was, relatively to the queſtion, a worthy anſwer, and took advantage of the common notion that a bean, after being ſkinned, is ſtill a bean, as before. More ſerious difficulties have ariſen from the attempt to ſeparate the *eſſential* from the *accidental*, particularly with regard to material objects. The Carteſians denied weight, hardneſs, &c. to be eſſential to matter, until at laſt they made it nothing but ſpace, and contended that a cubic foot of iron contained no more matter than a cubic foot of air.

The law, in criminal caſes, demands a degree of accuracy in the ſtatement of the *ſecundum quid* which many people think is abſurd: and it appears to me that the lawyers often help the popular miſapprehenſion, and give it excuſe, by confounding errors of things with errors of words, after the example of the world at large. Any error of any kind, provided it be ſmall in amount, paſſes for a miſtake in words only, by virtue of its ſmallneſs. By a miſtake in words, I mean the addition or omiſſion of words which, whatever they might do under another ſtate of things, do not, as matters ſtand, affect the meaning.

Take two inſtances, as follows;—Some years ago, a man was tried for ſtealing a ham, and was acquitted upon the ground that what was proved againſt him was that he had ſtolen a portion of a ham. Very recently, a man was convicted of perjury, 'in the year 1846,' and an objection (which the judge thought of importance enough to reſerve) was taken, on the ground that it ought to have been 'in the year *of our Lord* 1846.' There may, of courſe, be acknowledged rules, which, as long as they are rules, muſt be obeyed, and which may make the ſecond miſtake as neceſſarily vitiate an indictment as the firſt. But, in diſcuſſing the policy of the rules, it would ſeem to me that the two caſes are entirely different. In both, no doubt, the reſt of the indictment might, by implication, make good the meaning required: but there ſeems a great difference between allowing the remainder to correct an error, and allowing it to make good an inſufficiency (ſuppoſing the date, in the ſecond caſe, to be really inſufficient). In the ſecond caſe, the accuſed may ſee the omiſſion as well as another, and may conſider of his defence againſt every alternative: in the firſt, he may be actually led to appear in court with a defence not relevant to what will be brought againſt him. The ſecond may be a hardſhip, the firſt is an injuſtice. And this, even on the ſuppoſition that the reſt of the indictment is to be allowed in explanation: for we have no more right to ſuppoſe that the true parts will correct the erroneous ones, than that the erroneous parts will affect the conſtruction of the true ones. But there is good reaſon to think that the ſufficient deſcription of one ſentence may ſupply what is wanted in the inſufficient deſcription of another, when inſufficiency is all.

But, perhaps, it will be held to be the better rule, that the remainder of the indictment ſhould not be allowed in explanation. It will then be admitted by all that a material error, or a material inſufficiency, ſhould be allowed to nullify the charge. The difference between the law and common opinion entirely relates to what conſtitutes a material amount of one or the other. And here it is impoſſible to bring the two together: for the law muſt judge ſpecies, while the common opinion will never riſe above the caſe before it. In the two inſtances, which by many will be held equally abſurd, a great difference will be ſeen by any who will imagine the two deſcriptions, in each caſe, to be put before two different perſons. One is told that a man has ſtolen a ham; another that he has ſtolen a part of a ham. The firſt will think he has robbed a proviſion warehouſe, and is a deliberate thief: the ſecond may ſuppoſe that he has pilfered from a cook-ſhop, poſſibly from hunger. As things ſtand, the two deſcriptions may ſuggeſt different amounts of criminality, and different motives. But put the ſecond pair of deſcriptions in the ſame way. One perſon is told that a man perjured himſelf in the year 1846; and another, that he perjured himſelf in the year of our Lord

1846. As things ftand, there is no imaginable difference: for there is only one era from which we reckon. The two defcriptions mean the fame thing: nor can it even be faid that one is complete and the other incomplete; but only that one is lefs incomplete than the other. The next queftion might have been, what lord was meant, our Lord Jefus Chrift, or our Lord the King? both being phrafes of law. The anfwer will be, that the number 1846 leaves no doubt which was meant. A very good anfwer, certainly; but equally conclufive as to the fimple phrafe 'in the year 1846.' The firft cafe is one in which the two defcriptions have a real difference of meaning: it is not fo in the fecond.

3. The *petitio principii* is one of the logical terms which has almoft found its way into ordinary life. It is tranflated by the phrafe *begging the queftion*, that is, affuming the thing which is to be proved. This is alfo called *reafoning in a circle*, coming round, in the way of conclufion, to what has been already formally affumed, in a manner expreffed or implied. I fhall referve what I have to fay on the juftice of this tranflation, and take it for the prefent as good.

Every collective fet of premifes contains all its valid conclufions; and we may fairly fay that, fpeaking objectively of the premifes, the affumption of them is the affumption of the conclufion; though, ideally fpeaking, the prefence of the premifes in the mind is not neceffarily the prefence of the conclufion. But by this fallacy is meant the abfolute affumption of the fingle conclufion, or a mere equivalent to it, as a fingle premife. If the conclufion be 'Every X is Z' and if it be formally known that A and X are identical names, and alfo B and Z, then to affume 'Every A is B' as a premife in proving 'Every X is Z' would be a manifeft *petitio principii*, or begging of the queftion. But even this muft be faid hypothetically; it is fuppofed fully agreed between the difputants that the two identities are granted. Let it be otherwife, and there is no *petitio principii*: it is *then* fair to propound A)B, which, if difputed, is to be proved, and afterwards to reafon as in A)B + B)Z = A)Z, X)A + A)Z = X)Z. Strictly fpeaking, there is no formal *petitio principii* except when the very propofition to be proved, and not a mere fynonyme of it, is affumed. This of courfe, rarely occurs: fo that the fallacy to be guarded againſt is the aſſumption of that which is too nearly the ſame as the concluſion required. And then the fallacy is nothing diſtinct in itſelf: but merely amounts to putting forward and claiming to have granted that which ſhould not be granted. When this is done, it matters little as to the character of the fallacy, whether the undue claim be made for a propoſition which is nearer to, or further from, the concluſion to be proved. When proof is offered, the advancement of the concluſion in other words is of courſe not *petitio* principii: when proof is not offered, the aſſumption of that which (with other things proved) would prove the concluſion, is a fallacy of the ſame character in all caſes. There is an opponent fallacy to the *petitio principii* which, I ſuſpect, is of the more frequent occurrence: it is the habit of many to treat an advanced propoſition as a begging of the queſtion the moment they ſee that, if eſtabliſhed, it would eſtabliſh the queſtion. Before the advancer has more than ſtated his theſis, and before he has time to add that he propoſes to prove it, he is treated as a ſophiſt on his opponent's perception of the relevancy (if proved) of his firſt ſtep. Are there not perſons who think that to prove any previous propoſition, which neceſſarily leads to the concluſion adverſe to them, is taking an unfair advantage?

There is another caſe in which begging the queſtion may be unjuſtly imputed. It ſhould be remembered that *demonſtrative* inference is not the only kind of inference: there is *elucidatory* inference, *recapitulatory* inference, &c. A propoſition may have its aſſerted *explanation* preſented as a ſyllogiſm, the inference of which, as demonſtration, might well be called a reſult of *petitio principii.* Say 'it never could have been doubted that men would apply ſcience to the production of food.' If there ſhould be any heſitation about this, the explanation of *man* under the phraſe which is excluſively characteriſtic of him, *rational animal*, would remove it: the animal muſt have food, the rational being will have ſcience. But it would be begging the queſtion to aſſert that the ſyllogiſm of elucidation 'A rational animal is, &c.; man is, &c; therefore man is, &c.' is a demonſtration. And out of this ariſes the fallacy of preſuming that an author meant *demonſtration*, when he can only be fairly conſtrued to have attempted elucidation of what he ſuppoſed would, upon that elucidation, be *granted.* The forms of language are much the ſame in the two caſes.

It has been obſerved that Ariſtotle hardly ever uſes the phraſe ἀρχὴν αἰτεῖσθαι, *principium petere:* it is τὸ ἐξ ἀρχῆς and τὸ ἐν ἀρχῇ, that which is (ought to come) out of, or is in, the principle. By the word *principium* he diſtinctly means *that which can be known of itſelf.* He lays down five ways of *aſſuming* that which ought to come out of a ſelf-known principle, of which begging the queſtion is the firſt. The others are aſſuming the univerſal to prove the particular; aſſuming a particular to help to prove the univerſal; aſſuming all the particulars of which the univerſal may be compoſed; and aſſuming ſomething which obviouſly demonſtrates the concluſion.

Among the earlier modern writers, as far as I have ſeen them, there is ſome diverſity in their deſcription of the *petitio principii.* That the *principium* was meant to be the thing known of itſelf, the ἀρχή of Ariſtotle, as far as the introduction of the word is concerned, ſeems clear enough. Was it not then by a mere corruption that it was frequently confounded with the concluſion, the 'quod in *principio* quæſitum fuit?' Did not the ſame inaccuracy,* which confounds the τὸ ἐν ἀρχῇ of Ariſtotle with the ἀρχή itſelf, govern the change of the word? Moſt writers take the fallacy of the *petitio principii* as meaning that in which the concluſion is deduced either from itſelf, or from ſomething which requires proof more, or at leaſt as much, *ignotius aut æque ignotum.* But ſome, in their definitions, and ſtill more in their examples, ſupport the following meaning, which I ſtrongly ſuſpect to be the true derivation of the phraſe, however the *principium* and *quod in principio* might afterwards have been confounded with one another. The philoſophy of the time conſiſted in a large variety of general propoſitions (principles) deduced from authority, and ſuppoſed to be ultimately derived from intrinſic evidence, ſelf-known, or elſe by logical derivation from ſuch principles. Theſe were at the command of the diſputant, his opponent could not but admit each and all of them: the laws of diſputation demanded† the aſſent which the geometer requires for his poſtu-

* Sir W. Hamilton of Edinburgh (notes on Reid, p. 761,) ſays that *principium* is always uſed for that on which ſomething elſe depends.

† Does a traditional remnant of this convention ſtill linger in the not unfrequent notion that a diſputant is entitled to the conceſſion of his *principia?* We uſed to hear 'You muſt grant me my firſt principles, elſe I cannot

lates. Except when, now and then, literary ſociety was ſhaken to its very foundations by a diſpute which affected any of them, as a nominaliſt controverſy or the like moral earthquake. The moſt frequent ſyllogiſm was one which, having the form *Barbara*, had a *principium* for its major, and an *exemplum* for its minor: as in 'All men are mortal (*principium*); Socrates is a man (*exemplum*); therefore Socrates is mortal. The *petitio principii*, then, occurred, when any one, to prove his caſe, made it an example of a principle which was not among thoſe received, without offering to bring the former under the logical empire of the latter. And ſome writers define the fallacy as occurring *ſi contingat in ſyllogiſmo principium petere;* where by *principium* they mean the principle which generally occurs in the major premiſe, and by their inſtances they clearly ſhow that they mean to include nothing but the ſimple ſyllogiſm of principle and example. They would leave us to infer that if any one ſhould happen to conſtruct a ſyllogiſm in which *both* premiſes are principles, one or both not received, the inference, though denied by ſimple denial of one or both premiſes, would not be conſidered as technically the *petitio principii*, which with them was, as it were, *petitio principii exemplum continentis.*

It has often been aſſerted that all ſyllogiſm is a begging of the queſtion, or a *petitio principii* in the modern ſenſe, an aſſumption of the concluſion. That all premiſes do, when the argument is objectively conſidered, contain their concluſion, is beyond a doubt: and a writer on logic does but little who does not make his reader fully alive to this. But the phraſe, as applied to a good ſyllogiſm, is a miſapprehenſion of meaning: for its definition refers it to what is aſſumed *in one premiſe.* The moſt fallacious *pair* of premiſes, though expreſſly conſtructed to form a certain concluſion, without the leaſt reference to their truth, would not be aſſuming the queſtion, or *an* equivalent. But a further charge has been made againſt the ſyllogiſm, namely that very often the concluſion, ſo far from being deduced from the principle, is actually required to deduce it: that for inſtance, in 'All men are

argue.' Cardinal Richelieu's anſwer to his applicant's *il faut vivre*, namely, *Je n'en vois pas la néceſſité*, had ſomething of inhumanity in it: but, as applied to the *Mais, Monſieur, il faut ſe diſputer* of the preceding aſſumption, it would generally be quite the reverſe.

mortal; Plato is a man; therefore Plato is mortal' we do not know that Plato is mortal becauſe all men are mortal, but that we need to know that Plato is mortal, in order to know that it is really true that all men are mortal. There is much ingenuity in this argument: but I think a little conſideration, not of the ſyllogiſm, but of how we ſtand with reſpect to the ſyllogiſm, will anſwer it.

When we ſay that A is B, we do not merely mean that the thing called A is the thing called B: if we ſpoke of objects as objects, it would not matter under what name, and 'A is B' would be no other than 'B is B' and the very propoſition itſelf would be of its own nature a mere identity, an aſſertion that what is, is. It ſeems to me that between objects, thus viewed, there can neither be propoſitions nor ſyllogiſms. A may remind us of a thing as ſuggeſting one idea to our minds; B of the ſame thing as ſuggeſting another: and the propoſition 'A is B' then aſſerts that the two ſtates of our mind are from the ſame external ſource. Our logic, in wholly ſeparating names from objects, and dealing only with the former, makes a ſort of ſymbolic repreſentation of the diſtinction between ideas and objects.

Now the objection above ſtated to the ſyllogiſm appears to me to be founded upon thinking of the object, as if it had no names. Suppoſe all things marked, each with every name which can be applied to it. Undoubtedly then, each one marked *man* will have the mark *mortal* upon him, and ſome the mark *Plato*, it may be: and by the time all the marks are put on, and to a perſon who is ſuppoſed to be immediately cognizant of the ſimultaneous exiſtence of two or more marks on the ſame thing, it would be an abſurdity to attempt any ſyllogiſm at all. What coexiſtence of marks could there be which he muſt not be ſuppoſed to have noted in making the induction neceſſary for a univerſal propoſition. When he collected the elements of 'All men are mortal' he ſaw $\begin{matrix}\text{Plato}\\ \text{man}\end{matrix}$ among the reſt and ſet it down. But ſuppoſe that his knowledge is not acquired, as to different marks, all at once: but that each coincidence of marks is to be a ſeparate acquiſition to his mind. Then he does not know, by the time he has found out that 'All men are mortal' whether Plato be mortal or not. Plato may be a ſtatue, a dog, or a book written by a man of that name. *Plato* does not carry *man* with it: his major tells him nothing about Plato, until he has the minor, 'Plato is a man' and then, no doubt, he has absolutely acquired the conclusion 'Plato is mortal.' The whole objection tacitly assumes the superfluity of the minor; that is, tacitly assumes we know Plato to be a man, as soon as we know him to be Plato. Grant the minor to be superfluous, and no doubt we grant the necessity of connecting the major and the conclusion to be superfluous also. Grant any degree of necessity, or of want of necessity, to the minor, and the same is granted to the connection of the major and conclusion.

In the preceding case, the syllogism is looked upon as one of communication, by the authors of the objection; while at the same time it is tacitly assumed that the minor does not communicate: Plato, by virtue of our acquaintance with the name, is taken to be a man.

Moreover, it is to be noted that the proposition used in argument, whether to ourselves or to others, is very frequently not so much the mere attribution of one idea to another, as a declaration that *pro hac vice* the idea contained in the more extensive term is all that is wanted, and that the differences which constitute the species are not to the purpose. Or (page 234) it is the diminution of the comprehension which is necessary, and the increase of extension is only contingent. It is stripping the complex idea of the unnecessary parts, to prevent only what is requisite. Thus any one who will assert that, in the Mosaic account, no animal life whatever was destroyed by slaughter before the deluge, must be convinced by being reminded that an antediluvian (Cain) killed Abel who was a man and therefore an animal.

With the *petitio principii* may be classed (for it might also be referred to other fallacies) cases of the imperfect *dilemma*. Suppose we say 'Either M or N must be true: if M be true, Z is impossible; if N be true, Z is impossible; therefore Z is impossible.' Now if the disjunctive premise ought to have been 'either M or N or Z is true,' here would have been almost an express *petitio principii*. For example, say 'A body must either be in the state A or the state B; it cannot change in the state A; it cannot change in the state B; therefore, it cannot change at all.' Now, if the alternative A or B be necessary, the correct ſtatement may be 'A body muſt either be in the ſtate A, or in the ſtate B, or in the ſtate of tranſition from one to the other.' Of this kind is the celebrated ſophiſm of Diodorus Cronus, that motion is impoſſible, for all that a body does, it does either in the place in which it is, or in the place in which it is not, and it cannot move in the place in which it is, and certainly not in the place in which it is not. Now, motion is merely the name of the tranſition from the place in which it is (but will not be) to that in which it is not (but will be). It is reported that the inventor of this ſophiſm ſent for a ſurgeon to ſet his diſlocated ſhoulder, and was anſwered that his ſhoulder could not have been put out either in the place in which it was, or in the place in which it was not; and therefore, that it was not hurt at all.

4. The *ignoratio elenchi*, or *ignorance of the refutation*, is what we ſhould now call anſwering to the wrong point: or proving ſomething which is not contradictory of the thing aſſerted. It may be conſidered either as an error of form or of matter; and it is, of all the fallacies, that which has the wideſt range. Such, for inſtance, as the caſe of a writer I have read, who admits that certain evidence, if given at all, would prove a certain point; and admits that ſuch evidence has been given: but refuſes to admit the point as proved, becauſe the evidence was given in anſwer to objections, and in a ſecond pamphlet. The *pleadings* in our courts of law, previous to trial, are intended to produce, out of the varieties of ſtatement which are made by parties, the real points at iſſue; ſo that the defence may not be *ignoratio elenchi*, nor the caſe the counter-fallacy, which has no correlative name, but might be called *ignoratio concluſionis*. If a man were to ſue another for debt, for goods ſold and delivered, and if defendant were to reply that he had paid for the goods furniſhed, and plaintiff were to rejoin that he could find no record of that payment in his books; the fallacy would be palpably committed. The rejoinder, ſuppoſed true, ſhows that either defendant has not paid, or plaintiff keeps negligent accounts; and is a dilemma, one horn of which only contradicts the defence. It is plaintiff's buſineſs to prove the ſale, from what *is* in his books, not the abſence of payment from what *is not*; and it is then defendant's buſineſs to prove the payment by his vouchers.

It is commonly ſaid that no one can be required to prove a

negative, and often that no one *can* prove a negative. There is much confufion about this: for any one who proves a pofitive, proves an infinite number of negatives. Every thing that can be proved to be in St. Paul's Cathedral at any one moment is fairly proved *not* to be in more places than I can undertake to enumerate. What is meant is, that it is difficult, and may be impoffible, to prove a negative without proving a pofitive. Accordingly, when the two fides of the queftion confift of a pofitive and negative, the burden of proof is generally confidered to lie upon the perfon whofe intereft it is to eftablifh the pofitive. This being underftood, it is *ignoratio elenchi* to attempt to transfer the charge of proving the negative to the other party. But this rule is by no means without exception: there are many departures from it in the law, for example, though not under the moft logical phrafes. For inftance, a homicide, as fuch, is confidered by the law a murderer, unlefs, failing juftification, he can prove that he had no malice. Here, in the language of the law, the homicide, fuppofed unjuftifiable, is in itfelf a prefumption of malice, which the accufed is to rebut. It is not true, in point of fact, that fuch prefumption exifts on the mere cafe of homicide, independent of the manner of it: if the law will confult its own records, it will find that, for one homicide with malice of which it has had to take cognizance, there are dozens at leaft, done in heat of blood, and called manflaughters. But the cafe ftands thus;—the alternatives are few, fo that proving the negative of one, which the accufed is called on to do, can be done by proving the affirmative one out of a fmall number. There are but malice, heat of blood, mifadventure, infanity, &c. to which the action can be referred. Of thefe few things, it is eafier for the accufed to eftablifh fome one out of feveral, above all when motive is in queftion (of which only himfelf can be in poffeffion of the moft perfect knowledge) than it is for the profecutor to eftablifh a particular one. And the principle on which he is called on to eftablifh a negative (or rather another pofitive) is that the burden of proof fairly lies on the one to whom it will be by much the eafieft. The proof of a negative, then, being as eafy as, in fact identical with, the proof of one of the pofitive alternatives, fuch proof may, from the circumftances, lie upon a difputant, particularly when the number of the alternatives is few. But the *negative proof*, a very different thing, is of its own nature hardly attainable, and therefore hardly to be required: A book has been miſlaid; is it in one room or the other? If found in the ſecond room, there is proof of the negative as to the firſt: and almoſt any one who can read can be truſted to ſay, on his own knowledge, that in a certain room there is a certain book. But to give negative proof as to the firſt room, it muſt be made certain, firſt, that every book in the room has been found and examined, ſecondly, that it has been correctly examined. No one, in fact, can prove more than that he cannot find the book: whether the book be there or not, is another queſtion, to be ſettled by our opinion of the vigilance and competency of the ſearcher. Controverſialiſts conſtantly lay too much ſtreſs on their own negative proofs, on their *I cannot find*, even as to caſes in which it is palpably not their intereſt to find.

Somewhat akin to the preceding is the conſtant fallacy of controverſialiſts, conveyed in their ſtrong aſſertion of the reſults of their own arguments. Few can bear to admit that there is a queſtion for others to decide; and after ſumming up both ſides, to ſeparate the points which the reader is to pronounce upon. They muſt decide for him, and thus act both counſel and judge: probably becauſe their arguments are not ſo convincing to their own minds as they wiſh them to be to the reader's. They prove, at the utmoſt, their own conviction that they have the right ſide: but the thing to be proved is that ſuch conviction is well founded. They know the maxim *Si vis me flere, dolendum eſt primum ipſi tibi*, and think it will hold good of the reaſon, as well as of the feelings: as it will, to ſome. The conſequence is, that the deliberate reader ſuſpects them, and feels inclined rather to differ than agree: he will not dance to a writer who pipes too much. Juſt as "I'll tell you a capital thing," ſets the hearer upon avoiding laughter, and gives him notice to try; ſo 'I intend to give moſt unimpeachable proof,' puts the judicious reader upon looking for inadmiſſible aſſumptions, and he is ſeldom allowed by ſuch writers to look in vain. But, if the diſputant who begins by declaring his intention to be irreſiſtible, be ſuſpicious, the one who ends by announcing that he is ſo, is abſolutely ſelf-convicted. If it be very clear, why ſhould he ſay it? Does he tell his reader that he muſt remember to diſtinguiſh the black letters from the white paper, or does he print at the top of the book 'keep this ſide uppermoſt?' Theſe things (eſſential as they are) he really does leave to the reader: but he dares not truſt the latter to find out (though he ſays it is as clear as black and white) that his arguments are ſo ſtrong and ſo good, that nothing but wilful diſhoneſty, or hopeleſs prejudice, can reſiſt their force.

Another common form of the *ignoratio elenchi*, lies in attributing to the concluſion aſſerted ſome ultimate end or tendency. Thus, an argument in favour of checking the power of the Crown is called Jacobiniſm; of an increaſe of that power, abſolutiſm: though the argument propoſed may be ſound, independently of its propoſer's wiſhes. This is a caſe in which the reſult of the method is juſtifiable, though the method is wrong. Many readers will remember the advice given by an old judge to a young one, 'Give your judgments without reaſons; moſt likely your deciſions will be right; and it is juſt as likely that your reaſons will be wrong.' This advice ſhould be followed by many of thoſe who judge or decide arguments. The propoſer is of a known opinion, which gives him a ſtrong bias towards the concluſion of the argument. He is a witneſs (page 205), and the effect upon the mind of the receiver is to be that of the united argument and teſtimony. The teſtimony is, in the receiver's mind, of a low order; the propoſer is a radical, and the receiver is of opinion that a radical would pick a pocket: or elſe, perhaps, the propoſer is a tory, and the receiver is of the belief that a tory muſt have picked a pocket. Theſe opinions may be right or wrong; but they exiſt: and there is certainly no formal fallacy in admitting them, as affecting the teſtimony, to ſubtract from the probability of the truth of the concluſion. But there is a formal fallacy, a decided *ignoratio elenchi*, in throwing all the indiſpoſition to receive upon the invalidity of the argument.

There is a much more culpable form of the ſame ſpecies. If ſuch a concluſion were admitted, it would lead to ſuch and ſuch another concluſion, which is not to be admitted. In queſtions of abſolute demonſtration, this proceſs is ſound: if B be certainly falſe, and if it be the neceſſary conſequence of A, then A muſt alſo be falſe. But it is unſound when it takes the form, 'I believe B to be falſe; I believe it to follow from A; therefore I aſſume a right to diſbelieve A whatever evidence may be offered for it? This fallacy is ſufficiently expoſed in page 209. There is a tradition of a Cambridge profeſſor who was once aſked in a mathematical diſcuſſion 'I ſuppoſe you will admit that the whole is greater than its part,' and who anſwered, 'Not I, until I ſee what uſe you are going to make of it.' This was no doubt the extreme caſe; the more ordinary one ariſes in a great meaſure from the great fallacy of all, the determination to have a particular concluſion, and to find arguments for it. Obſerve a certain perſon who is led on by a wily opponent in converſation: nothing is preſented to him except what his reaſon fully concurs in, and no inference except what is indiſputable. At a ſudden turn of the argument, he ſees a favourite concluſion, which he cares more for than for all the reaſonings that ever were put together, upſet and broken to pieces. He conſiders himſelf an ill-uſed man, entrapped, ſwindled out of his lawful goods; and he therefore returns upon his ſteps, and finds out that ſome of the things which he admitted when he did not ſee their conſequences, are no longer admiſſible. Neither he nor the opponent has the leaſt idea of the nature of probable arguments, and of their oppoſition: both proceed as if the train of reaſoning were either demonſtration or nothing. The concluſion, formed perhaps upon teſtimony, which is more likely to be a guide to truth for the mind in queſtion than any appreciation of argument which that mind could make, muſt, according to the maxims of the age, be referred to argument, and argument only. The perpetual and wilful fallacy of that mind is the determination that all argument ſhall ſupport, and no argument ſhall ſhake, the concluſion. If there were only a diſtinct perception of another ſource of conviction, ſo ſtrong that ordinary argument can neither materially weaken, *nor materially confirm it*, there would be ſenſe in the concluſion; ſenſe, becauſe there is truth. Right or wrong, ſuch is the ſource of moſt convictions in, perhaps, moſt minds: ſuch ſource ought therefore to be acknowledged. It would be an excellent thing, if, in any diſputed matter, thoſe who are better ſatisfied by authority of the truth of one ſide of the concluſion than of the validity of argument in general, would avow it, keep their own ſide, and let others do the ſame. But here is the difficulty: the perſons who ſhould avow ſuch a ſtate of mind are as much diſpoſed to make converts as others: they do not like to debar themſelves from diſſemination of their opinions. Accordingly they propound their beſt arguments, be they what they may, as what ought to produce all the conviction which themſelves feel. On this point ſee page 194.

The whole claſs of *argumenta ad hominem*, having ſome reference to the particular perſon to whom the argument is addreſſed, will generally be found to partake of the fallacy in queſtion. Such are *recrimination* and *charge of inconſiſtency*, as, 'You cannot uſe this aſſertion, becauſe in ſuch another caſe you oppoſe it.' But if the original argument itſelf ſhould be a perſonal attack, then ſuch a retort as the preceding may be a valid defence.

In many ſuch *argumenta ad hominem*, it is not abſolutely the ſame argument which is turned againſt the propoſer, but one which is aſſerted to be like to it, or parallel to it. But *parallel caſes* are dangerous things, liable to be parallel in immaterial points, and divergent in material ones. A celebrated writer on logic aſſerts, that no one who eats meat ought to object to the occupation of a ſportſman on the ground of cruelty. The parallel will not exiſt until, for the perſon who eats meat, we ſubſtitute one who turns butcher for amuſement. There is, or was, a vulgar notion that butchers cannot ſit on a jury. Suppoſe that ſuch a law were propoſed, on the ground of the habits ariſing from continual infliction of death. Would it really be a counter-argument that men who eat meat have the ſame *animus* and are liable to acquire the ſame habits. It is contended (juſtly or not) that a deſire to *take life* for *ſport* is a cruel deſire; to anſwer that thoſe who eat fleſh from which life has been taken by others have therefore alſo cruel deſires, ought to be called arguing *a dicto ſecundum quid ad dictum ſecundum alterum quid*. The matter is clear enough. Cruelty of *intention* (the thing in queſtion) muſt be ſettled by our judgment of the circumſtance in which the ſport conſiſts. A perſon who ſeeks bodily exerciſe and the excitement of the chaſe, and who can acknowledge to himſelf that his object is gained on the birds which he miſſes, as well as upon thoſe which he hits, even if thoughtleſs, cannot be ſaid to act with cruelty of intention. But the ſportſman, as he calls himſelf, who collects his game in one place, merely that he may kill, without exerciſe, or feeling of ſkill, is either culpably thoughtleſs, or elſe a ſavage, who delights in the infliction of death. Let any man aſk himſelf, whether in the event of his being called upon to vote for a perfectly abſolute ſovereign, he would feel much concerned to inquire whether the candidate was or was not a ſportſman of the firſt kind: and then let him aſk himſelf the ſame queſtion with reſpect to the ſecond.

The moſt amuſing, and perhaps the moſt common, example of the *ignoratio elenchi*, is the taking exception to ſome part of an illuſtration which has nothing to do with the parallel. The word illuſtration (though it mean *throwing light* upon a thing) is uſually confined to that ſort of light which is derived from ſhowing a proceſs of difficulty employed upon an eaſier caſe. The firſt fallacy may be committed by the illuſtrator. He has before him the ſubject matter of the premiſes, their connexion in the proceſs of inference, and the reſult produced. Either may be illuſtrated; thus, if it be doubtful whether ſuch premiſes may be employed, the illuſtrator may throw away his mode of connexion, and chooſe another: if the proceſs of inference be doubtful, he may chooſe other premiſes: and ſo on. But he may illuſtrate the wrong point: and this is a fallacy very common to teachers and lecturers. The greateſt difficulty in the way of learners is not knowing exactly in what* their difficulty conſiſts; and they are apt to think that when *ſomething* is made clear, it muſt be *the* ſomething. I am of opinion that the examples given of ſyllogiſms in works of logic are examples of wrong illuſtration. The point in queſtion is the form, the object is to produce conviction of the form, of its neceſſary validity. If the ſtudent receive help from an example ſtated both in matter and form, the odds are that the help is derived from the plainneſs of the matter, and from his conviction of the matter of the concluſion. If this be the caſe, he has not got over his difficulty. Many learners are puzzled to ſee that 'Every Y is X' is not a neceſſary conſequence of 'Every X is Y.' If the want of con-

* Every learner, in every ſubject, ſhould accuſtom himſelf to endeavour to ſtate the point of difficulty *in writing*, whether he want to ſhow the reſult to another or not. I wiſh I had kept a record of the number of times which I have inſiſted on this being done, previouſly to undertaking the explanation, and of the proportion of them in which the writer has acknowledged that he ſaw his way as ſoon as he attempted to aſk the road in preciſe written language. That proportion is much more than one half. Truly ſaid Bacon, that writing makes an exact man.

nexion be eftablifhed by an inftance, as by appealing to their knowledge that every bird is not a goofe, though every goofe be a bird, their knowledge of the propofition is not logical. The right perception may, no doubt, be acquired by reflection on inftances: but the minds which are beft fatisfied by material inftances, are alfo thofe which give themfelves no further trouble.

The illuftration being fuppofed correct, there is more than one fallacious mode of oppofing it. Some perfons will difpute the very method of illuftration of form, in which the fame mode of inference is applied to eafier matter; but thefe are mere beginners, hardly even entitled to a name which fuppofes the poffibility of progrefs. Others will deny the analogy of the matter, and thefe there is no means of meeting: for illuftration is *ad hominem*, and the perception of it cannot be made purely and formally inferential: a denier of the force of an illuftration is inexpugnable as long as he only denies. But when he attempts more, when he indicates the point in which the illuftration fails, he very often falls into the error of attacking an immaterial point. If any one were to contend (as fome do) that it is unlawful to take the life of any animal, he might be afked what he would fay if Guy Faux had trained a pigeon to carry the match to the vault, would it have been lawful to fhoot the bird on its way or not? There are not a few who would think it an anfwer to fay that he could not have trained the pigeon, or that pigeons were not then trained to carry.

5. The *fallacia confequentis* (now very often called a *non fequitur*) is the fimple affirmation of a conclufion which does not follow from the premifes. If the fchoolmen had lived in our day, they would have joined with this the affirmation of logical form applied to that which wants it, a very common thing among us. A little time ago, either the editor or a large-type correfpondent (I forget which) of a newfpaper imputed to the clergy the maintenance of the 'logic' of the following as 'confecutive and without flaw.' This was hard on the clergy (particularly the Oxonians) for there was no middle term, neither of the concluding terms was in the premifes, and one negative premife gave a pofitive conclufion. It ran thus,

Epifcopacy is of Scripture origin.

The church of England is the only epiſcopal church in England,

Ergo, the church eſtabliſhed is the church that ſhould be ſupported.

Many caſes offend ſo ſlightly that the offence is not perceived. For inſtance 'knowledge gives power, power is deſirable, therefore knowledge is deſirable' is not a ſyllogiſm; there is no middle term. It is a ſorites, as follows, 'knowledge is a giver of power, the giver* of power is the giver of a deſirable thing, the giver of a deſirable thing is deſirable, therefore knowledge is deſirable.'

It ſhould be noted, however, that the copula 'gives' reſembles 'is greater than' (page 5) and is an admiſſible copula in inferences with no converſion, provided that 'A gives B and B gives C,' implies 'A gives C.' The ſame may be ſaid of the verbs to bring, to make, to lift, &c. And many of theſe verbs are, by the unſeen operation of their having the effect of *is* in inference, often ſupplanted by the latter verb in phraſeology. Thus we ſay 'murder *is* death to the perpetrator' where the copula is *brings*; 'two and two *are* four' the copula being 'have the value of' &c. But this practice may lead to fallacies, as above ſhown: which muſt be avoided by attention to the claſs of verbs which communicate their action or ſtate, ſuch as make, give, bring, lift, draw, rule, hold, &c. &c. All theſe verbs are applied to denote the cauſe of the ſeveral actions: ſo, to give that which gives, or to bring that which brings, is to give or to bring. The boy who was ſaid to rule the Greeks becauſe he ruled his mother, who ruled Alcibiades, who ruled the Athenians, who ruled the Greeks, would have been correctly ſaid ſo to do, if the matters of rule had been the ſame throughout.

6. The *non cauſa pro cauſa*. This is the miſtake of imagining neceſſary connexion where there is none, in the way of *cauſe*, conſidered in the wideſt ſenſe of the word. The idioms of language abound in it, that is, make their mere expreſſions of phenomena attribute them to apparent cauſes, without intent to aſſert real connexion. Thus we ſay that a tree *throws* a ſhadow,

* Becauſe power is deſirable. See page 115, as to this ſtep.

to deſcribe that it hinders the light. When the level of a billiard table is not good, the favoured pocket is ſaid to *draw* the balls. A particular caſe of this fallacy, which is often illuſtrated by the words *poſt hoc*, *ergo propter hoc*, is the concluſion that what follows in time follows as a conſequence. When things are ſeen together, there is frequently an aſſumption of neceſſary connexion. There is, of courſe, a preſumption of connexion : if A and B have never been ſeen apart, there is probability (the amount of which depends upon the number of inſtances obſerved) that the removal of one would be the removal of the other. It is when there is only one inſtance to proceed upon that the aſſumption falls under this fallacy ; were there but two, inductive probability might be ſaid to begin. The fallacy could then conſiſt only in eſtimating the probability too high.

As may be ſuppoſed, the *non cauſa pro cauſa* ariſes more often from mere ignorance than any other fallacy. To take the two inſtances that I happened to meet with neareſt to the time of writing this page ;—Walpole, remarking on the uniform practice among the old writing-maſters of putting their portraits at the beginning of their works, remarks that theſe men ſeem to think their profeſſion gives poſterity a particular intereſt in their features. Probably they did not think about it : the uſage of the day prevented any man from being chargeable with undue vanity who exhibited his phyſiognomy, and *moſt of the writing maſters were themſelves engravers*, and either did their own portraits, or more probably made uſe of their acquaintance with the more celebrated engravers for whom they did the under drudgery, to get themſelves done on eaſy terms. Again, Noble (in his continuation of Granger) remarks that Saunderſon had ſuch a profound knowledge of muſic, that he could diſtinguiſh the fifth part of a note. The author did not know, firſt, that any perſon who cannot diſtinguiſh leſs than the fifth part of a note to begin with, ſhould be bound over to keep the peace if he exhibit the leaſt intention of learning any muſical inſtrument in which intonation depends upon the ear ; and ſecondly, that if Saunderſon were not ſo gifted by nature, knowledge of muſic would no more have ſupplied the defect, than knowledge of optics would give him ſight.

The *fallacia plurium interrogationum* conſiſts in trying to get one anſwer to ſeveral queſtions in one. It is ſometimes uſed by barriſters in the examination of witneſſes, who endeavour to get *yes* or *no* to a complex queſtion which ought to be partly anſwered in each way, meaning to uſe the anſwer obtained, as for the whole, when they have got it for a part. An advocate is ſometimes guilty of the argument *à dicto ſecundum quid ad dictum ſimpliciter:* it is his buſineſs to do for his client all that his client might *honeſtly* do for himſelf. Is not the word in Italics frequently omitted? *Might* any man honeſtly try to do for himſelf all that counſel frequently try to do for him? We are often reminded of the two men who ſtole the leg of mutton; one could ſwear he had not got it, the other that he had not taken it. The counſel is doing his duty by his client; the client has left the matter to his counſel. Between the unexecuted intention of the client, and the unintended execution of the counſel, there may be a wrong done, and, if we are to believe the uſual maxims, no wrong doer. The anſwer of the owner of the leg of mutton is ſometimes to the point, 'Well, gentlemen, all I can ſay is, there is a rogue between you.' That a barriſter is able to put off his forenſic principles with his wig, nay more, that he becomes an upright and impartial judge in another wig, is curious, but certainly true.

The above were the forms of fallacy laid down as moſt eſſential to be ſtudied by thoſe who were in the habit of appealing to principles ſuppoſed to be univerſally admitted, and of throwing all deduction into ſyllogiſtic form. Modern diſcuſſions, more favourable, in ſeveral points, to the diſcovery of truth, are conducted without any conventional authority which can compel preciſion of ſtatement: and the neglect of formal logic occaſions the frequent occurrence of theſe offences againſt mere rules which the old enumeration of fallacies ſeems to have conſidered as ſufficiently guarded againſt by the rules themſelves, and ſufficiently deſcribed under one head, the *fallacia conſequentis.* For example, it would have been a childiſh miſtake, under the old ſyſtem, to have aſſerted the univerſal propoſition, meaning the particular one, becauſe the thing is true in moſt caſes. The rule was imperative: *not all* muſt be *ſome*, and even *all*, when not known to be *all*, was *ſome.* But in our day nothing is more common than to hear and read aſſertions made in all the form, and intended to have all the power, of univerſals, of which nothing can be ſaid except that moſt of the caſes are true. If a contradiction be aſſerted and proved by an inſtance, the anſwer is 'Oh! that is an *extreme caſe.*' But the aſſertion had been made of *all* caſes. It turns out that it was meant only for ordinary caſes; why it was not ſo ſtated muſt be referred to one of three cauſes;—a mind which wants the habit of preciſion which formal logic has a tendency to foſter, a deſire to give more ſtrength to a concluſion than honeſtly belongs to it, or a fallacy intended to have its chance of reception.

The application of the *extreme caſe* is very often the only teſt by which an ambiguous aſſumption can be dealt with: no wonder that the aſſumer ſhould dread and proteſt againſt a proceſs which is as powerful as the ſign of the croſs was once believed to be againſt evil ſpirits. Where anything is aſſerted which is true with exceptions, there is often great difficulty in forcing the aſſertor to attempt to lay down a canon by which to diſtinguiſh the rule from the exception. Every thing depends upon it: for the queſtion will always be whether the example belongs to the rule or the exception. When one caſe is brought forward which is certainly exception, the aſſertor will, in nine caſes out of ten, refuſe to ſee why it is brought forward. He will treat it as a fallacious argument againſt the rule, inſtead of admitting that it is a good reaſon why he ſhould define the method of diſtinguiſhing the exceptions: he will virtually, and perhaps abſolutely, demand that all which is certainly exception ſhall be kept back, ſimply that he may be able to aſſume that there is no occaſion to acknowledge the difficulty of the uncertain caſes.

The uſe of the extreme caſe, its deciſive effect in matters of demonſtration, may furniſh preſumption as to what it is likely to be in matters of aſſerted near approach. As in the following inſtance. It ſeems almoſt matter of courſe, when ſtated, to thoſe who have not ſtudied the ſubject of life contingencies, that the proper value of a life annuity is that of the annuity made certain during the average exiſtence of ſuch lives as that of the annuitant. That if, for example, perſons aged 22 live, one with another, 40 years, an office which receives from every ſuch perſon the preſent value of forty payments certain, will, without gain or loſs, in the long run, be able to pay the annuities. If this be (as was ftoutly contended by fome writers of the laft century) a univerfal truth, it will hold in this extreme cafe. Let there be two perfons, one of whom is certain to die within a year from the grant (and therefore never claims anything) and the other of whom is certain to live for ever. It is clear that the value of an annuity to both is o + the value of a perpetual annuity. But the *average life* of both is eternal: one perpetual duration makes the average of any fet in which it is, perpetual. Hence by the falfe rule the value is *two* perpetual annuities, or juft double of the truth.

We might fuppofe that moft perfons have no idea of a univerfal propofition: but ufe the language, never intending *all* to fignify more than *moft*. And in the fame manner principles are ftated broadly and generally, which the affertor is afterwards at liberty to deny under the phrafe that he does not *carry them fo far* as the inftance named. It would not do to avow that the principle is not always true: fo it is ftated to be *always true*, but not capable of being *carried* more than *a certain length*. Are not many perfons under fome confufion about the meaning of the word *general?* In fcience it always has the meaning of *univerfal:* and the fame in old Englifh. Thus the catechifm of the church of England afferts that there are two facraments which are *generally* neceffary to falvation: meaning neceffary for all of the *genus* in queftion, be it man, Chriftian, member of the church, or any other. But in modern and vernacular Englifh, *general* means only *ufual*, and generally means ufually.

A great deal of what is called evafion belongs to this head, or to that of the *ignoratio elenchi*, as the fophift anfwers. The advocates, for inftance, of the abfolute unlawfulnefs of war never tell, unlefs preffed, what they think of the cafe of refiftance to invafion. Is the country to be given up to the firft foreigner who choofes to come for it? Sometimes the extreme cafe comes into play: fometimes the affertion that no one will come; which is irrelevant as to the queftion what would be right if he did come.

Among amufing modern evafions are 'There is no *occafion* to confider that' and 'I do'nt confider it in that *point of view*.' Any one who watches the manner in which men defend their opinions will frequently fee 'A is B and B is C, therefore A is C' anfwered, not by denial of either premife, but by 'that is not the proper point of view' or 'I don't ſee it in that light.' This ſhould be called the confuſion between logic and perſpective.

The denial of one univerſal is often made to amount to, or to paſs into, the aſſertion of the oppoſite, or ſubcontrary, univerſal. This craving after general truths, the moſt manifeſt fault of the old logicians in their choice of premiſes, did not expire with them. Bacon ſays 'the mind delights in ſpringing up to the moſt general axioms, *that it may find reſt.*' Many perſons are deſirous of 'ſettled opinions,' which is well; unleſs by ſettled opinions they mean univerſal, as is often the caſe. That ſome are and ſome are not is no ſettlement: it makes every caſe require examination, to ſee under which it falls. And with the above we may couple the tendency to believe that refutation of an argument is proof of the falſehood of its concluſion, and that a falſe conſequence muſt be a falſe propoſition. Hence it ariſes that ſo many perſons dare not give up any argument in favour of a propoſition which they fully believe: they think they abandon the propoſition.

It ſometimes happens that an aſſertion is made, which it is difficult to ſuppoſe can be anything but a caſe of a univerſal propoſition: and yet the aſſertor takes care not to make his propoſition univerſal, but perſiſts in the particular caſe. A logician in our day has aſſerted that when Calvin ſays that *all* officers of the church ſhould be elected by the people, he muſt be underſtood as ſpeaking in reference to *deacons* only, becauſe the aſſertion is made in the chapter on deacons. If it had been roundly ſtated that all univerſal propoſitions are to have their univerſes limited by the headings of the works or chapters in which they occur—for inſtance, that the aſſertion that all men are mortal, occurring in a hiſtory of England, is to be taken as made of Engliſhmen only—there would have been at leaſt no ambiguity. But as it is, we are left to ſurmiſe whether this be meant, or whether the propoſition be to apply to Calvin only, or to Reformers only, or to men whoſe names begin with C, &c. The odds are that the *application* of a univerſal propoſition will be dictated by the heading of a chapter: but the extent to which a premiſe is *aſſerted as true* is not to be judged of by that to which it is *wanted for uſe:* and the leſs, the nearer we go to the day of the old logicians.

Wrong views of the quantity of a propoſition are as frequent as any fallacies. *Some*, meaning moſt, and *ſome*, meaning few, are frequently confounded. This is the neceſſary conſequence of the nature of human knowledge, in which we can but rarely form a definite idea of the proportion which the extent ſpoken of bears to the whole. It is part of the value of the mathematical theory of probabilities, that the mind is accuſtomed to the view of reſults drawn from perfectly definite ſuppoſed caſes; as uſeleſs, it may be, in themſelves, as many of the queſtions in a book of arithmetic, but nevertheleſs good for exerciſe. It is not ſurpriſing that fallacies about quantity ſhould be capable of moſt ſtriking expoſure in queſtions concerning meaſurable quantity, that is, in queſtions of mathematics: nor that there ſhould be claſſes of fallacy of which it is difficult to illuſtrate the detection by any other inſtances. What can be more clear, for example, to ordinary apprehenſions than the broad ſtatement that 'of things of the ſame kind, that which is ſometimes right muſt be better than that which is always wrong.' But a little conſideration will ſuggeſt that what is always wrong may be as good as that which is ſometimes right, if we do not know how to diſtinguiſh the caſes in which the latter *is* right: and alſo that what is not much wrong, generally, may be more uſeful than that which is moſtly very wrong, when it is not abſolutely right. A watch which does not go is right twice a day: but it is not ſo uſeful as one which does go, though very badly.

To give an account of all the fallacies which depend upon wrong notions of quantity would require much ſpace, and more aſſumption of mathematical knowledge in my reader than is conſiſtent with my plan. But I may mention the miſtaken uſe of abſolute terms and notions in queſtions of degree. There can be, a diſputant will ſay, but a right and a wrong; and if this be not right, it is wrong. Many perſons will announce that their watches are *quite right*, abſolutely at the true time, to a ſecond: and will end by giving the time which was ſhown when they looked, as being accurately that of the inſtant at which they announce it. The proverb *Fruſtra fit per plura, quod fieri poteſt per pauciora* contains an inaccuracy of degree: a bargain which coſts twenty ſhillings and is worth fifteen, is not twenty ſhillings loſt, but only five, though the vexation of the party overreached will ſeldom ſuffer him to ſee this.

Proverbs in general are liable to this miſtake. They are often uſed in exactly the ſame manner as the firſt principles of the old logicians. In fact, remembering that theſe firſt principles were bandied from mouth to mouth till they were perfectly *proverbial*, as we now call it, among the learned; and obſerving the application of our modern proverbs, as made by the maſs of thoſe who have not profited by mental diſcipline, we may ſee that the faults of the ſchoolmen are only thoſe of the ordinary human mind. It is hard indeed if there be a purpoſe which a proverb cannot be found to ſerve: it is a univerſal propoſition of no very definite meaning, ſanctioned by uſage, having the appearance of authority, and capable of ſtretching or contracting like Prince Ahmed's pavilion. One only is allowable—*In generalibus latet error:* this deſtroys all the reſt, and then, when cloſely looked at—commits ſuicide.

All miſtakes of probability are eſſentially miſtakes of quantity, the ſubſtitution of one amount of knowledge and belief for another. It is often difficult to convey a proper notion of the degree of force which is meant to be given; and ſtill more ſo to retain it throughout the whole of a diſcuſſion. A perſon begins by ſtating an explanation as poſſible, or probable enough to require conſideration, as the caſe may be. The forms of language by which we endeavour to expreſs different degrees of probability are eaſily interchanged; ſo that, without intentional diſhoneſty (but not always) the propoſition may be made to ſlide out of one degree into another. I am ſatisfied that many writers would ſhrink from ſetting down, in the margin, each time they make a certain aſſertion, the numerical degree of probability with which they think they are juſtified in preſenting it. Very often it happens that a concluſion produced from a balance of arguments, and *firſt preſented* with the appearance of confidence which might be repreſented by a claim of ſuch odds as four to one in its favour, is afterwards *uſed* as if it were a moral certainty. The writer who thus proceeds, would not do ſo if he were required to write $\frac{4}{5}$ in the margin every time he uſes that concluſion. This would prevent his falling into the error in which his partiſan readers are generally ſure to be more than ready to go with him, namely, turning all balances for, into demonſtration, and all balances againſt, into evidences of impoſſibility.

One of the great fallacies of evidence is the diſpoſition to dwell on the actual possibility of its being false: a possibility which must exist when it is not demonstrative. Counsel can bewilder juries in this way till they almost doubt their own senses. A man is shot, and another man, with a recently discharged pistol in his hand, is found hiding within fifty yards of the spot, and ten minutes of the time. It does not follow that the man so found committed the murder: and cases have happened, in which it has turned out that a person convicted upon evidence as strong as the above, has been afterwards found to be innocent. An astute defender makes these cases his prominent ones: he omits to mention that it is not one in a thousand against whom such evidence exists, except when guilty.

All the makers of systems who arrange the universe, square the circle, and so forth, not only comfort themselves by thinking of the neglect which Copernicus and other real discoverers met with for a time, but sometimes succeed in making followers. These last forget that for every true improvement which has been for some time unregarded, a thousand absurdities have met that fate permanently. It is not wise to toss up for a chance of being in advance of the age, by taking up at hazard one of the things which the age passes over. As little will it do to despise the usual track for attaining an object, because (as always happens) there are some who are gifted with energies to make a road for themselves. Dr. Johnson tells a story of a lady who seriously meditated leaving out the classics in her son's education, because she had heard Shakspeare knew little of them. Telford is a standing proof (it is supposed by some) that special training is not essential for an engineer.

The disposition to judge the prudence of an action by its result, contains a fallacy when it is applied to single instances only, or to few in number. That which, under the circumstances, is the prudent rule of conduct, may, nevertheless end in something as bad as could have resulted from want of circumspection. But upon dozens of instances, such a balance would appear in favour of prudence as would leave no doubt in favour of the rule of conduct, even in the instances in which it failed. The fallacy consists in judging from the result about the conduct of one who had only the previous circumstances to guide him. 'You acted unwisely, as is proved by the result,' is a paralogism, except when it implies 'You did, as it happens in this inſtance, take a courſe which did not lead to the deſired reſult:' Take a ſtrong caſe, and the abſurdity will be ſeen. A chemiſt makes up a preſcription wrongly, and his cuſtomer leaves him for another: this other, ſo it may happen, makes it up ſtill more wrongly, and poiſons the patient. Who would venture to ſay that he acted unwiſely, as is proved by the reſult, in leaving the tradeſman whom he knew to be careleſs, for another of whom he knew no harm. The only way in which blame can be imputed, is when it can be ſaid 'You acted unwiſely, in not finding out, as you might have done, that the reſult which has happened is the one which was likely to happen.' One reſult proves very little as to the ſuperior wiſdom of the courſe which produced it; ſeveral may give a preſumption of it, and the greater the number, the greater the preſumption.

So little is this thought of, that the common phraſe, 'I acted for the beſt,' meaning originally 'I acted in the manner which under the circumſtances, appeared likely to lead to the beſt reſults,' very often loſes its proper meaning, and is uſed as ſynonymous with 'I acted with good intentions.'

Theſe, and many other points, I can only ſlightly touch on: I will proceed to notice a few other cauſes of error.

And firſt, of equivocations of ſtyle. I have before referred to ſuch a phenomenon as the alteration of a good ſyllogiſm into a bad one, to make the ſentence *read better*. But nothing ever reads well (for a continuance) except the natural current of a writer's thought. I ſhould like it to be the law of letters, that every book ſhould have inſerted in it the printer's affidavit, ſetting forth the number of verbal eraſures in the manuſcript, fair copies being illegal. It would be worth at leaſt one review.

There is a wilful and deliberate equivocation, which it is ſuppoſed the age demands. It is the uſe of ſynonymes, or ſuppoſed ſynonymes, to prevent the ſame word from occurring twice in the ſame paſſage. So far is the neceſſity of this practice recognized, that there are few printing-offices in London, the *readers* of which do not *query* the ſecond introduction of any word which prominently appears twice. And then the author obeys the hint, ſtrikes out one of the offenders, ſticks in a dictionary equivalent, and would have been content if the printer's reader had done it for him. And ſo he writes a *good ſtyle*. To be ſure, he does not ſay what he meant, exactly; for ſynonymes are ſeldom or never logical equivalents: but what is that to elegance of expreſſion?

The demand for non-recurrence of words ariſes from the public (I beg its pardon) not knowing how to read. If, when a word occurs twice, the proper emphaſes were looked for, and obſerved, there would be nothing offenſive about the repetition. It is the reader who makes one and one into two, by giving both units equal value. Take this ſentence from Johnſon, (the firſt I happened to light on, in the preface to Shakſpeare), and read it firſt as follows:—"He therefore indulged his natural *diſpoſition*: and his *diſpoſition*, as Rymer has remarked, led him to comedy:" and then as follows—"He therefore indulged his *natural diſpoſition;* and *his* diſpoſition, as Rymer has remarked, led him to comedy." This reading is what the context requires, and the ill effect of the repetition is next to nothing. Take the next ſentence:—"In tragedy he often *writes*, with great appearance of toil and ſtudy, what is written *at laſt* with little felicity: but in his comic ſcenes he ſeems to produce, without *labour*, what *no* labour can improve." Theſe were the firſt inſtances I found, from a chance opening of the *Elegant Extracts*, purpoſely choſen as a miſcellany. The laws of thought generally dictate this rule, that the firſt occurrence of a word is the more emphatic of the two: the leſſon of experience is, that a writer who prevents recurrence by the uſe of the dictionary of ſynonymes, is a good ſtyle-maker for none but a bad reader, and may very poſſibly be a good arguer for none but a bad logician. Of courſe, I ſhould not deny that recurrence of both word and emphaſis is a defect, if it be frequent.

The confuſion between the means and the end, and putting one in the place of the other, is well enough known in morals: but there is a correſponding tendency to forget the diſtinction between the principle which is to be acted on, and the rule of action by which adherence to that principle is ſecured. A reference to the derived rule is in all reſpects as good as one to the firſt principle, between parties who underſtand both, and the connexion between them. But thoſe who underſtand the rule only, are apt to forget that a rule may or may not be the true expreſſion of a principle, according to the circumſtances in which it is propoſed to apply it. If, indeed, it were of univerſal application, thoſe who do and thoſe who do not underſtand the principle might be on the ſame footing as to ſecurity: but there are few ſuch rules.

The preceding caution may be applied in all departments of thought, in law and in logic, in morals and in arithmetic. It is impoſſible, for inſtance, to ſtate the rule of three in ſuch a manner as eaſily to include the caſes in which it ſhall apply, and exclude thoſe to which it does not. To ſay that it muſt be uſed where the fourth quantity, the one ſought, is to be a fourth proportional to the three which are given, though correct, ſtill leaves it open to inquiry what are the caſes in which this condition is to be ſatisfied: and many caſes might be, and are propoſed, in which the inquiry is not eaſy to a beginner. In law, there are not only rules, but rules for their application. To an unlearned ſpectator, particularly in the courts of equity, in which the advocate addreſſes a judge, and not a jury, the argument takes that technical form which makes many perſons think that the whole law is, at beſt, only arbitrary rule. It may be that ſome of thoſe who there addreſs the court can make nothing better of it: and juſt as there are arithmeticians, and good ones too, who are but the ſlaves, and never the maſters, of their proceſſes, ſo there may be advocates, and even judges, who have not one element of the legiſlator in them. But there are enough of a higher ſpecies.

The great art of uſing rules is to apply them in aid, and not in contravention, of the principles which they are intended to embody. A rule may have exceptions, it is ſaid; but this is hardly a correct ſtatement. A rule with exceptions is no rule, unleſs the exceptions be definite and determinable: in which caſe the exceptions are excluſions *by another rule.* The parallel is perfect between rules and propoſitions (page 143). Thus, 'All Europe, except Spain and Portugal' is a univerſal propoſition; but 'All the ſtates of Europe except two' is a particular one. A rule which applies to all ſtates except Spain and Portugal is a rule: but a rule which applies to all except two (unknown) is no rule. When it is ſtated, in ordinary language, that every rule is ſubject to exception, it is meant, for the moſt part, that the circumſtances under which adherence to the rule gains the object, are thoſe which moſt frequently occur, and that the circumſtances under which adherence to the rule would defeat the object are rare. If this were remembered, much confusion would often be saved. We want a word which shall so far express *rule*, that it shall imply that which will generally succeed, without the notion of *obligation* which accompanies that of *rule*, and which perpetually misleads. We want, in fact, the *rule nisi* of the courts, which is to be a rule unless cause be shown against it: and which will, in most cases, be ultimately made absolute, but is not absolute from the beginning.

The common mistake is, that the rule *nisi* is an absolute rule, and that therefore it may be substituted for its leading object or first principle, and that even the very words which express that object gained, may be taken as equally expressive of satisfaction of the rule, and *vice versâ*. For instance, it is commonly stated that the rule by which a discoverer is determined, is publication; that he who first publishes the discovery, is *to be held* the discoverer; one lapse more, and it is said that he *is* the discoverer; yet one more, and it will be said that the publication is the discovery. The very remarkable circumstances attending the recent discovery of the planet *Neptune*, involving points of peculiar interest and delicacy, have caused this rule to be much discussed, and have brought out every variety of statement of it. The thing to be determined is the *actual truth of the question*, the real history of the human mind with regard to it. No one has a right under any rule, no matter what its authority, nor by whom imposed, to substitute the thing which is not, for the thing which is, or the less probable for the more probable. If philosophers were to attempt, by a law of their own framing, to substitute the conventional result for the real one, the common sense of mankind would dispute their authority, and reverse their decision. The first rule (*nisi*) is undoubtedly that the first printer is the first publisher, the second, that the first publisher is the discoverer. These will, unless cause be shown against them, be made absolute in every case. A notion which is very prevalent, namely, that the first publisher has *therefore* the rights of the discoverer, is as incorrect as that the first *printer* is therefore the first *publisher*. To take the current language, one would suppose that printing one hundred copies would be held better than circulating one thousand in manuscript, and that even though the first publisher could be proved to have plagiarised, he has still the rights of discovery.

Juſt as (page 244) early notions make laws of literal interpretation ſuperſede thoſe of intended meaning, ſo, in the earlier ſtages of law, rules are often made to over-ride the principles on which they profeſs to be founded, and to defeat truth and common ſenſe. There is more excuſe here than there would be in a queſtion of ſcience, for peace and convenience are main objects of law, and it may be that rigid adherence to a rule, as a rule, at a certain avowed ſacrifice of truth and juſtice, may be the only practicable means of preventing a larger ſacrifice of both. In old times, the rule of affiliation, *Pater eſt quem nuptiæ demonſtrant*, was held ſo abſolutely, that the huſband of the mother would be the legal father, though the two had been confined in two different jails a hundred miles apart for twelve months preceding the birth of the child. The modern law has made this rule to be no more than it ought to be, namely, one which muſt hold unleſs the contrary be proved.

It is not uncommon, in diſputation, to fall into the fallacy of making out concluſions for others by ſupplying premiſes. One ſays that A is B; another will take for granted that he muſt believe B is C, and will therefore conſider him as maintaining that A is C. But it may be that the other party, maintaining that A is B, may, by denying that A is C, really intend to deny that B is C. In religious controverſy, nothing is more common than to repreſent ſects and individuals as *avowing* all that is eſteemed by thoſe who make the repreſentation to be what, upon their premiſes, they ought to avow. All parties ſeem more or leſs afraid of allowing their opponents to ſpeak for themſelves. Again, as to ſubjects in which men go in parties, it is not very uncommon to take one premiſe from ſome individuals of a party, another from others, and to fix the logical concluſion of the two upon the whole party: when perhaps the concluſion is denied by all, ſome of whom deny the firſt premiſe by affirming the ſecond, while the reſt deny the ſecond by affirming the firſt. Any ſect of Chriſtians might be made atheiſts by logical conſequence, if it were permitted to join together the premiſes of different ſections among them into one argument. This is a fallacy which, however common, could eaſily be avoided, and would be, if thoſe who uſe it cared for anything but victory. But there is another form of the ſame, which every one is ſubject to, and which it is not ſo eaſy to perceive. It is that of drawing upon our former ſelves for the premiſes which are to guide us for the time being. Concluſions remain in our minds long after the grounds on which they were formed are abandoned: and it may happen that one premiſe of an argument will ſtill have force, when the very reaſons on which the ſecond premiſe is now admitted are contradictory of thoſe which once induced us to admit the firſt. Thus many who have learnt to advocate the legal toleration of opinions which they ſtill believe, by force of education, to be abſolute crimes againſt ſociety, are logically the advocates of toleration of crime; whereas, the arguments which they have learned to think valid for the firſt premiſe, ought, if worth anything, to teach them to deny the ſecond. I have myſelf heard from one mouth in one converſation (of courſe not in one part of it) that all ſins againſt the Creator are ſins againſt ſociety, that all ſins againſt ſociety ought to be puniſhed by ſociety, that certain opinions then named are ſins againſt the Creator, and that it is the height of injuſtice to puniſh any one for his opinions.

In printed controverſy, the ſtatement of the oppoſite opinion or aſſertion may be made by deſcription without *citation* (by chapter or page), by deſcription with citation, or by *quotation* with or without deſcription. The firſt is not allowable. The preſumption is ſtrong that a perſon who oppoſes an opinion, imputes an error, or makes a charge, upon the writings of another, is bound at leaſt to cite, in a manner which cannot be miſtaken, the part of thoſe writings to which he refers. There are writers who refer deſcriptively and even commentatively, putting the reference of citation, and thus (as Bayle ſays Moréri conſtantly does) lead the reader to ſuppoſe that the words of their paraphraſe and comment are thoſe of the paſſage itſelf. I do not ſee that quotation is obligatory, though highly deſirable; but the reader muſt remember, when there is only citation, that it is not the author cited who ſpeaks, but the perſon who brings him forward. It is a man's own account of his own witneſs: with the advantage of an apparent offer of enabling the reader to go and verify the ſtatement for himſelf. If the citer be honeſt, the paſſage in queſtion exiſts: if judicious, it is to the effect ſtated. Conſequently, whenever the citer's honeſty or judgment is expreſsly in queſtion, no mere citation is admiſſible.

When citations are few they ought perhaps to be quotations: when they are many, it may be impracticable to make them so. But extensive citation ought to be encouraged. Lazy readers do not like it: they are not pleased to have a power of verification offered of which they do not mean to avail themselves; and they would rather, in case of being misled, have to throw the blame upon the author than upon their own non-acceptance of the offered means of verification. Accordingly, they express their disgust at "pages loaded with references." But the more diligent readers consider every citation as a boon. At the same time it is to be remembered that there are writers who, relying on the common disinclination to verify, add a large number of citations, and give the appearance of a strong body of authorities, which are often nothing to the purpose, and sometimes not taken from actual examination, but copied from other writers.

Perhaps the greatest and most dangerous vice of the day, in the matter of reference, is the practice of citing citations, and quoting quotations, as if they came from the original sources, instead of being only copies. It is in truth the reader's own fault if he be taken in by this, or by the false appearance of authority just alluded to; for it is in his own power to certify himself of the truth: though there may be difficulty when the citations are many, or when some of them are from very rare books. Honesty and policy both demand the express statement of every citation and quotation which is made through another source. If a person quote what he finds of Cicero in Bacon, it should be 'Cicero (cited by Bacon) says, &c.' It has happened often enough that a quoter has been convicted of altering his author, and has had no answer to make except that he took the passage from some previous quoter.

Quotations are frequently made with intentional omission and alteration. But no rule ought to be more inflexible than that all which is within the marks of quotation ought to be a literal transcript of the book quoted. Sometimes the omission is made because part of the sentence is unnecessary, *as the quoter thinks*. But this is just the point which he has no business to decide without letting his reader know that he *has* decided it, which is easily done by the recognized mark of omission (.....) If a person would quote the Æneid for the antiquity of Carthage, he has no businefs to write down, as from Virgil, 'Urbs antiqua fuit Carthago :' it fhould be 'Urbs antiqua fuit . . . . . Carthago,' if he decide upon omitting 'Tyrii tenuere coloni.' In this cafe, not only may the omiffion make the propofition appear more categorical than it is in the original, turning it from 'There was an old city, Carthage,' rather towards 'Carthage was an old city ;' but a reader may choofe to think that the omitted words qualify the epithet, or even offer proof deftructive of it. What if he fhould deny the antiquity of Tyre? The *omiffion* may (or may not) be right, but the omiffion without notice, or *fuppreffion*, is certainly wrong.

Moreover, it is dangerous to truth to fhorten without notice, inafmuch as thofe who quote the quotation will be apt to do the fame thing; that is, thinking they have the whole paffage, to fhorten it further. What this may end in, no one can predict: but miftakes have been brought about in this way quite as abfurd as any that ever were made. It may reafonably be fuppofed that many very ludicrous errors arife thus. A good many years ago, I fucceeded, by means of a fhortened quotation, put away until it was wanted, in arriving at, and publifhing, the conclufion that Archimedes was once fuppofed to have been an anceftor of Henry IV. of France. The real purport of the fentence was that he was fuppofed to have been an anceftor of the Sicilian martyr St. Lucia, on whofe day Henry IV. was born. It has happened that A has been faid to have afferted in a fecond book, that B related the death of C, when the truth is that A faid in the firft book that B died many years before C (See the *Companion to the Almanack* for 1846, page 27). I do not fpeak of omiffions made becaufe the part omitted would prove more than the quoter likes: this of courfe is fraud.

Unjuftifiable as unnoted omiffions may be, ftill more fo are additions and alterations. Writers have fometimes inferted gloffes of their own, into the text which they quote, either as addition or alteration. Explanatory *additions* may eafily be made within brackets [ ], which are underftood marks of fuch a thing: but alterations are intolerable. But why, the reader may afk, are fuch things infifted on? Is not the fimple rule, *Be honeft*, enough to include thefe and hundreds of things like them, without detail? To this I reply that within a twelvemonth before the time I write this, a clergyman, a man of high education and character both, publiſhed a ſermon in which he gave a verſe from the Bible within marks of quotation, in which he wilfully ſtruck out one word, and inſerted another, without notice: and his ſermon went through ſeveral editions, either without detection, or without that detection leading to ſucceſsful remonſtrance. I do not ſuppoſe there was diſhoneſty here; but rather the following reaſoning;—'I am ſure it was meant; therefore I may ſtate that it was ſaid.' Such reaſoning is one of the curſes of our literature.

There is one alteration within the marks of quotation which may at firſt ſeem reaſonable: it is alteration of grammar to bring the quoted phraſes into connected Engliſh with the quoter's context. As when a man ſays "I know" and another perſon, quoting him, ſays "He knows." But it is ſurely juſt as eaſy to put down He ſays "I know." There is often an alteration of emphaſis in this adaptation of grammar, and generally an introduction of irony: and it is the *premier pas* to ſomething worſe. As far as I have ſeen, thoſe who do it as a matter of courſe, are apt ſometimes to put their own paraphraſes under marks of quotation. A writer ſhould ſuit his own grammar to that of his quotation, and not the converſe.

Omiſſion of context, preceding or following the quotation, may alter its character entirely: and this is one of the moſt frequent of the fallacies of reference, both intentional and unintentional. The only way to inſure full confidence is to give the egg in its ſhell: that is, to begin at a point which clearly precedes the immediate ſubject of quotation, and to continue until the matter is as clearly paſt: to give a ſentence preceding and a ſentence following the matter quoted for its own ſake, diſtinguiſhing the latter. This is not always concluſive: becauſe the ſubject may be reſumed in a ſentence or two, or in another part of the book. But it will inform the reader, in moſt caſes, whether he is or is not likely to differ from the quoter as to the meaning of the part quoted. And this refers particularly to quotations of opinion: thoſe of fact may often be more briefly treated with ſafety.

In quoting ancient authors, in caſes where the text is not notorious, the various readings ſhould be given, eſpecially when it is an author whose text has an indifferent reputation for accuracy. Or if this cannot be done, the edition should be cited. Shameful things have occurred in controversy, by omission of a part of the ordinary text, which the quoter *chose to consider* as an interpolation, without choosing to consider that the reader ought to have liberty to judge for himself on that point.

Among the cases of indirect citation, should be included that in which a book is mentioned as existing, not on the authority of the writer's own eyes, but on that of a catalogue. The number of nonexisting books which are entered in catalogues and copied, as to their titles, into other works, is greater than any one who has not examined for himself would suppose possible. In those who know this, confidence is destroyed; and this sometimes affects questions of opinion. I am told that Dugald Stewart, who had a strong notion of the practical impossibility of presenting Euclid in a syllogistic form, never would believe that it had been done by Herlinus and Dasypodius. Such a work is entered in catalogues: but I must say that the state of catalogues is such that Stewart or any one else had full right to doubt of any work, upon no other than catalogue evidence. The work does exist, and I have a copy of it. But, seeing how matters stand, no one has a right to declare that an old book ever was written, without informing his reader on what sort of evidence he relies.

CHAPTER XIV.

On the Verbal Description of the Syllogism.

IN page 75, I have made a first attempt to express the relations of propositions in language which will make syllogisms capable of verbal description, and the inference of their conclusions matter of self-evidence. It is desirable that this should be more fully done, and I accordingly renew the attempt, with the best words of description which I can find or make. Any one who can suggest words which better convey the meaning to himself, will find it easy to substitute them for those which I have used.

The conditions to be ſatisfied are, that the words ſhould have as much imported meaning as poſſible, that every word and its contrary ſhould have the connexion of contrariety well marked, and that the verbal deſcriptions ſhould be capable of being eaſily formed from the ſymbolic notation. As may be ſuppoſed, theſe conditions are to ſome extent contradictory of each other: the ſacrifice of either to the others is then to be made to the moſt advantageous effect.

There are two ways in which it may be neceſſary to deſcribe the ſyllogiſm. Firſt, the one hitherto uſed throughout this work, in which one concluding term is referred to the other by the intervention of the middle term: what X is of Y, and what Y is of Z, determine what X is of Z. Secondly, that in which the two terms are referred to one another by compariſon of both with the middle term: what X and Z ſeverally are of Y determine what X is of Z.

In the firſt mode, the middle term is mentioned, and its deſcription is middle in the ſentence; while the reference term is underſtood in the predicate of each deſcription. Thus when we ſay 'a ſubcontrary of a ſupercontrary is a ſubidentical,' it is that a ſubcontrary of a *ſupercontrary (of* Z) is a ſubidentical (of Z); and the *ſupercontrary of* Z is the middle term.

In the ſecond mode, the middle term is underſtood in the ſubject, and the concluding terms in the predicate, of the deſcription of the ſyllogiſm. Thus when we ſay 'genus and ſpecies are genus and ſpecies,' it means that two terms which are ſeverally genus and ſpecies of the middle term (one entirely containing, the other entirely contained in, the middle term) are genus and ſpecies to one another (the firſt genus, the ſecond ſpecies).

Now it will be very eaſily ſeen, that the way to change the firſt deſcription into the ſecond is as follows. Say the deſcription runs thus, 'P of Q is R.' If Q be its own correlative, as happens when Y and Z are convertibly connected, then 'P of Q' merely becomes 'P and Q:' but if Q have another, Q^0, for its correlative, then 'P of Q' becomes 'P and Q^0.' Again, if R be its own correlative, its plural takes its place: but if R have R^0 for its correlative, it becomes 'R and R^0.' Thus 'ſubcontrary of ſupercontrary is ſubidentical' of the firſt mode, becomes 'ſubcontrary and ſupercontrary are ſubidentical and ſuperidentical' meaning that C_1 and C' of the middle term are D_1 and D' of each other. But 'ſubcontrary of ſuperidentical is ſubcontrary' becomes 'ſubcontrary and ſubidentical are ſubcontraries.'

I need hardly ſay that 'P of Q is R' with reſpect to X in terms of Z, muſt be read 'Q^0 of P^0 is R^0' with reſpect to Z in terms of X. This rule we have already uſed.

It is thus ſhown that it is only neceſſary to dwell on the firſt mode; and now ariſes the queſtion what words are to be employed in deſcribing the eight ſtandard propoſitions. After a good deal of conſideration, I prefer to denote the univerſal relations by poſitive terms, and their contrary particulars by the correſponding negative ones: not without full perception of the ſacrifice which enſues of the firſt condition above mentioned to the third.

The words *genus* and *ſpecies* immediately ſuggeſt themſelves to denote the relation of Y to X and X to Y in X)Y. Theſe are to be underſtood as employed up to their limit; or the genus and ſpecies may be coextenſive. For two names which have nothing in common, as in X.Y, I propoſe to ſay that they are *externals* of each other. And for two names which have nothing out of one or the other, as in x.y, that they are *complements* of each other. Remember that complemental does not mean *only juſt complemental* (which is contrary), but may be contrary or ſupercontrary.

In X:Y, I call X a *non-ſpecies* of Y, and Y a *non-genus* of X. Theſe words have not as much as I could wiſh of imported meaning, nor are there any poſitive terms which I can propoſe to ſupply their places. They appear as ſynonymous with *not entirely contained in* and *not containing the whole.* In XY, let X and Y be *non-externals*; and in xy, let X and Y be *non-complements*. Accordingly, in deſcribing what X is with reſpect to Y, we have as follows, ſhowing the ſubſtitutions which occur in reading the ſyllogiſtic ſymbols into this language.

A_1, ſpecies	O_1, non-ſpecies.
A', genus	O', non-genus.
E_1, external	I_1, non-external.
E', complement	I', non-complement.

If we conſider *genus* and *complement* as *larger terms*, and *ſpecies* and *external* as *ſmaller ones*, and if we put down each univerſal followed by its two weakened particulars, writing firſt that which is of the ſame accent, we have

Univerſal.	Firſt weakened form.	Second weakened form.
A' Genus	I' non-complement	$I_{,}$ non-external.
$A_{,}$ Species	$I_{,}$ non-external	I' non-complement.
E' Complement	O' non-genus	$O_{,}$ non-ſpecies.
$E_{,}$ External	$O_{,}$ non-ſpecies	O' non-genus.

Thus it appears that the primary weakened form of a larger name contains a larger name, and of a ſmaller a ſmaller: and the contrary for the ſecondary forms. The words primary and ſecondary do not refer to importance, but only to order of derivation: thus $A_{,}$ was in our table X)Y, weakened into XY, before it became y)x, weakened into yx or xy.

The rules for forming particular ſyllogiſms by weakening univerſal premiſes may now be repeated. In a univerſal ſyllogiſm, ſubſtitute for the *firſt* premiſe and for the concluſion their *primary* weakened forms, or for the *ſecond* premiſe and for the concluſion their *ſecondary* weakened forms. In a ſtrengthened ſyllogiſm, ſubſtitute for the *firſt* premiſe its *ſecondary* form, or for the *ſecond* premiſe its *primary* form.

I now write down the whole body of ſyllogiſms, that the reader may exerciſe himſelf in the independent comprehenſion of their meaning, and in aſſent to their inferences; deducing the particular ſyllogiſms from the univerſals only.

Univerſal and particular Syllogiſms.

Symbol.	Deſcription of X with reſpect to Z.
$A_{,}A_{,}A_{,}$	Species of ſpecies is ſpecies.
$I_{,}A_{,}I_{,}$	Non-external of ſpecies is non-external.
$A_{,}I'I'$	Species of non-complement is non-complement.
$A'A'A'$	Genus of genus is genus.
$I'A'I'$	Non-complement of genus is non-complement.
$A'I_{,}I_{,}$	Genus of non-external is non-external.
$A_{,}E_{,}E_{,}$	Species of external is external.
$I_{,}E_{,}O_{,}$	Non-external of external is non-ſpecies.
$A_{,}O'O'$	Species of non-genus is non-genus.

A'E'E'	Genus of complement is complement.
I'E'O'	Non-complement of complement is non-genus.
A'O,O,	Genus of non-ſpecies is non-ſpecies.
E,A'E,	External of genus is external.
O,A'O,	Non-ſpecies of genus is non-ſpecies.
E,I,O'	External of non-external is non-genus.
E'A,E'	Complement of ſpecies is complement.
O'A,O'	Non-genus of ſpecies is non-genus.
E'I'O,	Complement of non-complement is non-ſpecies.
E,E'A,	External of complement is ſpecies.
O,E'I,	Non-ſpecies of complement is non-external.
E,O,I'	External of non-ſpecies is non-complement.
E'E,A'	Complement of external is genus.
O'E,I'	Non-genus of external is non-complement.
E'O'I,	Complement of non-genus is non-external.

Strengthened Syllogiſms.

A,A'I'	Species of genus is non-complement.
A'A,I,	Genus of ſpecies is non-external.
A,E'O'	Species of complement is non-genus.
A'E,O,	Genus of external is non-ſpecies.
E,A,O'	External of ſpecies is non-genus.
E'A'O,	Complement of genus is non-ſpecies.
E,E,I'	External of external is non-complement.
E'E'I,	Complement of complement is non-external.

No perſon could propoſe to himſelf a better exerciſe in the acquiſition of command over language, than practiſing the demonſtrations of theſe relations, or more properly their reduction into ſpecific ſhowing, as to the matter of the inference, in what its extent conſiſts. For inſtance, ‘ the complement of a non-complement is a non-ſpecies’ : How, and by how much ? The non-complement leaves ſomething which is neither in the term underſtood, nor in that non-complement. This, the complement of that non-complement muſt fill up : and by this then, at leaſt, the complement of the non-complement is not in the term underſtood, of which it is therefore ſo far non-ſpecies.

In the preceding view, I have particularly conſidered the connexion between contrary forms, and the adaptation of language to that connexion. But in the firſt derivation of the ſimple ſyllogiſms (page 88) the univerſals were related, not to their contraries, but to their particular concomitants. I now proceed to the conſideration of this view, and to the juſtification, on ſelf-evident principles, of the aſſertion that there is a real and ſtriking affinity between the univerſal ſyllogiſm and its concomitants, as $A_{\prime}A_{\prime}A_{\prime}$ and $O'A_{\prime}O'$, $E'E_{\prime}A'$ and $E'I'O_{\prime}$, &c.

The complex propoſitions $D_{\prime}$, D', and $C_{\prime}$ contain each a univerſal which, in common language, is generally confounded with it, and a particular, the exiſtence of which is therefore for the moſt part ſuppoſed in thought to accompany the univerſal. The remaining univerſal, E', is differently circumſtanced: if we ſay that X and Y complete the univerſe, we ſhould generally mean that they only juſt complete it, and ſhould not think of the ſuper-contrary relation, or of their overcompleting it. To be contained but not to fill; to contain with room to ſpare, or to *overfill;* to exclude and be excluded without completion; and to exclude and be excluded with completion (or to complete and be completed without incluſion);—are our moſt uſual ideas of the relations of the extent of names.

The reduction of the complex propoſition to the ſimple univerſal, when done by removal of the concomitant *particular*, is in all caſes a lowering of the quantity, by the removal of an exceſs, as follows:—

$D_{\prime}$ means that X is contained in Y, and more is contained.
D' means that X contains Y, and contains more.
$C_{\prime}$ means that X excludes Y, and excludes more.
C' means that X completes Y, and *more than completes.

Drop the ſecond clauſes, and $D_{\prime}$, &c. are reduced to $A_{\prime}$, &c. Drop the firſt clauſes, and it would ſeem as if we had ſtill the

* The alteration of grammar here ſeen is in deference to the word *complete*, the beſt I can get. In this propoſition, the verb refers to *the univerſe*, and it is X(joins in completing the univerſe)Y and joins in completing more (than the univerſe).

complex propoſitions; for *more* will contain its tacit reference to that which it *is more than.* Let this tacit reference be dropped, and then we have, inſtead of the whole complex propoſition, only its particular. And this abandonment is actually made in common language, by what would be called perhaps a lax, but is a very logical, uſe of the word *more.* 'There are more than fiſh on the dry land,' would be perfectly intelligible, and not as implying that there were any fiſh: 'he was actuated by more than the motive, &c.' very often means 'other than the motive' &c.

Now, in the complex ſyllogiſm, as we have ſeen (page 81), the exceſſive part of the concluſion (whence comes its ſecond clauſe, its additive *more*) is the ſum of the exceſſive parts of the premiſes. If one of the complex premiſes be deprived of its aſſertion of exceſs, or lowered into a ſimple univerſal, the concluſion ſtill remains, though not *à fortiori*, neceſſarily. This being done, *the valid exceſs of the concluſion depends upon the exceſs of the remaining premiſe;* and the concomitant particular ſyllogiſm, conſidered as part of the mixed complex ſyllogiſm, is the expreſſion of this, without the reſt. Finally, the exceſs may be uſed in the lax, or non-correlative, ſenſe, and then the concomitant ſyllogiſm ſtands by itſelf.

For example, $O_1A'O_1$ may be read thus:—Conſider O_1 as concomitant of A' in D'. 'X contains more than [ſomething that is not in] Y; Z contains X; therefore, Z contains more than [ſomething that is not in] Y.' If 'more than Y' mean 'Y and more,' this would be $D'A'D'$. Again, $O'E_1I'$ is 'more than X [ſomething not X] is contained in Y; Y excludes Z; therefore, X excludes more than Z [ſomething not in Z].' If 'more than X' were 'X and more,' &c.: this would be $D_1E_1C_1$. And ſo on for other caſes.

I now proceed to what I may call the *quantitative* deſcription of the ſyllogiſm: by which I mean the expreſſion of its caſes in terms of the quantities *only* of its names and propoſitions, leaving the alternative of affirmation and negation to be ſettled by the law of theſe quantities. My reaſon for the preſentation of the ſyſtem in ſo many different points of view will be obvious enough: that which claims to be complete, muſt ſhow itſelf to contain juſt the ſame, and no more, as to reſults, whatever may be the principle which is choſen as the baſis of conſtruction.

Every propofition, in fpeaking of two names, fpeaks of their contraries, and (page 63) of the four terms, two direct and two contrary, two are univerfal and two are particular. Since univerfal and particular are themfelves properly contraries, (for 'Every X' is 'Xs, *known* to be all' and 'Some Xs' are 'Xs, *not known* to be all') let us fignify the univerfal and particular forms of the *propofition* by V and v. Again, fpeaking of a *name*, let its mode of entry, univerfal and particular, be denoted by T and t. Writing down V(or v) applied to T(or t), T(or t) we can make eight varieties, which give us the eight ftandard forms applied to one order, fay XY; as follows:—

$A_1 = V(Tt)$	$A' = V(tT)$	$E_1 = V(TT)$	$E' = V(tt)$
$O' = v(Tt)$	$O_1 = v(tT)$	$I' = v(TT)$	$I_1 = v(tt)$

Thus I' or xy, may be defcribed as the particular in which both terms are univerfal: for X and Y are both univerfal in xy, or x:Y, or y:X. And v(TT) defcribes it thus.

If, underftanding the order to be XY, YZ, XZ, we write down any three propofitions, we make an attempt at a fyllogifm, valid or not, as the cafe may be: as in

$$V(Tt).v(tt).V(tT) \text{ or } VvV(Tt,tt,tT)$$

which muft be A_1I_1A'. It will affift the memory to obferve that fub-fymbols have VT or vt at the beginning, fuper-fymbols vT or Vt. Alfo, that affirmatives have an even number of capitals (*none** or *two*) and negatives an odd number (*one* or *three*). A univerfal and its particular concomitant have the fame entries of T and t, and contranominals have inverted modes of entry of thefe letters. The convertibles have T in both places, or t: the inconvertibles have T and t.

Firft, it is unneceffary to write down the term-letters of the conclufion, for they muft be taken from the premifes, in every cafe in which the conclufion is the ftrongeft that can be drawn from the premifes; and our fyftem has no others (nor, indeed,

* The reader muft here follow the mathematician in confidering o as an even number.

has the Ariſtotelian any other except *Bramantip*). Thus, TT,tt being the term letters of the premiſes, ſtrike out the ſecond T and the firſt t, which refer to the middle term, and Tt muſt belong to the concluſion. To prove this, obſerve that we know that t in the premiſe cannot give T in the concluſion : therefore T cannot give t; for if, the term being Z, T gave t, then, putting z properly in its place, t would give T, which it cannot. Again, we know that the valid forms, as to propoſitions, are VVV, VVv, vVv, Vvv; ſo that v occurring once only, muſt come third, and V muſt come in the firſt pair. Further, in the four term letters of the premiſes, VVV, vVv, Vvv, require Tt, or tT, to come in the middle, while VVv alone requires TT, or tt. Obſerve theſe laws, and every formation which can take place under them leads to a valid ſyllogiſm. Putting dots to repreſent a blank place, we form the eight univerſal ſyllogiſms by filling up the blanks in VVV(. . t,T . .) and VVV(. . T,t . .); the eight ſtrengthened ſyllogiſms from VVv(. . T,T . .) and VVv(. . t,t . .); the eight particulars which begin with a univerſal from Vvv(. . t,T . .) and Vvv(. . T,t . .); and the eight particulars which begin with a particular from vVv(. . t,T . .) and vVv(. . T,t . .). And, under the rules juſt given, we have no other caſes.

Taking the preceding as a baſis, we might make the rules of accentuation follow from it. For, ſince the firſt blank in our ſymbol, and the firſt concluding term, muſt agree, and ſince accents depend only on the firſt two letters in the ſymbol of a propoſition, we may proceed as follows. Let K and L, each of them, mean T or t, as the caſe may be, but with the proviſo that what it means in either place it ſhall mean in the other. Then, in VVV(KT,tL,KL) and in vVv(KT,tL,KL), in which ſymbols of concluſion are introduced, we ſee that the firſt and third accents muſt agree, which is part of the direct rule. As to the firſt and ſecond accents, they agree in the firſt inſtance above, if K be t, which puts an even number of capitals in the firſt ſymbol VKT, or an affirmative propoſition at the commencement : they differ if K be T, which puts a negative propoſition firſt. In the ſecond inſtance, they agree if K be T, which puts an affirmative firſt, &c. I leave it to the reader to deduce the other caſes of this rule, the inverſe rule, and alſo that premiſes give an affirmative, or a negative, conclusion, according as they have like or unlike signs. And thus it will appear, that the symbolic rules given in chapter V, are really expressions of the general rules of quantity.

It will be observed that the concomitant syllogisms of a universal have the same term letters as that universal, and only change VVV into Vvv, or vVv. Also, that the inverted syllogisms of page 96 only invert the order of all the term-letters, and the letters of the premises, when different.

Thus, $E_{\iota}A'E_{\iota}$ being VVV(TT,tT), its concomitants I'A'I' and $E_{\iota}O_{\iota}I'$, are vVv(TT,tT) and Vvv(TT,tT). But the inverted form $A_{\iota}E_{\iota}E_{\iota}$ is VVV(Tt,TT). Contranominals have different quantities in all the term-letters. The weakened forms of a universal change the first premise letter and the first term letter, or the second of both. Thus, $E_{\iota}E'A_{\iota}$ being VVV(TT,tt), its weakened forms, $O_{\iota}E'I_{\iota}$ and $E_{\iota}O_{\iota}I'$, are vVv(tT,tt) and Vvv(TT,tT).

The forms of the numerical syllogism (page 161) may be recovered by few and easy rules, in which the premises as they stand determine the conclusion, as follows:—Let ξ be designated as the number of X, and ξ' as that of x; and so on. Let a term of the conclusion be called *direct* when it is in the premise, and *inverse* when its contrary is in the premise. Then,

1. In every case, the conclusion has the sum of the quantities mentioned in the premises, as part of the expression of its quantity.

2. For every inverse term in the conclusion, the number of its direct term appears in the quantity of the conclusion, subtracted. Thus, x in a premise, with X in the conclusion, must have $-\xi'$ in the concluding quantity. But the direct terms of the conclusion never introduce anything into the concluding number.

3. When the entrances of the middle term are similar (YY, or yy), the terms of the two forms of conclusion are both direct and both inverse, with subtraction of the number of the middle term in the former, addition of the number of its contrary in the latter. Thus, yy gives $-\eta'$ in the direct, $+\eta$ in the inverse form.

4. When the entrances of the middle term are dissimilar (Yy,

or yY), each form of concluſion has one direct and one inverſe term; and no number from the middle term enters the concluding quantity.

Thus, the concluſions from $m\mathrm{xY}+n\mathrm{YZ}$ are immediately written down as

$$(m+n-\eta)\mathrm{xZ} \text{ and } (m+n+\eta'-\xi'-\zeta)\mathrm{Xz}:$$

while thoſe from $m\mathrm{xY}+n'\mathrm{yz}$ are at once

$$(m+n'-\xi')\mathrm{Xz} \text{ and } (m+n'-\zeta')\mathrm{xZ}$$

There are relations exiſting between the forms of the ſyllogiſm which I have not conſidered. For inſtance, the deſcription of X with reſpect to Z being that it is a ſpecies, ($A_{\prime}$), the deſcription of its contrary, x, is that it is a ſupercontrary, (E'). If then we give the name of *contradeſcriptives* to $A_{\prime}$ and E' we find that A' and $E_{\prime}$, $I_{\prime}$ and O', I' and $O_{\prime}$, are alſo contradeſcriptives. The arrangement of ſyllogiſms by contradeſcriptives, and the laws of connexion thence reſulting, will be an eaſy exerciſe for the ſtudent.

APPENDIX.

I.

Account of a Controverſy between the Author of this Work and Sir William Hamilton of Edinburgh; and final reply to the latter.

THIS appendix contains an account of a controverſy in which ſome of the matters treated in the preceding work involved me with Sir William Hamilton, Profeſſor of Logic and Metaphyſics in the Univerſity of Edinburgh. It has produced four publications (to which I ſhall refer as I, II, III, IV) namely:

I. 'Statement in anſwer to an aſſertion made by Sir William Hamilton, Bart. . . . . by Auguſtus De Morgan, ... (London, octavo, R. and E. Taylor, pp. 16, publiſhed April 30, 1847.)

II. 'A letter to Auguſtus De Morgan, Eſq. . . . . on his claim to an independent rediſcovery of a new principle in the theory of ſyllogiſm. From Sir William Hamilton, Bart. Subjoined, the whole previous correſpondence, and a poſtſcript in anſwer to Profeſſor De Morgan's "Statement,"' (London and Edinburgh, octavo, Longman and Co., Maclachlan and Co. pp. 44, excluſive of 'Proſpectus' hereinafter mentioned: received by me May 22, 1847.)

III. Letter from me to Sir W. Hamilton, dated May 24, publiſhed in the *Athenæum* Journal of May 29.

IV. Letter from Sir W. Hamilton to me, dated June 2, publiſhed in the ſame Journal of June 5.

There are two queſtions involved, one concerning my character, the other purely literary. The former ſtands thus. March 13, Sir W. Hamilton informed me by letter that (the Italics are his own words) *to* him *it is manifeſt that* for a certain *principle* I was *wholly indebted to* his *information,* and *that if* I *ſhould give* it *forth as a ſpeculation of* my *own* (which I had done to himſelf, and meant to do, as he knew, and have ſince done, in print) I ſhould, even *though recognizing always* his *priority, be guilty both of an injurious breach of confidence towards* him *and of falſe dealing towards the public.* This *hypothetical charge,* and *derogatory ſuppoſition of which he may formerly have ſurmiſed the poſſibility* (ſuch are his ſubſequent qualifications of it) is unreſervedly retracted at the *beginning* and *end* of II: but it is frequently inſinuated in the *middle,* by propoſing things as difficult to be explained otherwiſe, by hint that *others* may believe it, by *hopes* that they will not, by charges of falſehood, &c. &c. For the *formal* charge is ſubſtituted imputation of lapſe of memory, intellectual confuſion, &c. The following is the programme of the firſt intended argument, (II. p. 4.)

'I confeſs, that, for a time, I regarded your pretenſion, as an attempt 'at plagiariſm, cool as it was contemptible.

'From this view, feeling, information, reflection turned me; and I 'now, Sir, tender you my ſincere apology, for admitting, though founded 'on your own ſtatements, an opinion ſo derogatory of one, otherwiſe ſo 'well entitled to reſpect.

'In itſelf, this view was, to me, painful and revolting.—The cha-'racter, too, which you bear among your friends, I found to be wholly 'incompatible with a ſuppoſition ſo odious. You are repreſented as an 'active and able man, profound in Mathematics, curious in Logic, wholly 'incapable of intentional deceit, but not incapable of chronological miſ-'takes. Your habitual confuſion of times is, indeed, remarkable, even 'from our correſpondence. Your dates are there, not unfrequently of 'the wrong month, and not always, even of the right year. With much 'acuteneſs, your works ſhow you deficient in architectonic power, the 'concomitant of lucid thinking; and, that you are not guiltleſs of intel-'lectual raſhneſs is ſufficiently manifeſt, from your pretention to advance 'Logic, without having even maſtered its principles.'

With regard to the ſubſequent inſinuation of a retracted charge, my explanation (believing as I do, that Sir W. Hamilton always ſpeaks ſubjective truth) is that his mind inſenſibly fell back to its old bias as he felt that the ſubſtitute for his charge wanted ſtrength: my concluſion is, that it is unneceſſary henceforward to notice any thing he may ſay or write on my character: and my determination is to act accordingly.

Sir W. Hamilton's pamphlet contains about a ſcore and a half of quotations, on which hang ſundry jokes and ſneers, ſome of them at mathematicians in general, and myſelf as one of the body. On theſe I ſhall only ſay that my notions of the common ſenſe of controverſy, and my determination to perſiſt, generally, in the tone of reſpect to my opponent's learning and character which I have hitherto preſerved, would, were there nothing elſe, prevent my adopting the habit of which they are ſpecimens. But as no man willingly ſtands an unreturned fire of facetiæ without deſiring to prove that his forbearance does not ariſe from want of ammunition, I will permit myſelf (diſclaiming the *animus* under which ſuch things are uſually written) juſt to ſhow that quotation, application, alluſion, ſneer, joke, and fling at an opponent's ſtudies, are all among the weapons which I could have employed, if I had thought them worthy of my antagoniſt, or of thoſe whom I want to convince.

I might, for inſtance, have written ſomething like the following;—

Among the aſſets of the old logicians, diſcovered when the ſchools were ſwept out, there was found, as is well known, the queſtion *Utrum chimæra bombinans in vacuo poſſet comedere ſecundas intentiones:* a very good title, as Curll would have ſaid, wanting nothing but a treatiſe written to it. Now whether it be *comĕdere,* or whether the ſchoolmen invented *comēdere,* Sir W. Hamilton, on whom their mantle has fallen, has written the treatiſe, and ſucceſsfully maintained the affirmative. His notion that his communication could give any hint, is clearly and aptly described by *chimæra*, his style by *bombinans*, his proof by *vacuum*: and the *second intentions*, above noticed, chewed up and given forth with his first ones, are a practical example of the possibility of the Q.E.I. He, or rather the bombinating chimæra which has personified itself in his form, as the οὖλος ὄνειρος did in that of Nestor, is thus both retractor and detractor. But though the transition from slops to solids generally indicates convalescence, yet, as here made manifest, the passage from liquid to dental may be only the growing weakness, the periscence, of the case.

I assert the following documents to be all that are relevant with respect to the literary part of the controversy. They are given at the end of this appendix.

A is an extract from a communication of mine to the Cambridge Philosophical Society, made before I received any communication whatsoever from Sir W. Hamilton. I assert it to contain a distinct announcement and use of the principle of *quantification of the middle term*, be that middle term subject *or predicate*. On this point the reader is to judge.

B is a communication from Sir W. Hamilton to me. The reader is to judge first, whether it contain anything which is intelligible with respect to any system of syllogism; secondly, whether, if it should so contain anything, that something would have been information to me who had written A, on some matter afterwards found in C.

C is the relevant part of an addition made by me to A, when the latter came before me in proof. The reader is to judge first, whether C contain anything more than an application of A; secondly, if so, whether that something more is derived from anything intelligibly hinted at in B.

The only bare fact on which Sir W. Hamilton and myself are at issue is this. I assert and maintain that the matter of C was written in my possession before I received B: Sir W. Hamilton holds me mistaken, and thinks he can prove from the correspondence that in this point my memory has failed. This I continue to treat as irrelevant: for we are both agreed that the *corpus delicti*, if *delictum* there be, lies in C containing something not substantially contained in A, but sufficiently hinted at in B. Any reader who thinks that C does contain something suggested by B which is not in A, may declare against the correctness of my memory; any one who thinks the contrary, will hold it of no consequence whether my memory on the disputed fact be good or bad. With the first reader I have no case: with the second I have all I think worth caring about.

Sir. W. Hamilton maintains my letters to be essential parts of the case. They *may* become so, as soon as it is *pointed out* what C contains which is hinted at in B, and not contained in substance or principle, in A. When Sir W. Hamilton *points out*, by citation from C, what he alleges to have been taken, and by citation from B, what he thinks it has been taken from, and when I thereupon fail to produce equivalent knowledge from A or elſe to expoſe the irrelevance of his citation from B—then thoſe letters may become of importance. This he has not done, though ſpecially challenged to do ſo: and when I come to diſcuſs III and IV, it ſhall appear that he admits he has not done it.

I now give the beſt account I can of the origin of the diſpute, premiſing, that up to this 3d of September, 1847, I do not abſolutely know what the ſyſtem is which I am charged with appropriating. There is *a* ſyſtem which I think is moſt probably the thing in queſtion: but a ſyſtem containing a defect of ſo glaring a character, that I will not attribute it to Sir W. Hamilton, who deſcribes his own as "adequately teſted and matured" until he expreſſly claims it, or until I have the moſt indubitable proof.

In the common, or Ariſtotelian propoſition, the quantities of the ſubject and predicate are determined, the firſt *by expreſſion* or *implication*, the ſecond *by the nature of the copula* (ſee page 57 of this work). And the only quantities conſidered are *all* and *ſome*; the latter meaning anything that *not none* may mean, ſome, it may be all but not known to be all, perhaps not more than one. The matter contained in A ſuggeſted itſelf to me in the ſummer of 1846, and was forwarded to Cambridge with the reſt of the memoir on the 4th of October.

I will now introduce Sir W. Hamilton's deſcription of the various kinds of quantity (II p. 31, 32).

'Your "Statement" is chiefly plauſible from a wretched confuſion 'of diſtinct things. This confuſion, with which you delude yourſelf, 'and many of your readers, is of two independent ſchemes of logical 'quantification; the one, aſſerting *an increaſe* in the expreſſly *quantified* '*terms*, the other, *a minuter diviſion of the forms of quantification itſelf*. 'To diſintricate this entanglement, we have ſimply to conſider, in their 'contraſts, the *three* following ſchemes of quantification:—

'The *firſt* ſcheme is that which logically—confines all expreſſed 'quantity to the *Subject*, preſuming the *Predicate* to be taken—in *ne*-'*gative* propoſitions, always determinately in its *greateſt* and *leaſt* ex-'tenſion (univerſally and ſingularly), in *affirmative* propoſitions, always 'indeterminately in *ſome part* of its extenſion (particularly).

'The *ſecond* ſcheme is that which logically—extends the expreſſion 'of quantity to *both* the propoſitional terms, and allows the *Predicate* to 'be of *any quantity*, in propoſitions of *either quality*. This not only 'ſupplies a capital defect, but affords a principle on which Logic ob-'tains a new and general development.

'The *third* ſcheme is that which logically—admits *more expreſſed* '*quantities* than a determinately leaſt or greateſt extenſion (quantity ſin-'gular and univerſal), and an indeterminately partial extenſion (quantity 'particular.) This, though it corrects, *perhaps*, an omiſſion, yields no 'principle for a general logical development.

'The firſt doctrine is the common or Ariſtotelic; the ſecond is mine; 'and in the third—in ſo far as you have gone, and apart from the con-'ſideration of right or wrong—I do not queſtion your originality.

'Now, the ſecond and third ſchemes are both oppoſed to the firſt, 'but in different reſpects; conſequently the ſecond and third may, each 'of them, combine with itſelf, *either* the whole other, *or* that part of 'the firſt to which it is not itſelf oppoſed. More is impoſſible.

'Let the following be noted:*—*Your* OLD *view (that in the body of 'the Cambridge Memoir) is a combination of the* THIRD *ſcheme of quan- 'tification with the* FIRST; *your* NEW *view (that in its Addition) is a 'combination of the* THIRD *ſcheme of quantification with the* SECOND: *and 'the confuſion, of which you are* NOW *guilty, is the recent and uniform, 'and perverſe identification, in your* PRESENT "*Statement,*" *of the* SECOND '*ſcheme with the* THIRD.

'Before, however, proceeding to comment on your confuſion of the 'ſecond and third ſchemes, I may alſo relieve a confuſion in the term '*definite* and its reverſe, *indefinite*, as applied to logical quantification.

'In the *firſt*, common, or Ariſtotelic meaning, *definite*, or more pre- 'ciſely *predefinite* (διοριστὸς, προσδιοριστὸς,) is equivalent to *expreſſed*, '*overt*, or, more proximately, to *deſignate* and *pre-deſignate*; in this 'ſenſe, *definite quantity* denotes *expreſſed*, in oppoſition to merely *under- 'ſtood, quantity*.

'In the *ſecond* meaning, that which I have always uſed, (and certain 'ancients, I find, were before me,) *definite* is equivalent to determinately '*marked out*; a ſenſe in which *definite quantity* is *extenſion undivided 'or indiviſible, univerſal* or *ſingular* (this including any collected plu- 'rality of individuals) as oppoſed to *particular quantity*.

'In the *third* meaning, which you have uſurped, *definite* is equivalent 'to *numerically ſpecified*; and in this ſenſe, a *definite* is an *arithmetically 'articulate* quantity, as oppoſed to one arithmetically inarticulate.— 'This your meaning of the word I did not, before the appearance of 'your "Statement," apprehend; for of courſe I preſumed you to uſe it 'in its firſt or common meaning, from which you never hint that you 'conſciouſly intend to deviate.'

Three ſchemes of quantity are here mentioned.

Firſt, the ordinary one.

Secondly, that in which the *ordinary quantities*, *all* and *ſome*, are applied in every way to both ſubject and predicate.

Thirdly, that in which numerically definite quantity is applied to ſubject or predicate or both: the eſſential diſtinction of this caſe is *numerical* definiteneſs: it really contains the ſecond ſyſtem, when numerical quantity is algebraically expreſſed. Of theſe, it appears, Sir W. Hamilton claims the ſecond, or rather, the application of ſuch a ſcheme to the ſyllogiſm. What then is it? I ſuppoſe it to be the following. My order of reference is XY.

* Let the following alſo be noted:—My *old* view (that in the body of the Cambridge paper) is entirely on the *firſt* ſcheme, except in one *digreſſive* ſection and one ſubſequent paragraph (from both of which A is quoted) in which the *ſecond* and *third* are combined: my next view (that in the addition) is alſo a combination of the *ſecond* and *third* ſchemes: and my "Statement" contained alſo a uniform, but not recent, identification of the ſame *ſecond* and *third* ſchemes, which I never ſeparated in thought until I ſaw this paragraph. Any one who can form an opinion of the way in which the ſubject would preſent itſelf to the mind of a mathematician, will ſee that the ſecond ſcheme would preſent itſelf concomitantly with, and as an eſſential part of, the *algebraical* form of the third. A. De M.

All X is all Y means that X and Y are identical: it is my D. *All X is some* Y is A_1. *Some X is all Y* is A'. *Some* X is *some* Y is I_1. As to negative propositions, *All X is not all Y* is E_1. *Some X is not all Y* is O_1. *All X is not some Y* is O'. *Some X is not some Y* is true of all pairs of terms one of which is plural. In its indefinite form, it is what I have in Chapter VIII. called *spurious*.

The propositions of this system are then the complex D, or A_1+A', the six Aristotelian forms A_1, A', E_1, O_1, O', I_1, and the spurious form, which may be called U. In looking over (Sept. 5) Sir W. Hamilton's pamphlet, I happened to light on the assertion (incidentally made) that his system gives *thirty-six* valid moods in each figure. On examining the preceding system, I find this to be the case. I should not have published the results, had not Sir W. Hamilton made it necessary for me to comment on them. I shall denote the proposition U, or 'Some Xs are not some Ys' by X :: Y; and I shall, supposing each case to be formed in the first figure, then transpose it into my own notation.

1. There are *fifteen* forms in which D enters. Whenever D is either of the premises, the other premise and conclusion agree. Thus we have A_1DA_1, DUU, &c. &c.

2. *Fifteen* Aristotelian forms $A_1A_1A_1$, $A'A'A'$; $A_1E_1E_1$, $E_1A'E_1$; $A_1O'O'$, $O_1A'O_1$; $A'O_1O_1$, $O'A_1O'$; $A'I_1I_1$, $I_1A_1I_1$; E_1I_1O', $I_1E_1O_1$; $A'A_1I_1$; $A'E_1O_1$, E_1A_1O'.

3. *Six* more U syllogisms $A'O'U$, O_1A_1U; $A'UU$, UA_1U; $I_1O'U$, O_1I_1U.

The two things to be considered are;—the introduction of the identical proposition; and that of the spurious one, as I call it.

It is, I suppose, a fundamental rule of all formal logic, that every proposition must have its denial, its contradiction. Now D has no simple contradiction in this system: that O' and O_1 both contradict it (and also E_1) is true: but the mere contradiction is the disjunction 'O' or O_1'. A person who can show that one or the other of these is true, has demonstratively contradicted D, even though it could be proved impossible to determine which of the two it is.

The proposition U is usually spurious. But if we introduce it, we must introduce its contradictory also. Now if either X or Y be plural names, it must be true: consequently, the contradiction of U is 'X and Y are singular names, and X is Y.' When a syllogism having the premise U is introduced, either that premise may be contradicted, or it may not. If it may, there is no form to do it in: if it may not, then it is a spurious proposition, and cannot, by combination with others, prove anything but a like spurious conclusion.

Let X :: Y denote 'Some Xs are not some Ys,' and X_1Y_1 denote 'there is but one X and one Y, and X is Y.' Then either X :: Y or X_1Y_1 must be true, and one only. A logical system which admits one and not the other, which contains an assertion incapable of contradiction without going out of the system, can hardly be said to be "adequately tested and matured," and is not self-complete. The proposition X_1Y_1 includes in itself the conditions of D, and is a kind of *singular form* of D.

I presume, from the number of Sir W. Hamilton's moods, *thirty-six*, as above obtained, that the contradiction neither of D nor of U finds a place. Admit them, and the contradiction of U alone (call it V) demands sixteen new moods in each figure. I will now proceed.

In my publication, speaking now of (A) what was sent to Cambridge before I communicated with Sir W. Hamilton, I had no quantification intermediate between the ordinary one, and the numerical one applied to either subject or predicate, as wanted in the *canon of the middle term* there given. Look at the last of the seven syllogisms in the *second extract*, where *both the predicates*, being of the middle term are quantified, and the condition of validity is quantitatively stated. But for 'Y_1+Y_2 less than 1' should be read 'y_1+y_2 greater than 1.' The equivalence of this to 'Y_1+Y_2 less than 1' is a mistake. In the *first extract*, the general canon is given which is afterwards used in C.

Up to the time when Sir W. Hamilton published his letter in reply to my statement, (II), I never had separated the idea of his second scheme of quantification from that of the third.

Thus then we stood on October 3, when I sent my paper to Cambridge. Sir W. Hamilton had been teaching the application of the ordinary quantities to both subject and predicate: I had arrived at the algebraical representation of the numerical quantification of terms, whether subject or predicate matters not, as long as they were middle terms.

1846, *October* 6. My communication (containing A) was in the hands of Dr. Whewell (as he informs me) for transmission to the Cambridge Society: I never saw it again till the next February. *October* 7, Sir W. Hamilton wrote to me, in answer to an application of mine on the *history* of the syllogism, further informing me that he taught an extension and simplification of its *theory*, which he offered to communicate. *November* 2, (the offer having been accepted) Sir W. Hamilton forwarded the communication B, which I give entire; consisting of a letter, and the *Requisites* which he had furnished to his students, for a prize Essay. *December* 28, he wrote again, forwarding a printed *Prospectus* of his intended work on logic. This is not material; for, on receiving it, I thought certain, what from the previous communication I had thought possible, that Sir W. Hamilton was in possession of the theory of numerically definite syllogisms (but this was a mistake of mine, as will presently appear). I accordingly, to preserve my own rights, immediately forwarded (as will presently be stated more in detail) an identifying description of the sheets of paper on which my numerical theory was written, and an account of both my systems (in letters dated *December* 31, 1846, and *January* 1, 1847). Of this, Sir W. Hamilton (who has published both letters) is my witness. 1847, *February* 27, I dated the addition to the proof sheet of my Cambridge paper, which was despatched to Cambridge the next day. This addition contains C, which itself contains (in substance) all that part of my letter of *January* 1 which refers to the disputed point. *March* 13, Sir W. Hamilton wrote the letter containing the charge of plagiarism; having been for two months prevented by illness from resuming the subject.

All ſubſequent correſpondence referred to proceedings, and not to the ſubject matter of the charge.

Many days before the middle of October, I had applied the ſyſtem of quantification in the manner ſhewn in C. Sir W. Hamilton thinks my memory has failed here: I know better. My memory does not depend upon a date, but upon the opening of the Univerſity College Seſſion, which takes place in the middle of October. But it matters nothing, for the notion of the complete quantification of a predicate, *when wanted becauſe it is the middle term,* will prove the poſſeſſion of that proceſs as well as quantification *in all caſes whether wanted or not.* On receiving B, I looked with curioſity at 2°, on which, in fact, Sir W. Hamilton grounds his declaration of having made a communication. He demands of his pupils,

'The reaſons why common language makes an *ellipſis* of the *expreſſed quantity*, frequently of the *ſubject*, and more frequently of the *predicate*, though both have always their quantities in thought.'

On looking at this, and ſeeing mention of the quantities which the terms have in thought, *in common language,* I took it for granted that the common quantities were ſpoken of: namely, that of the ſubject from the tenor of the propoſition, that of the predicate from the nature of the copula. I never ſhould have imagined that in the common language of common people, there were any other quantities, even if, in their minds, the *predicate* have theſe. Had this been all, I ſhould have paſſed it over, as referring to common quantities, and making common people a little more of logicians, as to the predicate, than I have found them to be. That this *common language* meant the language of any ſcientific ſyſtem, I had not the leaſt idea: ſtill leſs that it referred to the language of the writer's own unprinted ſyſtem, current only between himſelf and his hearers. And, though I gained a ſuſpicion that Sir W. Hamilton might have (which he had not) adopted numerical quantification, it was not from this paſſage, which by itſelf was nothing, but from what is now coming, which made this paſſage ambiguous.

On looking further into B, (which ſee) I found that Sir William's ſyſtem, whatever it might be, noted defects in the *converſion of propoſitions,* and a *general canon of ſyllogiſm.* Now I had two ſyſtems, each of which had its own way of adding to the converſions, and each its own canon of ſyllogiſm. In my firſt ſyſtem (which has now grown into Chapter V) the permanent introduction of the *contranominals* is a completion of converſion: and the reduction, by the remarks in pages 96, &c. of all ſyllogiſms to univerſal affirmative premiſes, was the canon of ſyllogiſm. In the ſecond, ſeen in A and C, which has grown into Chapter VIII, there is the univerſality of ſimple converſion, and the canon of the middle term. Sir W. Hamilton may deny (I believe he does) that theſe are canons: let it be ſo; but I took them for canons, and thought of them when I ſaw the word *canon* in his ſummary. And then the queſtion was, had Sir W. Hamilton one of theſe ſyſtems, or a third one? I had been throughout our correſpondence well pleaſed with the idea that I had hit upon ſomething in common with Sir W. Hamilton; and in my anſwer to communication B I ſaid,

'I am not at all clear that I ſhall *not* have to claim only ſecondary 'originality on ſeveral points. When I ſee "defects of the common 'doctrine of converſion" and a "ſupreme canon" of categorical ſyllo-'giſm, I muſt wait for further information. .... I think I may yet be 'able to flatter myſelf that I have followed you in ſome points unknow-'ingly.'

The reader will obſerve that this inſtructive communication is ſuppoſed to tell me, that in my thoughts the predicate has all kinds of quantity: though in truth *both have their quantities* is not Engliſh for *either may have any one of two ſpecies of quantity*. Sir W. Hamilton has expreſſed (perhaps) the dictum which is to have taught me new quantification, in terms of that new quantification unknown. By *both have quantities* he ſeems to aſſert that he meant *both have all quantities*. That both have their quantities, is true in the common ſyſtem: theſe words, which expreſs a truth of the common ſyſtem, Sir W. Hamilton declares to be a ſure mode of communicating the difference between his ſyſtem and the common one. This may do in his own lecture room, in which he has the *arbitrium et jus et norma loquendi* in his own hands. A diſtinctively unmeaning phraſe may, in virtue of his explanations, paſs current between him and his pupils: and a private bank, of courſe, muſt receive its own notes. But they are not lawful tender anywhere: nor good tender out of the neighbourhood.

I ſhall now proceed to the letters in the *Athenæum* (III and IV). Theſe contain the iſſues raiſed by the pamphlets: my ſhort letter contains the ſtrength of my caſe: I am to preſume that my opponent's letter contains the ſtrength of his anſwer, and I think it does ſo. At leaſt I can ſee nothing ſtronger in his pamphlet.

MR. DE MORGAN.

I take this mode of acknowledging the receipt of your printed letters to me. I promiſed you an anſwer, if you would bring forward the grounds of your aſſertion that I had acted with breach of confidence and falſe dealing. But you now admit that your grounds are no grounds; you declare your conviction that (though chargeable with confuſion, want of memory, &c. &c.) I have acted with good faith; and you offer a proper retraction and apology. You ſtate in various places and manners, that though you are ſatisfied of my integrity, all may not be ſo; and, thereupon, you call for an anſwer. But I think that others will be

SIR W. HAMILTON.

In reply to your letter in the laſt number of the *Athenæum*:—you were not wrong to abandon your promiſe "of trying the ſtrength of my poſition;" for never was there a weaker pretenſion than that, by you, ſo ſuicidally maintained. You would, indeed, have been quite right had you never hazarded a ſecond word; for every additional ſentence you have written is another miſ-ſtatement, calling, ſometimes, for another correction.

quite ſatisfied with your own anſwer to your own charge.

There is nothing left which I care to diſcuſs with you. Our views of logic, their coincidences, their differences, their firſt dates, my memory, &c. I am content to leave to thoſe who will read my ſtatement and your letters, with two remarks.

There is no ſtrength in an abandoned poſition. My pamphlet was publiſhed in defence of my own *character:* when Sir Wm. Hamilton retracted his charge of breach of confidence and falſe dealing, there was nothing to which I ſtood engaged, nothing I cared to write ſeparate pamphlets on, eſpecially when the approach of this preſent publication was conſidered. Any one who reads page 9 of my pamphlet, in which the promiſe was made, will ſee that it has reference to what I there call "the infamy which would attach to any one who had deſerved the terms he uſed for the conduct he deſcribed." I certainly forgot to ſay "unleſs you retract:" but as he had already refuſed to retract (though he had propoſed to *ſuſpend* the charge, provided I would then undergo an examination) it did not enter into my head to provide for ſuch a contingency. The aſſertions about weakneſs, misſtatement, &c. are for the reader's judgment. I did not, in this letter, allude expreſſly to Sir W. Hamilton's various inſinuations that the old charge might be true: both becauſe, at the firſt hurried reading, I did not become aware of their extent; and alſo, becauſe I wiſhed to take time before I made up my mind as to the way of treating what I ſaw of them.

Mr. De Morgan.

1. As ſoon as the queſtion of character was diſpoſed of, it was your buſineſs to ſhow that my *Addition*,* written after I communicated with you, contained ſome principle not contained in my *Memoir*,† written before I communicated with‡ you. This you do not do. You aſſert, and you deſcribe, and you ſum up; but you do not quote,—except a few words, which are not in that part of my *Memoir* which I declared to contain the principles uſed in my *Addition*.

Sir W. Hamilton.

You do not deny, that your correſpondence aſſerts a claim to the principle communicated to you by me; but you complain that I have not ſhown that your *Addition* involves a new doctrine, uncontained in *that part!* [from the overt contradictions of its *other* parts I had] of your Memoir which you declared to contain the principles uſed in your *Addition*. And this you can ſay, when I explicitly ſtated that "*throughout the whole paper* (the Memoir) not only is there much in contradiction—there

* Here given in C. † Here given in A, ſo far as relevant.
‡ Sir W. Hamilton's part of this is B.

is abſolutely *nothing* in (more then fortuitous) conformity with the theory of a quantified predicate" (L. p. 34). This, too, you can ſay whilſt before your eyes, *unanſwered*, there was lying "my *formal requeſt*, that you would point out *any paſſage* of your *previous writings* in which this doctrine (that aſſerted in your '*Statement*,' of a quantification of the middle term, be it ſubject or predicate) is contained" (Ibid)—for *I could find none; and none has by you been indicated.*

I do deny, in one ſenſe, that my correſpondence aſſerts a claim to the principle communicated by Sir W. Hamilton: for I deny that he *communicated any principle.* I preſume of courſe that the Proſpectus and letter ſent on the 28th of December are out of the queſtion: ſince I gave the ſyſtem on which the charge was made by return of poſt. Sir W. Hamilton has very properly confined himſelf, in his pamphlet, to his communication (B) of *November* 2, as containing the communication which he aſſerts me to have uſed. Let the reader look through it and aſk himſelf what *new principle* is communicated, and where.

Sir W. Hamilton aſſerts that he has ſhown my *Addition* to contain a new doctrine, not contained in *one definite part* of my *memoir*, by the contradictions of its *other parts*. Let P, Q, R, be parts of a memoir; and S an addition. By ſhowing that P and Q contradict one another, Sir W. Hamilton thinks he ſhows that S contains a doctrine not involved in R. The fact is, that all my memoir except '*Section* iii. *On the quantity of propoſitions*' and one other paragraph (from both which A is taken) belongs to the ſyſtem of Chapter V. in this work: while Section iii., the other paragraph, and the *addition*, belong to Chapter VIII. Let the reader take notice that Sir W. Hamilton (who, by the way, ſeems to conſider 'I explicitly ſtated' as a ſufficient anſwer to 'you have not ſhown') *does* find *ſomething* in my memoir *in conformity with the theory of a quantified predicate.* He ſays it is *fortuitous*: but it did not ſeem to him requiſite to bring it forward, and point out its *fortuitouſneſs.* This point is for the reader to judge of. "How dare you," he ſays, "rob me of my quantified predicate." "Good Sir," I anſwer, "I had it before I knew you." "What if you had," he replies, "it is enough if I inform you that it was only by accident."

Sir W. Hamilton cannot find either in the *memoir* or the *addition* (he ſays here only in the *previous writings*, but in his pamphlet (p. 34) he ſtates it of both *memoir* and *addition*), any thing about the doctrine of quantification of the middle term, whether it be ſubject or predicate, which doctrine he ſays *is repugnant to all that is there taught.* It is true that in the next ſentence he refers to *previous writings*, as cited. I will therefore conclude that Sir William included the *addition* by miſtake, and meant the *memoir* only. Whether my Section iii. (A) is or is not full of quantification of the middle term, without reference to whether that middle term be ſubject or predicate, I am quite content to leave to the reader. Sir W. Hamilton ſays he cannot find it. This I believe, and wonder at: but it does not follow that it is not there. Let the reader look.

Again, when Sir W. Hamilton aſſerted that C contains ſomething which I got from him, and which is therefore not in A, I repeat that he ought to have pointed out *what it is*. His aſſertion that he *cannot find* it in A neither proves that it *is not* in A, nor that it *is* in C.

This is the pinch which obliged him to write forty-four pages of *accuſation* in anſwer to ſixteen of *defence*: and this is the point on which the queſtion will finally turn. I am tediouſly often obliged to bring the whole matter to its A B C; but what elſe can I do with an opponent who writes an *ignoratio elenchi* of forty-four pages long.

Sir W. Hamilton is not good at finding. Immediately after what he has quoted from himſelf as above, comes the following paſſage;—

'In regard to your third aſſertion, that "*perfectly definite quantification deſtroys the neceſſity of diſtinguiſhing ſubject and predicate;*" this 'is altogether a miſtake. It is not "definite quantification," (in whatever ſenſe the word *definite* be employed), but the quantification of *both the terms* which "deſtroys the neceſſity of diſtinguiſhing ſubject and 'predicate;" and this by ſhowing, that propoſitions are merely *equations*, 'and enabling us to convert them *all—ſimply*.'

I now quote from myſelf. Of the two ſentences now coming, Sir W. Hamilton quotes the firſt, *omits the ſecond*, which ſhows that my phraſe '*perfectly* definite' means *definite in both terms*, and then makes the preceding remark.

'In fact, perfectly definite quantification deſtroys the neceſſity of diſ'tinguiſhing ſubject and predicate. To ſay that ſome 20 Xs out of 50, 'are all to be found among 70 Ys, or that 20 out of 50 Xs are 20 out of '70 Ys, is preciſely the ſame thing as ſaying that 20 out of 70 Ys are '20 out of 50 Xs.'

In a writer of whom diſhoneſt intention might be concluded, we ſhould know how to explain the omiſſion of the ſecond ſentence. But there is no diſhoneſty in Sir W. Hamilton: the omiſſion muſt be referred to the ſame diſpoſition which prevents him from ſeeing quantification of the middle term in A. What I take that diſpoſition to be, matters nothing to my reader. Perhaps this ſentence alone will enable ſome to detect that I had not any idea of the ſecond ſyſtem of quantifition independently of the third.

Mr. De Morgan.

2. All the alleged inconſiſtencies which you find in my letters, &c. will not help you till you have

Sir W. Hamilton.

You ſay, that my expoſure of your inconſiſtencies is unavailing, except "I ſhow that my commudone this:—and even then, you will have to ſhow that your communication was intelligible.

In glancing over my letters and the maſs of notes which you have written on them, I ſee that I have ſeveral times uſed inaccurate language, as people do in hurried letters. Still more often you have miſunderſtood me. If my occaſional inaccuracy and your occaſional miſunderſtanding ſhould be held to furniſh ſome excuſe for you when you precipitately charged me with diſhonourable conduct, I ſhall be better pleaſed than not.

nication was intelligible." You forget that it is for *you* to explain how, having "*ſubſcribed to,*" as having "*rightly underſtood,*" twenty-two ſentences of my proſpectus (L. pp. 19, 16), you could ſubſequently declare that communication to be *unintelligible!!* (L. p. 59). I have now no doubt, however, that you then "ſubſcribed to" more ſentences than, by you, were "rightly underſtood." Indeed, had you only betimes avowed that all you had "ſubſcribed to, as rightly underſtood," was to you really unintelligible, and that the repetition of my doctrine was in your mouth mere empty ſound, two pamphlets might have eaſily been ſpared.

Firſt, the *proſpectus* is not the "communication." The communication is that of November 2 (B). Let the reader look at it, and ſee whether it be intelligible communication of new principle.

In my pamphlet I have ſeveral times ſpoken of *the* communication, though there were two. This was natural enough, inaſmuch as there was *one* communication (that of Nov. 2), on which the charge was made againſt which that pamphlet was a defence. Sir Wm. Hamilton has never ventured to maintain that I derived anything from the communication of Dec. 28, containing the proſpectus, to which I replied on the evening I received it, as preſently mentioned. But he makes, in various places of which the above is one, a mixture of the two communications.

Secondly, I have looked carefully at pages 19 and 16 of Sir Wm. Hamilton's letter, and at all the reſt of our correſpondence, without finding that I have ever admitted that I ſubſcribed to any part of the proſpectus as by me "rightly underſtood." Page 59 is no doubt a miſprint for 39. I have neither found, nor have I the ſlighteſt remembrance of, any ſubſcription of mine to any thing Sir Wm. Hamilton ever wrote as "rightly underſtood."

I repeat the account given in my pamphlet of the manner in which I ſubſcribed to this proſpectus;—

'The next communication is dated *Dec.* 28, and conſiſted of—1. A 'letter. 2. A printed proſpectus of Sir William Hamilton's intended 'work on logic. Nothing turns on this, for the ſimple reaſon that my 'anſwer contained the moſt expreſs and formal proof that, come by it 'how I might, I was *then* in the moſt complete *written* poſſeſſion of all 'I have ſince publiſhed.... The proſpectus which accompanied this letter 'is very full on the *refults* which Sir William Hamilton can produce 'from his principles; but gives nothing, I think, certainly nothing intel- 'ligible to me, on thofe principles themfelves.

'As foon as I faw thefe *refults*, I inftantly faw that many of them 'agreed with my own. I had then no doubt that we poffeffed fomething 'in common; and I faid fo very diftinctly in my reply. As the reader 'will prefently fee, this firft impreffion has not been confirmed. Feeling 'it now time to fecure whatever of independent difcovery might belong 'to me, I anfwered Sir William Hamilton in two letters, dated Decem- 'ber 31 and January 1. In thefe letters—

'1. I returned the printed profpectus with the refults underlined 'which *my* fyftem would produce.

'2. I ftated that I had a fyftem written on certain fheets of paper, 'which I defcribed as to number, fize, &c., adding the head words of 'each page. I felt inclined to get the fignature of fome good witnefs put 'upon thefe papers; but at the fame time I felt reluctant that Sir Wil- 'liam Hamilton fhould fee, if it ever became neceffary to produce thefe 'papers, that I had been taking *precautions* againft him. I therefore de- 'termined to make himfelf my witnefs.

'3. I ftated diftinctly the firft principles of *both* my fyftems, and the 'fyllogiftic formulæ to which they lead.'

Thirdly, I fubftantiate the above, fo far as the fubfcription is concerned, by quoting two paffages from Sir W. Hamilton's publication of my letter of December 31.

'I received your obliging communication *this morning* and am now 'fully fatisfied that I have, in one of my views of fyllogifm, arrived at 'your views in fubftance, or fomething fo like them, that I could fub- 'fcribe in my own fenfe to a great part of your paper ...... This 'chapter [meaning the one on the fheets of paper above referred to] 'I might exprefs in your words wherever they are underlined in the 'profpectus which I return, hoping you will fend another.'

Where are thofe words "rightly underftood" which Sir W. Hamilton attributes to me three times in one paragraph?

He muft have been quoting from memory. Seeing *his refults*, I found they were alfo *my refults*; fo I told him that I could "fubfcribe" (and I cannot find I have ufed this word more than once, and it is in page 19 referred to by Sir William) "in my own fenfe to a great part of" his "paper." If words can fpeak meaning, I here tell him that I fubfcribe in *my own* fenfe, leaving it to the future to fhow whether I fubfcribe in *his*, that is, whether I *underftand him rightly*.

[I was reading this for the prefs, when I found out the words which, applied in one fenfe hypothetically to *one* of his refults, Sir W. Hamilton has transferred in a different fenfe to all. One of his refults, fpeaking of the moods, is the eftablifhment of 'Their numerical *equality* under all the figures,' the Italics being his. I could not make out the Englifh of this. The others I *underftood* in the grammatical fenfe. For example, 'The abrogation of the fpecial laws of fyllogifm' is intelligible: I did not know whether my fenfe of thefe words, that is to fay, my abrogation of thofe laws, was the fame as Sir W. Hamilton's; ftill that he did abrogate certain laws was clear. But numerical equality of moods I could only underftand as referring to the numerical quantities which I fuppofed (the reader will remember that I fent back the profpectus by the next poft, and had little time to look at it) Sir W. Hamilton's fyftem to contain. It means, I find, that there are the fame number of moods in all figures: but to attribute *numerical equality* to different things is a mode of faying that there is *the fame number of them in different fets* to which I was unaccuftomed. Having however, as I thought, divined what the Englifh of this might mean, I underlined it, adding (as Sir W. Hamilton ftates in one of the foot-notes, which I never remarked till now) thefe words, "If I underftand this rightly I may underline it I think.". I meant, "If I can make out *the words*." This *underftand rightly*, Sir W. Hamilton actually takes from this fentence, joins it to my "fubfcription" mentioned in *another document*, and reprefents me as declaring that I have "*fubfcribed to as rightly underftood*" *twenty-two fentences*, &c., and himfelf as quoting from one paffage.]

But, had I *betimes* avowed my non-underftanding, two pamphlets might have been fpared. Where are we now? I did avow my not underftanding the firft communication, and my fubfcribing to the fecond *in my own fenfe*. To which Sir W. Hamilton fubfequently anfwered to the effect that I fpoke falfe, that I did underftand the *firft*, *for* that I had fent him, in letters written immediately after the fecond was received, his "fundamental doctrine" and "many of its moft important confequences." What have I been contending for all along, except that the doctrine of his *firft* communication *was* to me mere empty found, and that all I was able to produce when I received the *fecond*, was my own? But Sir. W. Hamilton actually gives me a right to fay, with reference to the *fecond*, the more developed and more intelligible communication, that I did not underftand it, infifts upon my faying it, and reproaches me for not faying it. Well then, to ufe a Scottifh phrafe, the lefs I lie when I fay I did not underftand the *firft*, which is the point at iffue. So that, as to the matter of our controverfy, Sir W. Hamilton admits that there was (fortuitous he calls it) entrance of the theory of the quantified predicate in my writings prior to his communications; and as to the conduct of it, he admits that I *did not* underftand his communication; and in the face of fact, reproaches me with maintaining that I *did* till after the pamphlets were written: when it was of the effence of my ftatement, firft, that I did not underftand, fecondly that neither I nor any one elfe could have underftood, fave only the pupils to whom the requifites were addreffed.

Mr. De Morgan.	Sir W. Hamilton.
Your copious and flafhing criticifms on my intellect (by which you avenge yourfelf for the retraction of your afperfion on my integrity), I will profit by fo far as I	I difregard your mifreprefentation that "I avenge myfelf for the retraction of my afperfion on your integrity by my copious and flafhing criticifms on your intellect."

discover them to be true: the rest shall amuse me;—and the whole will be good for the printer. Take one retort from me on the same terms. You have much skill in forming new words; and, as is fair, you put your own image and superscription on your own coinage. I think you have got into the habit of assuming the same authority over that already existing portion of our language which is commonly said to belong to the Queen—and that you need an interpreter. If I can arrive at your meaning by the time I write the preface to my work on logic, I will state your claim, accompanied by your own words; if not, I can still state your own words. Till then, I have nothing more to say.

When your (excusable) irritation has subsided, you will see that I *could only* secure you from a verdict of plagiarism by bringing you in as suffering under an illusion. What, however, is all in all;—my criticisms will not, I think, be found untrue.

If guilty of lese majesty by reference to the Queen's English, have I not my accuser as abettor? For you not only passed my mintages (*quantify* and *quantification*) as current coin; but, in borrowing, actually "thanked me for the words" (L. p. 22). However, my verbal innovations are, at least, not elementary blunders. I do not, for example, confound a *term* with a *proposition*, the *middle* with the *conclusion* of a syllogism.

Sir W. Hamilton unconsciously adapts his language to a very true supposition, namely, that he has, in his pamphlet, made himself the jury in this case. He is unfortunate about the mintage. I say to him 'You make new words well, but I am afraid you alter the old ones.' To which he replies 'Why, you thanked me for my new words.' So I did, and so I do again: but what has that to do with the *lese majesty* part of my insinuation.

Sir W. Hamilton says that I have somewhere (where he does *not* say) used *term* for *proposition*, *middle* for *conclusion*, *collectively* for *distributively*. This may be; such slips of the pen are common enough. He sets them down as blunders of ignorance. I am not afraid the reader will follow him. He ought to have said where they occur, that is, when he first mentioned them, in his pamphlet. Till I put these letters together, I was satisfied, on Sir Wm. Hamilton's statement, that I had done all these enormities: but now, after the case of "*rightly understood*" which I have just had to discuss, I do not feel so well satisfied.

Sir W. Hamilton.

Finally, I beg leave to remind you.—There is now evidence in your possession that for *seven years*, at least, the doctrine of a quantified predicate has been puclickly taught by me; whilst, on your part, there is a counter assertion or innuendo, which, as you cannot prove, it concerns *your character* formally to annul.

I never denied that Sir W. Hamilton had taught *a* doctrine of the quantified predicate. By the time I wrote my pamphlet, I was pretty ſure that it was not the ſame as mine. Sir W. Hamilton's anſwer confirmed me in this, as appears in page 300.

I now come to mention a part of the diſcuſſion which I ſhould perhaps have omitted, if I had not pledged myſelf in my pamphlet to give an account of a certain offer which I there made to Sir William Hamilton, in the event of that offer not being accepted. It is a curious inſtance of that diſpoſition to hold a correſpondent or an opponent capable of ſolving enigmas, and bound to do it, which appears in his preſuming that (ſee B, paragraph 2°) an obſcure reference to what is done in *common language* would enable me to gueſs at the *uncommon language* of his ſyſtem and his lectures. I inſert it, alſo, as a ſpecimen of the various miſunderſtandings and miſapprehenſions which Sir W. Hamilton imputes to me, referring to a matter which readers will ſeparately comprehend. Had I ſpace or inclination to deal with them all, I believe I could ſerve them all in the ſame way.

Oct. 7, 1846, I learnt from Sir Wm. Hamilton that his doctrine had obtained conſiderable publicity through the *notes and eſſays* of his ſtudents. In my reply, referring to this ſyſtem, and to his offer of communicating it, I aſked if he had a pupil whom he could truſt with the communication; the anſwer was B, preſently given. But, *Dec.* 28, in ſending the proſpectus, Sir W. Hamilton informed me that, before forwarding it (the firſt communication in which that he had other than Ariſtotelian quantification was intelligibly announced) he had waited for a reply from Mr. ———. 'That gentleman' continues Sir W. Hamilton, in words ſome of which I place in Italics 'was a pupil of 'mine ſix years ago, and obtained one of the higheſt honours of the claſs; 'he was therefore *fully competent to afford you information*, which I 'begged him to do, in regard to my logical doctrines as they were taught 'ſo far back. I knew him to be a graduate of your College, and he tells 'me that he was for three years a pupil of your own. If you are ſtill 'intereſted in the matter, you can therefore obtain from him as an 'acquaintance, what information you wiſh, more agreeably than from a 'ſtranger. When he *attended me*, beſides the twofold wholes in which 'the ſyllogiſm proceeds, *the quantification of the predicate*, and the effect 'of that on the doctrine of converſion, on the doctrine of ſyllogiſtic 'moods, on the ſpecial ſyllogiſtic rules, &c., were *topics diſcuſſed*, and 'partly *given out for exerciſes. They were, in fact, then mere common-'place.*'

Jan. 13, 1847, Mr. ——— called on me at Univerſity College, after an evening lecture of mine, put his *notes* into my hands, and has ſince ſtated (in which I have no doubt he is correct, though I do not remember it) that he informed me he *was doubtful whether they contained exactly what I wanted*, and that he would gladly furniſh any additional information. Now I conceived, as I thought it was intended by Sir W. Hamilton I ſhould do, that the notes of one of the beſt ſtudents, even if not exactly what I wanted, were ſure to contain *ſomething* of the *mere commonplace* (by which I took to be meant the ordinary matter of the lectures) which was *diſcuſſed*, and *given out as exerciſes* to thoſe who attended. But in theſe notes I found nothing on quantification (I had now this key word, which did not appear in the main communication B) differing from what is uſual; and after expreſſing this in my pamphlet, I proceeded as follows:

'But if there really be anything in which Sir William Hamilton has 'preceded me, I ſhall be, of all men except himſelf, moſt intereſted in 'his having his full rights. And I make him this offer, and will take his 'acceptance of it as reparation in full for his ſuſpicions and aſſertions. 'With the conſent of the gentleman to whom theſe notes belong, which 'I am ſure will not be refuſed to our joint application, I will forward to 'him a copy of their table of contents, having more than a hundred and 'fifty headings. From theſe Sir William Hamilton ſhall ſelect thoſe 'which are, in his opinion, ſure to contain proof of his priority on any 'point which I have inveſtigated. Of theſe I will have copies made and 'ſent to him: and will print in the work on Logic which I am preparing '(and in ſome one part of it) the parts which he ſhall ſelect as fit to 'prove (or to ſhow that he could prove, let him call it as he likes) his 'caſe, or the germs of his caſe (as he pleaſes, again). Provided always, 'that the matter ſhall not run beyond ſome eight or a dozen octavo pages 'of ſmall print. And I on my part propoſe that I ſhall be allowed to 'print, to one-half the amount ſelected by Sir William Hamilton, of ad'ditional extract: but if this be refuſed I will not inſiſt on it. With this 'I will put a heading fully deſcriptive of the reaſon and meaning of the 'inſertion, and ſuch diſtinct reference and account at the beginning of 'the preface as ſhall be ſure to call the reader's attention to it. So that 'my book ſhall eſtabliſh the claim, if it can be eſtabliſhed from the notes 'of one of the beſt ſtudents. If this offer be not accepted, an account of 'it will take the place of any other reſult. If Sir William Hamilton, or 'any one elſe, can propoſe anything to make this offer fairer, I ſhall pro'bably not be found indiſpoſed to accept the addition. And though, I 'will frankly ſay, my preſent conviction is that the acceptance of the 'offer would alone cauſe my work to knock Sir William Hamilton's 'aſſertions to atoms, yet I will pledge myſelf, in any caſe, to abide by it.

Had our places in this diſcuſſion been changed, I ſhould have taken care that no reader of my anſwer ſhould have been left in ignorance of ſo fair an offer on the part of my opponent: more eſpecially if that opponent had been accuſed by me of fraud and falſehood, in a manner which I felt obliged formally to retract. But Sir Wm. Hamilton does not notice *the offer*, even by an alluſion: and refers to the notes in the following way:

'In regard to Mr ——— and his Notes, I beg leave to ſay, that in 'my relative letters, neither to that gentleman nor to you, did I ever 'refer to his *Notes* of my lectures, but excluſively to his *perſonal information* in regard to them. And for a ſufficient reaſon. The Paragraphs 'on Logic dictated to, and taken down by, my ſtudents, on which I after'wards prelect, were written ſo far back as the year 1837, and prior to 'many of my new views, and to the *whole* doctrine of a quantified predi'cate. Theſe views, as developed, were, and are, introduced in a great 'measure as corrections of the common doctrine; in the older Notes 'especially, they may, therefore, not appear in the dictated and numbered 'Paragraphs at all; whilst, frequently, (particularly at first,) they were 'given out as data, on which, previous to farther comment, the students 'were called on or excited to write expository Essays. I distinctly recol-'lect, that in the Session during which Mr. ——— attended my course of 'Logic (1840-1) it was required, on the hypothesis of a quantified pre-'dicate,—to state in detail, the valid moods of each syllogistic figure; and 'I, further, distinctly recollect, that Mr.——— was one of those who 'essayed this problem. If wrong on this point, I shall admit that my 'memory is as treacherous as yours. It was, indeed, quite natural, that 'Mr. ——— should give, and that you should receive, his Notes; but, 'of course, you could have sought or obtained no personal information 'from him, in reference to the point in question, without mentioning the 'fact. .... Were it, however, requisite to give proof from *Notes* of so 'manifest a fact, I doubt not that scores of students would be willing to 'place theirs at my disposal.

On the appearance of Sir W. Hamilton's pamphlet, Mr. ——— wrote him a very straightforward letter, of which he sent me a copy, with permission to both of us to use it. The general tenor is that Sir W. Hamilton is correct in his statements of what he had taught (which statements I never impugned as to fact; I did not know what they meant). On the point in question Mr. ——— says (the Italics are mine);—

'During the Session in which I attended your lectures (1840 and '1841) your new system, based on the thorough going quantification of 'the predicate (the second of the three systems mentioned in page 31 of 'your published letter) and its consequences in making all propositions 'simply convertible &c. *was not developed by you in your ordinary series 'of Lectures. I believe it was not touched upon in them, but it was partly 'explained to the class verbally,** and then given out as a subject for Es-'says.* When the Essays were given in they were read aloud in the class, 'and commented upon by you, and in so doing you fully explained the 'system as "a full extension and thereby a complete simplification of the 'syllogistic theory."

'These facts which were strongly fixed in my memory, because I 'believe on that occasion I happened to be the only Essayist who had 'rightly apprehended and worked out the thesis, will account for the 'circumstance that my notes, which were originally taken in shorthand, 'although containing a full Report of all your ordinary Lectures, are 'completely silent on the subject."

The reader may find out, if he can, where Sir W. Hamilton referred to *personal information* as distinguished from *notes*, or to his teaching of his new system, as a matter distinct from that of his ordinary lectures: and must judge what his success is in saying what he means

* I think this should be *extempore:* meaning that Sir W. Hamilton usually reads his lectures.

to ſay. And he may find out further, how I was to gueſs that the *mere commonplace* of the *topics diſcuſſed* in Sir William's teaching was to come, after an interval of ſix years, from his old pupil's *perſonal information*, and not from the full and (as I found them) excellent *notes* which he made at the time.

I ſhould add that Mr. ———, ſubſequently to the printed controverſy, anſwered every query which I put to him on Sir W. Hamilton's ſyſtem, but did not feel juſtified (as in a like caſe I ſhould not either) in anſwering poſitively as to the minute details of it, after laying it by for years.

I have mentioned one or two inſtances in which, as ſeems to me, Sir W. Hamilton has a ſtrange idea of the ſenſe of his own words: I will now take one of the caſes in which he has dealt as ſtrangely with mine. The way in which we uſe language, is one of the means which the reader has, for forming his judgment on the whole of this diſpute: and he muſt decide which of us is incapable of giving to the phraſes of the other their proper ſignification.

When I returned to Sir W. Hamilton his proſpectus, with thoſe parts underlined which I could interpret in my own ſenſe, the more important parts relating to logical mood and figure were not thus underlined. In the accompanying letter, I uſed theſe words, 'To mood and figure, I have attended but little; what I get on theſe points will be from your hint, or from your book.' The whole letter was on what I had done in the way of inveſtigation, not of elementary reading: and I may ſafely ſay that it is clear I meant that I had not made mood and figure, as conſtituent parts of a theory of ſyllogiſm, ſubjects of *inveſtigation*, with a view to new properties. But Sir W. Hamilton, in two places, makes me avow ignorance of the ordinary ſyſtem of mood and figure. In a foot-note to the above, he ſays, "And yet, though con-'feſſedly to ſeek in the very alphabet of the ſcience, Mr. De Morgan 'would be a logical inventor! What is here acknowledged in terms, is 'ſufficiently manifeſted from miſtakes."* And in his pamphlet (II. p. 9), he repreſents me as 'no proficient—no thorough ſtudent,—in the ſcience;' and refers to this paragraph of mine as the ground of the aſſertion. It would have been ſtrange, if, avowing ignorance of the ordinary doctrine of mood and figure, I had ſaid that what I ſhould get on theſe points muſt be from Sir W. Hamilton's hint or unpubliſhed book, when any ordinary treatiſe would have given it: ſo ſtrange, that this clauſe ought, I think, to have ſuggeſted the obvious meaning. Is Sir W. Hamilton's interpretation a *fair* one? I do not doubt that he meant it to be fair. What I aſk is, has he the power to read fairly as well as the will?

The two preceding caſes (that of the *notes* and that of the *avowed ignorance*) are ſpecimens of Sir W. Hamilton's *give and take*, of the

* Sir W. Hamilton ſhould have cited a few: but when he declares I have made elementary blunders, he does not give ſo much as a reference. The plan is a ſafe one.

manner in which he expects to be underftood, and of that in which he claims a right to underftand. They are alfo, of courfe, fpecimens of my own.

In (A), the fymbols A, E, I, O, are the A_1, E_1, I_1, O_1, of this work: and a, e, i, o, are the A', E', I', O'.

(A) *From the paper as fent to Cambridge before I had any communication whatfoever from Sir William Hamilton (without any corrections).*

SECTION III. *On the quantity of propofitions.*

"The logical ufe of the word *fome*, as merely 'more than none,' needs no further explanation. Exact knowledge of the extent of a propofition would confift in knowing, for inftance in 'fome Xs are not Ys' both what proportion of the Xs are fpoken of, and what proportion exifts between the whole number of Xs and of Ys. The want of this information compels us to divide the exponents of our proportion into 0, more than 0 not neceffarily 1, and 1. An algebraift learns to confider the diftinction between 0 and quantity as identical, for many purpofes, with that between one quantity and another: the logician muft (all writers imply) keep the diftinction between 0 and *a*, however fmall *a* may be, as facred as that between 0 and $1-a$; there being but the fame form for the two cafes. We fhall now fee that this matter has not been fully examined.

"Inference muft confift in bringing each two things which are to be compared into comparifon with a third. Many comparifons may be made at once, but there muft be this procefs in every one. When the comparifon is that of identity, of *is* or *is not*, it can only be in its ultimate or individual cafe, one of the two following:—'This X is a Y, this Z is the very fame Y, therefore this X is this Z; or elfe 'This X is a Y, this Z is not the very fame Y, therefore this X is not this Z.' And collectively, it muft be either 'Each of thefe Xs is a Y; each of thefe Ys is a Z; therefore each of thefe Xs is a Z;' or elfe 'Each of thefe Xs is a Y, no one of thefe Ys is a Z, therefore no one of thefe Xs is a Z.'

"All that is effential then to a fyllogifm is that its premifes fhall mention a number of Ys, of each of which they fhall affirm either that it is both X and Z, or that it is one and is not the other. The premifes may mention more: but it is enough that this much can be picked out; and it is in this laft procefs that inference confifts.

"Ariftotle noticed but one way of being fure that the fame Ys are fpoken of in both premifes; namely, by fpeaking of all of them in one at leaft. But this is only a cafe of the rule: for all that is neceffary is *that more* Ys *in number than there exift feparate* Ys *fhall be fpoken of in both premifes together*. Having to make $m+n$ greater than unity, when neither m nor n is fo, he admitted only that cafe in which one of the two m or n, is unity and the other is anything except 0. Here then are two fyllogifms which ought to have appeared, but do not,

Moſt of the Ys are Xs	Moſt of the Ys are Xs
Moſt of the Ys are Zs	Moſt of the Ys are not Zs
∴ Some Xs are Zs	∴ Some of the Xs are not Zs

And inſtead of moſt, or $\frac{1}{2}+a$, of the Ys, may be ſubſtituted any two fractions which have a ſum greater than unity. If theſe fractions be m and n, then the middle term is *at leaſt* the fraction $m+n-1$ of the Ys. It is not really even neceſſary that all the Ys ſhould enter in one premiſs or the other: for more than the fraction $m+n-1$ of the whole may be repeated twice.

"And in truth it is this mode of ſyllogiſing that we are frequently obliged to have recourſe to; perhaps more often than not in our univerſal ſyllogiſms. '*All* men are capable of ſome inſtruction; all who are capable of any inſtruction can learn to diſtinguiſh their right and left hands by name; therefore all men can learn to do ſo.' Let the word *all* in theſe two caſes mean only *all but one*, and the books on logic tell us with one voice that the ſyllogiſm has particular premiſes, and *no concluſion can be drawn.* But in fact idiots are capable of no inſtruction, many are deaf and dumb, ſome are without hands: and yet *a* concluſion is admiſſible. Here m and n are each very near to unity, and $m+n-1$ is therefore near to unity. Some will ſay that this is a probable concluſion: that in the caſe of any one perſon it means there is the chance m that he can receive inſtruction, and n that one ſo gifted can be made to name his right and left hand: therefore $m\times n$ (very near unity) is the chance that this man can learn ſo much.

"But I cannot ſee how in this inſtance the probability is anything but another ſort of inference from the demonſtrable concluſion of the ſyllogiſm, which muſt exiſt under the premiſes given. Beſides which, even if we admit the ſyllogiſm as only probable with regard to any one man, it is abſolute and demonſtrative in regard to the propoſition with which it concludes.

"But this is not the only caſe in which the middle term need not enter univerſally: this however is matter for the next Section. I now go on to another point."

Extract II.

"I now take the two caſes in which particular premiſes may give a concluſion: namely

$$I_{II} \qquad XY+XY=XZ \qquad\qquad XY+Y:Z=X:Z \qquad O_{IO}$$

on the ſuppoſition that the Ys mentioned in both premiſes are in number more than all the Ys. If Y_1 and Y_2 ſtand for the fractions of the whole number of Ys mentioned or implied in the two premiſes, and y_1 and y_2 for the fractions of the *ys* implied or mentioned, we ſhall by a repetition of the proceſs on YX+YZ=XZ (the other being obtained in the courſe of the proceſs) arrive at the following reſults or their counterparts: remembering that Y_1+Y_2 is greater or leſs than 1, according as y_1+y_2 is leſs or greater.

Deſignation.	Syllogiſm.	Condition of its exiſtence.
I_{II}	$YX+YZ=XZ$	Y_1+Y_2 greater than 1
O_{Io}	$YX+Y{:}Z=XZ$	....
i_{oo}	$Y{:}X+Y{:}Z=xz$	....
O_{oi}	$X{:}Y+yz=X{:}Z$	Y_1+Y_2 leſs than 1
i_{ii}	$yx+yz=xz$	....
O_{oi}	$X{:}Y+yz=X{:}Z$	....
I_{oo}	$X{:}Y+Z{:}Y=XZ$	....

(B) *Communication received on the 4th or 5th of November from Sir William Hamilton, being the pretext for his charge that I have, with injurious breach of confidence towards himſelf, and falſe dealing towards the public, appropriated his "Fundamental Doctrine of Syllogiſm" privately communicated to me: and, after the retraction of that charge, noticed in pages 297, 8, for the aſſertion that I have done the ſame thing unconſciouſly.*

"16, Great King Street,
November 2nd, 1846.

"Dear Sir,—I have been longer than I anticipated in anſwering your laſt letter. I now ſend you a copy of the requiſites for the prize Eſſay, which I gave out to my ſtudents at the cloſe of laſt ſeſſion. It will ſhow you the nature of my doctrine of ſyllogiſm, in one of its halves. The other, which is not there touched on, regards the two wholes, or quantities in which a ſyllogiſm is caſt. I had intended ſending you a copy of a more articulate ſtatement which I meant, at any rate, to have drawn up; but I have not as yet been able to write this. I will ſend it when it is done. From what you ſtate of your ſyſtem having 'little in common with the old one,' and from the contents of your Firſt Notions, we ſhall not, I find, at all interfere, for my doctrine is ſimply *that of Ariſtotle, fully developed.*

It will give me great pleaſure if I can be of any uſe, in your inveſtigations concerning the hiſtory of Logical doctrines. I have paid great attention to this ſubject, on which I found, that I could obtain little or no information from the profeſſed hiſtorians of Logic; and my collection of Logical books is probably the moſt complete in this country. But, as I mentioned to you in my former letter, it is only in ſubordinate matters that in *abſtract* Logic there has been any progreſs.

"I remain, dear Sir, very truly yours,

"W. Hamilton."

Eſſay on the new Analytic of Logical Forms.

Without wiſhing to preſcribe any definite order, it is required that there ſhould be ſtated in the Eſſay,—

1°. What Logic *poſtulates* as a condition of its applicability.

2°. The reaſons why common language makes an *ellipſis* of the *expreſſed quantity*—frequently of the *ſubject*, and more frequently of the *predicate*, though both have always their quantities in thought. [*This paragraph is the one on which Sir W. Hamilton principally relies*].

3°. *Converſion of propoſitions*—on the *common* doctrine.

4°. Defects of this.

5°. *Figure* and *Mood* of Categorical ſyllogiſm, and Reduction,—on *common* doctrine (General ſtatement).

6°. Defects of this (General ſtatement).

7°. The one *ſupreme Canon* of Categorical Syllogiſms.

8°. The evolution, from this canon, of all the *ſpecies* of Syllogiſm.

9°. The evolution, from this canon, of all the *general laws* of categorical Syllogiſms.

10°. The error of the *ſpecial laws* for the ſeveral Figures of Categorical Syllogiſm.

11°. *How many Figures* are there.

12°. What are the *Canons* of the *ſeveral Figures.*

13°. *How many moods* are there in all the Figures: ſhowing in concrete examples, through all the Moods, the *uneſſential* variation which Figure makes in a ſyllogiſm.

(Thoſe which follow 13° were wrong numbered.)

15°. What relation do the Figures hold to *extenſion* and *comprehenſion.*

16°. Why have the *ſecond* and *third* Figures *no determinate major* and *minor* premiſes and *two* indifferent concluſions: while the *firſt* Figure has a *determinate major and minor premiſe*, and a ſingle *proximate concluſion.*

17°. What relation do the Figures hold to *Deduction* and *Induction.*

N.B. This Eſſay open for competition to all ſtudents of the claſs of Logic and Metaphyſics during the laſt or during the enſuing ſeſſion.

April 15th, 1846.

(C) *Extract from the Addition to my Paper, taken, as can be ſhown, from the papers which I gave the means of identifying in January laſt, and which papers (though I hold it immaterial) I aſſert to have been written before I received any logical communication from Sir William Hamilton. (To be compared with the extracts given in A).*

"Since this paper was written, I found that the whole theory of the ſyllogiſm might be deduced from the conſideration of propoſitions in a form in which *definite quantity* of aſſertion is given both to the ſubject and the predicate of a propoſition. I had committed this view to paper, when I learned from Sir William Hamilton of Edinburgh, that he had for ſome time paſt publicly taught a theory of the ſyllogiſm differing in detail and extent from that of Ariſtotle. From the proſpectus of an intended work on logic, which Sir William Hamilton has recently iſſued, at the end of his edition of Reid, as well as from information conveyed to me by himſelf in general terms, I ſhould ſuppoſe it will be found that I have been more or leſs anticipated in the view juſt alluded to. To what extent this has been the caſe, I cannot now aſcertain; but the book of which the proſpectus juſt named is an announcement, will ſettle that queſtion. From the extraordinary extent of its author's learning in the hiſtory of philoſophy, and the acuteneſs of his written articles on the ſubject, all who are intereſted in logic will look for its appearance with more than common intereſt.

"The footing upon which we ſhould be glad to put propoſitions, if our knowledge were minute enough, is the following. We ſhould ſtate how many individuals there are under the names which are the ſubject and predicate, and of how many of each we mean to ſpeak. Thus, inſtead of 'Some Xs are Ys,' it would be, 'Every one of *a* ſpecified Xs is one or other of *b* ſpecified Ys.' And the negative form would be as in 'No one of *a* ſpecified Xs is any one of *b* ſpecified Ys.' If propoſitions be ſtated in this way, the conditions of inference are as follows. Let the *effective number* of a propoſition be the number of mentioned caſes of the *ſubject*, if it be an affirmative propoſition, or of the *middle term*, if it be a negative propoſition. Thus, in 'Each one of 50 Xs is one or other of 70 Ys,' is a propoſition, the effective number of which is always 50. But 'No one of 50 Xs is any one of 70 Ys' is a propoſition, the effective number of which is 50 or 70, according as X or Y is the middle term of the ſyllogiſm in which it is to be uſed. Then two propoſitions, each of two terms, and having one term in common, admit an inference when 1. They are not both negative. 2. The ſum of the effective numbers of the two premiſes is greater than the whole number of exiſting caſes of the middle term. And the exceſs of that ſum above the number of caſes of the middle term is the number of the caſes in the affirmative premiſs which are the ſubjects of inference. Thus, if there be 100 Ys, and we can ſay that each of 50 Xs is one or other of 80 Ys, and that no one of 20 Zs is any one of 60 Ys;—the effective numbers are 50 and 60. And 50+60 exceeding 100 by 10, there are 10 Xs, of which we may affirm that no one of them is any one of 20 Zs mentioned.

"The following brief ſummary will enable the reader to obſerve the complete deduction of all the Ariſtotelian forms, and the various modes of inference from *ſpecific particulars*, of which a ſhort account has already been given.

"Let *a* be the whole number of Xs; and *t* the number ſpecified in the premiſs. Let *c* be the whole number of Zs; and *w* the number ſpecified in the premiſs. Let *b* be the whole number of Ys; and *u* and *v* the numbers ſpecified in the premiſes of *x* and *z*. Let X_tY_u denote that each of *t* Xs is affirmed to be one out of *u* Ys: and $X_t : Y_u$ that each of *t* Xs is denied to be any one out of *u* Ys. Let $X_{m,n}$ ſignify *m*

Xs taken out of a larger ſpecified number n; and ſo on. Then the five poſſible ſyllogiſms, on the condition that no contraries are to enter either premiſes or concluſion, are as follows :—

1. $X_tY_u+Z_wY_v \;= X_{t+w-b,\,t}\,Z_w = Z_{t+w-b,\,w}X_t.$
2. $X_tY_u+Y_vZ_w \;= X_{t+v-b,\,t}\,Z_w = Z_{t+v-b,\,w}X_t.$
3. $Y_uX_t+Y_vZ_w \;= X_{u+v-b,\,t}\,Z_w = Z_{u+v-b,\,w}X_t.$
4. $X_tY_u+Z_w{:}Y_v = X_{t+v-b,\,t}:Z_w.$
5. $Y_uX_t+Z_w{:}Y_v = X_{u+v-b,\,t}:Z_w.$

"The condition of inference expreſſes itſelf; in the $X_{m,\,t}$ of the concluſion, m muſt neither be 0 nor negative. The firſt caſe gives no Ariſtotelian ſyllogiſm; the middle term never entering univerſally (of neceſſity) into any of its forms, under any degree of ſpecification which the uſual modes of ſpeaking allow. The other caſes divide the old ſyllogiſms among themſelves in the following manner: they are written ſo as to ſhow that there is ſometimes a little difference of amount of ſpecification between the reſults of different figures, which changes in the reduction from one figure to another. The Roman numerals mark the figures.

2.	$t=a,\ v=b$	$Y)Z_w+X)Y_u=X)Z_{u,\,w}$	*Barbara* I.
	$t=a,\ v=b$	$X)Y_u+Y)Z_w=Z_{a,\,w}X$	*Bramantip* IV.
	$t<a,\ v=b$	$Y)Z_w+X_tY_u=X^tZ_{u,\,w}$	*Darii* I.
	$t<a,\ v=b$	$X_tY_u+Y)Z_w=Z_{t,\,w}X_t.$	*Dimaris* IV.
3.	$u=b,\ v=b$	$Y)X_t+Y)Z_w=Z_{b,\,w}X_{b,\,t}$	*Darapti* III.
	$u<b,\ v\quad b$	$Y_uX_t+Y)Z_w=Z_{u,\,w}X_{u,\,t}$	*Diſamis* III.
	$u=b,\ v<b$	$Y)X_t+Y_v)Z_w=Z_{v,\,w}X_{v,\,t}$	*Datiſi* III.
4.	$t=a,\ v=b,\ w=c$	$Y\,.\,Z+X)Y_u=X\,.\,Z$	*Celarent* I.
	$t=a,\ v=b,\ w=c$	$Z\,.\,Y+X)Y_u=X\,.\,Z$	*Ceſare* II.
	$t=a,\ v=b,\ w=c$	$X)Y_u+Z\,.\,Y=Z\,.\,X$	*Cameſtres* II.
	$t=a,\ v=b,\ w=c$	$X)Y_u+Y\,.\,Z=Z\,.\,X$	*Camenes* IV.
	$v=b,\ w=c$	$Y\,.\,Z=X_tY_u=X_t:Z$	*Ferio* I.
	$v=b,\ w=c$	$Z\,.\,Y+X_tY_u=X_t:Z$	*Feſtino* II.
	$t=a,\ v=b,$	$X)Y_u+Z_w:Y=Z_w:X$	*Baroko* II.
5.	$u=b,\ v=b,\ w=c$	$Y\,.\,Z+Y)X^t=X_{b,\,t}:Z$	*Felapton* III.
	$u=b,\ v=b,\ w=c$	$Z\,.\,Y+Y)X_t=X_{b,\,t}:Z$	*Feſapo* IV.
	$v=b,\ w=c$	$Y\,.\,Z+Y_uX_t=X_{u,\,t}:Z$	*Feriſo* III.
	$v=b,\ w=c$	$Z\,.\,Y+Y_uX^t=X_{u,\,t}:Z$	*Freſiſon* IV.
	$u=b,\quad w=c$	$Y_v:Z+Y)X^t=X_{v,\,t}:Z$	*Bokardo* III.

I conclude by ſubmitting to the reader what I began with, namely, that until Sir William Hamilton produces ſomething from C, intelligibly hinted at in B, and neither ſubſtantially contained in the matter, nor immediately deducible from the principles, of A, he has no right whatever to assert that I have borrowed from him consciously or unconsciously. I have not found any person who thinks that such a thing can be produced : and I leave every reader to form his own opinion whether it can be done or not.

APPENDIX II.

On some forms of inference differing from those of the Aristotelians.

I THINK it desirable to state all I know of any attempt to deal with the forms of inference otherwise than in the Aristotelian method. Since the time of Wallis, three well known mathematicians have written on the subject, Euler, Lambert, and Gergonne : there may have been others, but I have not met with them.

Euler's 'Lettres à une Princesse d'Allemagne sur quelques sujets de Physique et de Philosophie' (3 vols. 8vo. Petersburg 1768-1772, according to Fuss) contain the representation of the syllogism by *sensible terms*, namely, areas. There was a Paris edition by Condorcet and Lacroix, in 1787, as is stated by Dr Henry Hunter, who published an English translation from it and from the original edition, London; 1795, 2 vols. 8vo. Euler makes use of circles to represent the terms. In a tract published (or completed) in 1831, in the Library of Useful Knowledge, under the name of 'the Study and Difficulties of Mathematics' I fell upon this method before I knew what Euler had done, using, for distinction, squares, circles, and triangles, as in Chapter I. of this work. The author of the "Outlines" presently mentioned, has what I consider a very happy improvement on Euler. The proposition 'some X is Y,' is represented by the latter as the circle of X, partly inside and partly outside the Y. The author of the "Outlines" puts a broken segment of the circle of X inside the circle of Y, leaving it unsettled whether the rest of the circle is united to the broken piece, or transferred elsewhere.*

But Euler had been preceded in the publication of this idea by Lambert, in his 'Neues Organon, &c.' Leipzig, 1764, 2 vols. 8vo. In this work, the terms are represented by lines, and identical extents by parts of the lines vertically under one another, as in page 79. The whole notion is represented by continuous line, the part left indefinite in particular propositions by dotted line. Some of the contranominal forms are more distinctly mentioned than is usual, but there is no introduction that I can find of any form of inference which is not Aristotelian.

* I should say that Euler does not use the numerical, but the magnitudinal notion, (see page 48 of this work).

In the ſeventh volume of the *Annales de Mathématiques* (Niſmes, 1816 and 1817, 4to.) there is a paper by the editor, M. Gergonne, entitled *Eſſai de dialectique rationelle.* I did not ſee this paper, nor Lambert's work, until after my memoir in the Tranſactions of the Cambridge Society had been publiſhed. The ſecond would have given me no hint: the firſt might have done ſo. There is the idea, and ſome formal uſe, of a complex propoſition: but the diviſion is erroneous. The ſubidentical, identical, and ſuperidentical forms are there; theſe are not eaſily miſſed: the others which Gergonne uſes are, the *complete excluſion* (the *contrary or ſubcontrary* of my ſyſtem, which, disjunctively, are only the common univerſal negative) and *partial incluſion with partial excluſion* (the *complex particular; or ſupercontrary,* of mine). The uſe of contraries is expreſſly* forbidden, the old converſion by contrapoſition formally declared *falſe,* and the particular propoſition aſſerted to be incapable of being made univerſal. But M. Gergonne's complex propoſitions, ſuch as they are, are uſed in a manner reſembling that in chapter V, of this work, though requiring a ſeparate *tâtonnement* for many things the analogues of which appear as connected reſults of my ſyſtem. Accordingly, I am bound to attribute to M. Gergonne the firſt publication of the idea of *a* complex ſyllogiſm, and of the compariſon of the ſimple one with it. But numerical ſtatement is not hinted at.

Sir William Hamilton's ſyſtem dates, as to its publication in lectures, from 1841, as far as has yet appeared. What I have to ſay of it will be found in another appendix.

In 1842, there was publiſhed anonymouſly 'Outline of the laws of thought'; London and Oxford (Pickering, and Graham) *octavo in twos* (ſmall). The author is the Rev. Wm. Thomſon, tutor of Queen's College, Oxford. It is a very acute work, and learned. The ſyſtem of propoſitions is extendedby the introduction of both the common quantifications of the predicate into the *affirmatives only,* which introduces the propoſitions U and Y, as the author calls them, or "All Xs are all Ys," and "Some Xs are all Ys."

The memoir in the Cambridge Tranſactions in which I gave the firſt account of what has ſince grown into Chapters IV, V, VIII, and X, of this work, is deſcribed as to date in the preceding appendix. With reference to the ſubject of chapter V, I may note the following defects of that memoir: 1. That only one arrangement of X and Z as premiſes being taken, only half the ſyſtem is given, and many correlative arrangements are not obtained (ſee page 140). 2. That owing to my not ſeeing diſtinctly that each univerſal propoſition has *two* weakened forms, the ſyllogiſms $A_1A'I'$ and $E'E'I_1$ are conſidered as a claſs apart. 3. That much of the power of forming eaſy rules is not gained, by the order of reference being made XY, ZY, XZ, inſtead of XY, YZ, XZ. The former appears at firſt the more natural order, and is certainly

* I am told that ſome works on logic uſed in the Iriſh colleges formally announce that the truth of the [ordinary] laws of ſyllogiſm depends upon the excluſion of contraries: but I have not met with any of them.

more eaſily deſcribed; namely, to refer each of the concluding terms to the middle term, with which both are compared. I obſerve, ſince, that M. Gergonne adopts this laſt order of reference: but the other is by an immenſe deal more convenient in its reſults, as I think I have ſhown.

With reſpect to the numerical quantification, what I did in the *Memoir* and *Addition* is given in full in the preceding appendix. Sir William Hamilton, who diſtinctly renounces all claim to the "arithmetically articulate" ſyſtem, and doubts whether it afford any baſis for a logical developement, ſtates that he had formerly obtained the "ultratotal quantification" (page 317) and thrown it away as a cumbrous and uſeleſs ſubtlety, without publiſhing it, as I underſtand, in any way. To his reply, he appends a note which I think it deſirable to republiſh at length, as a document in the hiſtory of this ſpeculation, and that I may make that hiſtory complete (II. p. 41).

'I have avoided, in the previous letter and poſtſcript, all details in 'regard to the *third* ſcheme of quantification (p. 32); becauſe that ſcheme 'except in ſo far as it is confounded with the *ſecond*, has no bearing in 'the controverſy; and I admit that whatever Mr. De Morgan has 'therein accompliſhed, he has accompliſhed independently of me. Further, I ſhall not deny him any claim of priority to whatever he may 'have ſtated in our correſpondence, in reference to this third ſcheme. 'Finally, I ſhall acknowledge, for I think it not improbable, that his 'ſyllogiſm (p. 19) ſuggeſted a reconſideration, on my ſickbed, of a certain former ſpeculation, in regard to the ultratotal quantification of the 'middle term in both premiſes together;—a ſpeculation determined by 'the vacillation of the logicians, touching the predeſignations *more, moſt,* '&c. but which I had laid, aſide, as a uſeleſs and cumbrous ſubtlety.

'Ariſtotle, followed by the logicians, did not introduce into his doc'trine of ſyllogiſm, any quantification between the abſolutely univerſal 'and the merely particular predeſignations, for valid reaſons.—1°, Such 'quantifications were of no value or application in the one whole (the 'univerſal, potential, logical), or, as I would amplify it, in the two cor'relative and counter wholes (the logical,— and the formal, actual, 'metaphyſical,) with which Logic is converſant. For all that is out of 'claſſification, all that has no reference to genus and ſpecies, is out of 'Logic, indeed out of Philoſophy; for Philoſophy tends always to the 'univerſal and neceſſary. Thus the higheſt canons of deductive reaſon'ing, the *dicta de Omni et de Nullo,* were founded on, and for, the 'procedure from the univerſal whole to the ſubject parts; whilſt, con'verſely, the principle of inductive reaſoning was eſtabliſhed on, and for, 'the (real or preſumed) collection of all the ſubject parts as conſtituting 'the univerſal whole.—2°, The integrate or mathematical whole, on 'the contrary, (whether continuous or diſcrete) the philoſophers con'temned. For whilſt, as Ariſtotle obſerves, in mathematics genus and 'ſpecies are of no account; it is, almoſt excluſively, in the mathemati'cal whole, that quantities are compared together, through a middle 'term, in neither premiſe, equal to the whole. But this reaſoning, in 'which the middle term is never univerſal, and the concluſion always 'particular, is,—as vague, partial, and contingent,—of little or no value 'in philoſophy. It was accordingly ignored in Logic; and the prede-'ſignations *more moſt*, &c., as I have ſaid, referred, to univerſal, or, '(as was moſt common) to particular, or to neither, quantity. This 'diſcrepancy among Logicians long ago attracted my attention; and I 'ſaw, at once, that the poſſibility of inference conſidered abſolutely, 'depended, excluſively on the quantifications of the middle term, in both 'premiſes, being, together, more than its poſſible totality—its diſtribution, 'in any one. At the ſame time I was impreſſed—1°, with the almoſt 'utter inutility of ſuch reaſoning, in a philoſophical relation: and 2°, 'alarmed with the load of valid moods which its recognition in Logic 'would introduce. The mere quantification of the predicate, under the 'two pure quantities of *definite* and *indefinite*, and the two qualities of '*affirmative* and *negative*, gives (abſtractly) in each figure, *thirty ſix* 'valid moods; which, (if my preſent calculation be correct,) would be 'multiplied, by the introduction of the two hybrid or ambiguous quan-'tifications of *a majority* and *a half*, to the fearful amount of *four hun-'dred and eighty* valid moods for each figure. Though not, at the 'time, fully aware of the ſtrength of theſe objections, they however 'prevented me from breaking down the old limitation; but as my ſu-'preme canon of Syllogiſm proceeds on the mere formal poſſibility of 'reaſoning, it of courſe comprehends all the legitimate forms of quanti-'fication. It is;—*What worſt relation of ſubject and predicate, ſubſiſts 'between either of two terms and a common third term, with which one, 'at leaſt, is poſitively related;—that relation ſubſiſts between the two 'terms themſelves:* in other words;—*In as far as two notions both 'agree, or one agreeing, the other diſagrees, with a common third notion: '—in ſo far, thoſe notions agree or diſagree with each other.* This canon 'applies, and proximately, to all categorical ſyllogiſms,—in extenſion 'and comprehenſion,—affirmative and negative,—and of any figure. It 'determines all the varieties of ſuch ſyllogiſms; is developed into all 'their general, and ſuperſedes all their ſpecial, laws. In ſhort, without 'violating this canon, no categorical reaſoning can, formally, be wrong. 'Now, this canon ſuppoſes that the two extremes are compared together, 'through the *ſame common middle;* and this cannot but be, if the 'middle, whether, ſubject or predicate, in both its quantifications to-'gether, exceed its totality, though not taken in that totality in either 'premiſe.

'But, as I have ſtated, I was moved to the reconſideration of this 'whole matter; and it may have been Mr. De Morgan's ſyllogiſm in 'our correſpondence (p. 19), which gave the ſuggeſtion. The reſult 'was the opinion, that theſe two quantifications ſhould be taken into 'account by Logic, as authentic forms, but then relegated, as of little 'uſe in practice, and cumbering the ſcience with a ſuperfluous maſs of 'moods. As to Mr. De Morgan's ſtatement in our correſpondence (p. '21) of the principle on which (by his later ſyſtem) ſuch ſyllogiſms 'proceed, this, to uſe his own expreſſion, "I did not comprehend at 'all;" nor do I now,* having, to ſpeak with the Rabbis, "reſerved it 'for the advent of Elias." I ſaw however, that, be it what it might, 'it had no analogy with mine; indeed, even from the fuller expoſition 'of his doctrines, contained in the body of the Cambridge Memoir and 'its Addition, which I afterwards received, I can find no indication 'of his having generaliſed either, 1° *the comprehenſive principle of all 'inference, that the two quantifications of the middle term, ſhould, together, exceed it as a ſingle whole; or*, 2°, *under a non-diſtributed 'middle, the* TWO *excluſive forms of its quantification.* On receipt, 'however, of Mr. De Morgan's Cambridge Memoir, I ſaw, or thought 'I ſaw, in the body of the paper, on his *old* view, ſome manifeſtation of 'a leſs erroneous doctrine upon this point, than that afterwards contained 'in his Letters and Addition, upon his *new*. Accordingly, to obviate 'all miſconſtruction, I wrote immediately the following letter,† of which 'an account has been previouſly given (p. 26, note).

EDINBURGH, 30*th March*, 1847.

'Your paper read to the Society I have curſorily peruſed; but though 'oppoſed to many of its doctrines, I admire the ingenuity which charac-

* The paſſages which Sir William Hamilton does not underſtand, are the following, and alſo that relating to the effective terms, in C of the preceding appendix.

"Now ſuppoſe propoſitions in which the quantitative part of the preceding is made more definite. Say that

$X_t \quad Y_u$		and $X_t : Y_u$
	mean	
Every one of t Xs is one or other of u Ys		No one of t Xs is any one of u Ys

Let the *effective number* of caſes in a propoſition be the number which it makes effective in inference. Then the effective number in a poſitive propoſition is the number of caſes of the *ſubject*.

The effective number in a negative propoſition is the number of caſes of the *middle term*.

And the criterion of inference being poſſible, is that the ſum of the effective numbers of the two premiſes (not both negative) is greater than the *whole number of caſes* of the *middle term*.

And the exceſs is the number of caſes involved in the inference, of all which are mentioned in the concluſion-term (or terms) of the poſitive premiſs (or premiſes).

For inſtance, let b be the whole number of Ys in exiſtence: I aſk whether we can infer anything from

	effective number	
$X_t \quad Y_u$		t
$Z_w : Y_v$	....	v

Anſwer, if t + v be greater than b, we can infer

$$X_{t+v-b} : Z_w$$

Or, if each of t Xs be one or other of u Ys, and no one of w Zs be any one of v Ys, then if t and v together are more in number than there are Ys, we may infer that no one of t + v − b Xs is any one of the w Zs juſt ſpoken of."

† This letter (the firſt paragraph of which is omitted, as not relevant to *this* appendix,) was addreſſed to me, and was ſent open to my friend Dr. Sharpey, to be delivered to me. Dr. Sharpey refuſed to deliver (and, as it happened, I was as much prepared to refuſe to receive) any thing on the literary ſubject matter of the controverſy which did not contain a retraction of Sir W. Hamilton's then ſubſiſting charge againſt me. Accordingly, I never ſaw it till it appeared in print.

'terises it throughout. On one point, I find we coincide, in principle, 'at least, against logicians in general. They have referred the quantify-'ing predesignations *plurimi*, and the like, to the most opposite heads; 'some making them universal,—some, particular,—and some between 'both; (for you are not correct in saying, (p. 6), that logicians are 'unanimous in regarding them as particular, [though most do]). This 'confliction attracted my attention; and a little consideration showed 'me, that besides the quantification of the pure quantities, *universal* and '*particular*, (which I call *definite* and *indefinite*,) there are two others of 'these, mixed and half developed, which ought to be taken into account 'by the logician, as affording valid inference; but which, without scien-'tific error, cannot be referred either to *universal*, (definite,) or to *par-'ticular*, (indefinite) quantity, far less left to vacillate ambiguously be-'tween these. I accordingly introduced them into my modification, in 'English doggerel, of "*Asserit A*," *&c.*, which [in the original cast] I 'formerly said was at your service; and as it affords a brief view of my 'doctrine on this point, I may now quote it.

'A, it affirms *of this*, *that*, *all*,*
Whilst E denies of *any*;
I, it affirms, whilst O denies,
Of *some* (or few or many).

'Thus A affirms, as E denies,
And definitely either;
Thus I affirms, as O denies,
And definitely neither.

'A *half*, left semi-definite,
Is worthy of its score;
U, then, affirms, as Y denies,
This, neither less nor more.

'Indefinito-definites,
To UI, YO, last we come;
And that affirms, and this denies,
Of *more*, *most*, (half plus some).

'"The rule of the logicians, that the middle term should be once at '"least distributed [or indistributable,] (*i.e.* taken universally or singu-'"larly, = definitely,) is untrue. For it is sufficient, if, in both the '"premises together, its quantification be more than its quantity as a '"definite whole. (Ultratotal)" - - - - - - - "It is enough for a '"valid syllogism, that the two extreme notions should (or should not), '"of necessity, partially coincide in the third or middle notion; and '"this is necessarily shown to be the case, if the one extreme coincide

* Better: 'A, it affirms of *this*, *these*, *all*.'

'" with the middle, to the extent of a half, (dimidiate quantification);
'" and the other, to the extent of aught more than a half, (ultradimi-
'" diate quantification). - - - - -

'" The firſt and higheſt quantification of the middle term (. .) is '" ſufficient not only in combination with itſelf, but with any of all the '" three inferior. The ſecond (. ,) ſuffices, in combination with the '" higheſt, with itſelf, and with the third, but not with the loweſt. '" The third (.) ſuffices, in combination with either of the higher, but '" not with itſelf, far leſs with the loweſt. The fourth and loweſt (,) '" ſuffices only in combination with the higheſt." [1. Definite; '" 2. Indefinito-definite; 3. Semi-definite; 4. Indefinite.]"'

Of the effect of this new ſyſtem of quantification in amplifying the 'ſyllogiſtic moods, (which in all the figures remain the ſame,) I ſay no- 'thing. It ſhould be noted, however, that the letters A, E, &c. do not 'mark the quantification [and qualification] of *propoſitions*, (as of old) 'but of *propoſitional terms*. The ſentences within inverted commas are 'taken from notes for the " Eſſay towards," &c.

'Before concluding, I ought to apologiſe, in the circumſtances, for 'the details, that have inſenſibly lengthened out, of a part of my doc- 'trine, which I have found, to a certain extent, coincident with what 'appears in your paper. I was anxious, however, that you and others 'ſhould have no grounds for ſurmiſing, that I borrowed any thing from 'my predeceſſors without due acknowledgment.—On ſecond thoughts, 'however, I deem it more proper to make this communication through 'a third party.'

The diſcuſſion between Sir William Hamilton and myſelf called a very able third party into the field, who addreſſed the following letter to the editor of the *Athenæum*, in which journal it was publiſhed, June 19, 1847.

'Sir,—As two great logical innovations—the one due to Sir William 'Hamilton, the other due to Mr. De Morgan—uſed in conjunction, have 'led me to the ſimpleſt and moſt general formulæ of ſyllogiſm that ever 'have been given (formulæ which correct a ſerious miſtake into which 'both Sir William Hamilton and Mr. De Morgan have fallen), I think 'it will gratify thoſe intereſted in logical ſcience if you would give them 'publicity through your columns.

'n^{I}, n^{II}, n^{III}, &c. are any numbers. When placed before a term, as '$n^{II}xs$, n^{II} marks the total number of the claſs x; placed before a pro- 'poſition, it marks the number of things of which we mean to ſpeak. 'Thus, n^{I}, of $n^{II}xs$ are of $n^{III}ys$, means that a number of things n are 'alleged to have both the characteriſtics x and y; and are to the whole 'claſs of xs as n^{I} to n^{II}, and to the whole claſs of ys as n^{I} to n^{III}: ſimi- 'larly with the negative propoſition n^{I} of $n^{II}xs$ are not of $n^{III}ys$, n^{I} 'things being here ſaid to have the characteriſtic x, and to want the 'characteriſtic y. It is clear, from the nature of a propoſition, that in 'affirmatives, n^{I} can never be greater than the leaſt extenſive of the 'terms, and in negatives never greater than the number of the claſs 'whoſe characteriſtic it is ſaid to have. But within theſe limits the pro-

'portion n^{I} to n^{II} may be wholly undetermined; we then mark it with 'the word *ſome*,—we call this, with Sir William Hamilton, indefinite 'quantity. It may be perfectly determined; as of equality when we mark 'it with *all, every*, or, following Mr. De Morgan, any other arithmeti-'cal proportion—as a half. (Sir William Hamilton has erred in calling 'a half, ſemi-definite; it is thoroughly definite). All this we call defi-'nite quantity. Laſtly, the indefinitude may be reduced within limits '—indefinito-definite, as *moſt*, &c.

'The firſt formula contains all ſyllogiſms with an affirmative conclu-'ſion, without any exception.

'I. n^{I} of $n^{II}xs$ are of $n^{III}ys$
n^{IV} of $n^{V}zs$ are of $n^{III}ys$
$(n^{I}+n^{IV}-n^{III})$ of $n^{II}xs$ are of $n^{V}zs$

'As Sir William Hamilton's principle takes away all diſtinction of 'ſubject and predicate in affirmative propoſitions, it will be ſeen that, by 'varying the proportions of the ſymbols, n^{I}, &c., every poſſible affirma-'tive logical inference, in whatever mood or figure, emerges.

'The ſyllogiſms with negative queſtions or concluſions, are not ſo 'ſimple. They fall into two diviſions, according as, in the negative 'premiſs, the things ſpoken of have the characteriſtic of the extreme, or 'of the middle; and from each of theſe, *two* concluſions, not *one*, are 'drawn, according as the things to be ſpoken of in the concluſion have 'the characteriſtic of the extreme in the affirmative premiſs, or of that 'in the negative premiſs.

'II. n^{I} of n xs are of $n^{III}ys$
'n^{IV} of $n^{V}zs$ are not of $n^{III}ys$ concludes;
'doubly 1° $(n^{I}+n^{IV}-n^{V})$ of $n^{II}xs$ are not of $n^{V}zs$
'2° $(n^{I}+n^{IV}-n^{II})$ of $n^{V}zs$ are not of $n^{II}xs$.

'It is to this formula I referred as correcting a ſerious error into which 'Sir William Hamilton and Mr. De Morgan have fallen—of holding, as 'a general principle of all inference, that the two quantifications of the 'middle term ſhould exceed it as a whole; for this ſyllogiſm proceeds 'wholly irreſpective of the total quantity of the middle, which is excluded 'from our ſymbolic concluſion.

'III. n^{I} of $n^{II}xs$ are of $n^{III}ys$
'n^{IV} of $n^{III}ys$ are not of $n^{V}zs$ concludes; alſo,
'doubly 1° $(n^{IV}+n^{I}-n^{III})$ of $n^{II}xs$ are not of $n^{V}zs$
'2° $(n^{IV}+n^{I}+n^{V}-n^{III}-n^{II})$ of $n^{V}zs$ are not of $n^{II}xs$.

'Such are the three ſymbolical formulæ of every poſſible logical infe-'rence. I have the demonſtrations that theſe are in all their extent valid, 'and are the only poſſible forms; but it is ſufficient to give here the re-'ſults.

'It will ſurpriſe no one who conſiders that the negative propoſition is 'not converted in the ſame ſenſe as the affirmative, that the negative 'ſyllogiſtic formulæ are not reducible to one. For the rule of negative 'conversion changes the things spoken of, and is as follows: n^{I} of $n^{II}xs$ 'are not of $n^{III}ys$; converts $(n^{III} + n^{I} - n^{II})$ of $n^{III}ys$ are not of $n^{II}xs$. The 'consequence of a form universally true, $(n^{III} - n^{II})$ of $n^{III}ys$ are not of $n^{II}xs$. 'As to the two conclusions, they are but the converse of each other.

'It will not be difficult to interpret these, by $n^{I} = n^{II}$ as *every* or n^{I} : 'n^{II} indefinite = *some*, &c. The usual Aristotelic forms will be seen to 'be derived from them. Thus the mood Cesare, and the corresponding 'indirect mood (or, if you will, the mood of the fourth figure, call it at 'another time Celantes or Cadere at will, but let it be Celantes for the 'nonce), come forth from the third formula.

'$n^{IV} = n^{III}$ gives no y is z . . . $n^{IV} : n^{V}$ indefinite
'$n = n^{II}$ every z is y . . . $n^{I} : n^{III}$ indefinite.

'Hence in Cesare, no x is z from our first,
'and in Celantes, no z is x from our second conclusion, and so of all 'the others.

'I owe it to Sir William Hamilton and Mr. De Morgan to say that 'without their improvements I could not have advanced one step. Mr. De 'Morgan has even attempted a like reduction to general formulæ, and has 'failed, chiefly through a misapprehension of Sir William Hamilton's prin-'ciple of quantified predicate. He has introduced a superfluous quantity, '—one logically useless, or worse than useless, as the result has shown. 'This confusion explains his errors. Had it not been for this circum-'stance, I should not have had the honour of presenting these formulæ 'to logicians.

'Permit me to add what I think also of some value. I am not of those 'who think with Sir William Hamilton that the syllogism always pro-'ceeds in the two counter wholes of intension and extension—that it 'must always be an involution or evolution in respect of classification. 'This is, no doubt, true in the most important reasonings of science; but 'it is not scientifically accurate to assert this universally.

'Quality, which is the comprehensive element, is of three kinds—not 'two, as heretofore affirmed; for since Kant, the division of affirmatives 'into analytic and synthetic, or (as Sir William Hamilton wishes) expli-'cative and ampliative, has been established. James Bernouilli has puz-'zled himself to reduce these two to the same form, but without success; 'for that contains an immediate relation of part to whole, and only a re-'mote one of part to part, while this contains an immediate relation of 'part to part, and remote of part to whole. These, as distinct kinds of 'quality, are erroneously elided in language. As the words *ampliative* 'and *restrictive* are generally opposed in logic, perhaps we might replace 'the old division of propositions, according to quality, into affirmative and 'negative—by one into *Explicative*, *Ampliative*, and *Restrictive*.

'Where, then, both premises are ampliative, the syllogism proceeds 'purely by force of extension. There is neither involution nor evolution '—neither induction nor deduction—but a passage or transition from one 'mark to another, or from class to class. Of this kind are all singular, 'or, as Ramus calls them, proper syllogisms. Let us call this new 'clafs of fyllogifms traductive, to contraft it with the inductive and de-'ductive.

'The ufe of thefe in philofophy as independent modes of inference will 'eafily appear. When we collect the fcattered fragments of our know-'ledge into unity of fcience, we ufe *induction and inductive fyllogifm;* 'when we apply the principles of fcience to fpecial events of things, we 'ufe *deduction and deductive fyllogifm;* but when, abandoning one fcheme 'of claffification, we transfer our knowledge *directly* to another, we ufe '*traduction and traductive fyllogifm.* Thus, in political fcience, what 'has been predicated by hiftorians of men claffed geographically is tranf-'ferred to men claffed according to conftitutions of government by tra-'duction. This laft efcapes Sir William Hamilton's rule, and never 'concludes through a comprehenfive containing and contained.

'I fhall not add, at prefent, any attempt to prove *à priori* the exclufive 'validity of fyllogiftic inference.

'I admit that I ought not, without good ground, to diffent from a ma-'tured opinion of Sir William Hamilton in any part of philofophy, ftill 'more in logic; but I obey the force of demonftration,—and, as Ludo-'vicus Vives faid in refpect to Ariftotle, *Verecundè diffentio.*

'Yours, &c.

'James Broun.

'*Temple, June* 9, 1847.'

My reply to this confifted in forwarding, on the fame 19th of June, to the editor of the *Athenæum*, a fummary of the refults of chapter VIII, then written. This fummary appeared on the 26th: I do not infert it, becaufe the chapter in queftion is a better anfwer; and though the publication faved my rights, the republication is unneceffary. Mr. Broun's three forms are the firft (without the contranominal), the ninth, and the eleventh, of page 161. Mr. Broun was wrong in deducing from the two latter forms that the principle of the middle term was erroneous: for in thefe very forms the two quantifications exceed the whole: being the whole (in premife one) plus fome (in the other). As to the fuperfluous quantity, it only becomes fuperfluous when fuch quantifications are introduced as diftinguifh fpurious from admiffible propofitions: fee pages 145, 146, in which it is fhown that the forms are correct.

Nothing but clofe comparifon, and that after practice, would detect the accordance of the two fymbolic modes of expreffion in pages 145 and 161. I am not therefore furprifed that Mr. Broun fhould, having obtained cafes of that in page 161, pronounce that in page 145 erroneous.

In the anfwer which I made, I promifed to ftate diftinctly how much of the chapter was written before Mr. Broun's letter appeared. This I now do. With the exception of pages 145, 146, the matter of which is moftly from my Cambridge Memoir, the whole of it was then written, excepting fuch verbal alterations and occafional introduction of fentences, as take place at the prefs, or at the laft reading of the manufcript. I had thought that there would be no neceſſity to introduce thoſe pages, except ſlightly, and in anſwer to certain objections which ſeemed likely to occur. The examination which the aſſertion that they are erroneous made me give my previous forms, pointed out the deſirableneſs of introducing them as they now ſtand.

September 17, 1847. I had finiſhed the preceding appendix, when I became aware of the exiſtence of the 'Commentationes Philoſophicæ Selectiores' of Godfrey Ploucquet, of Tubingen, Utrecht, 1781, quarto. The laſt title (p. 561) is 'De Arte Characteriſtica. Subjicitur Methodus calculandi in logicis, ab auctore inventa. 1763.' I find by a catalogue* that this *methodus calculandi* had been previouſly publiſhed in 1773, at Tubingen, at the end of a work entitled 'Principia de Subſtantiis et Phænomenis:' alſo that the 'Methodus demonſtrandi directé omnes ſyllogiſmorum ſpecies' of the ſame author (which is probably the thing I am going to deſcribe) was publiſhed at Tubingen in 1763. From the title of a work which, I am informed, exiſts, namely, 'Sammlung der Schriften welche von logiſchen Calcul des Prof. Ploucquet betreffen' Tubingen, 1773, one would ſuppoſe that this ſyſtem had obtained great local currency. I give a ſhort account of it: premiſing that Ploucquet appears to have been a well informed mathematician, much given to pure ſpeculation on mental ſubjects.

The *calculus* (a term which Ploucquet uſes in as wide a ſenſe as I do when I call the contents of Chapter V. a part of the calculus of inference) conſiſts in the invention of a ſimple notation, and the mechanical ſubſtitution, in one premiſe, of an identical equivalent to the middle term therein contained, taken from the other premiſe (this laſt being one in which the middle term is univerſal). There is neither uſe of contraries, nor numerical definition: but there is every variety of quantity of the predicate which can be produced by ſimple converſion of the ordinary forms. A term uſed univerſally is denoted by the capital letter; particularly, by the ſmall letter: affirmation by juxtapoſition; negation, by interpoſing ⊲. Thus X)Y is Xy; X.Y is X⊳Y; XY is xy; X:Y is x⊳Y. The following is a complete ſpecimen:

Sint præ- Pm
miſſæ s⊳M

Calculo: s⊳mP quoddam s non eſt P

Omnis ducatus eſt aureus
Quædam moneta non eſt aurea.

Da
m⊳A

Calc. m⊳aD. ſeu m⊳D, quædam moneta non eſt ducatus.

As Ploucquet ſeems to think that this actual application of the *calculus* to concrete inſtances, by aid of their initial letters, is a material part of

* The *ſecond edition* of Mr. Blakey's 'Eſſay on Logic' recently publiſhed, contains a catalogue of upwards of a thouſand works on logic, briefly titled.

his ſyſtem, I have inſerted the caſe entire. The rationale of the ſyſtem conſiſts in that ſubſtitution of identicals for each other, which I underſtand Sir William Hamilton (with perfect truth) to employ in every caſe. Thus we have in the above 'Some of the Ss are not any Ms, are not thoſe Ms which make up all the Ps, are not therefore any Ps.' This demand for identical ſubſtitutes requires both kinds of quantity for every predicate, and Ploucquet uſes them accordingly, as far as wanted to eſtabliſh the Ariſtotelian ſyllogiſms. Sir W. Hamilton goes further, and invents ſyllogiſms for all the kinds of quantity. Thus Ploucquet uſes mP or 'ſome Ms are all the Ps' and P ⊳ m or 'all Ps are not ſome of the Ms;' but not MP or p ⊳ m.

At the ſame time with the knowledge of Ploucquet I obtained that of the work of a follower and extender, M. W. Drobitſch, author of 'Neue Darſtellung der Logik . . . Nebſt einen logiſch-mathematiſchen Anhange,' Leipzig, 1836, octavo. As far as the ſymbolic part is concerned, Mr. Drobitſch begins by a convention which would reconcile any one to the ſound, not merely of *Barbara* and *Celarent*, but even of *Baroko* and *Freſiſon*. He makes S and P the ſubject and predicate of the concluſion and M the middle term; and puts the Ariſtotelian vowel between them: thus S)P is SAP, and P:S is POS. Hence his premiſes may be *map ſam* or *mop ſam;* and one of his ſyllogiſms is *mepſamſep*. In the *algebraical* part, he uſes large and ſmall letters for the univerſal and particular, or for the whole and part extent of a term. He alſo introduces the ſigns = and < to ſignify identity and (what I call) ſubidentity. This uſe of the mathematical ſigns involves an extenſion, which is made by all thoſe who ſignify the identity of X and Y by X=Y. The mathematician thinks of extent as quantity only: the logician includes both quantity and poſition. Thus when the former ſays that five feet are leſs than ſeven feet, he means any five feet, be they part of the ſeven feet or not: the latter, when he ſays that X is a name of leſs extent than Y, means not only that the former *can be* contained in the latter, but *that it is*. To make negative propoſitions, Mr. Drobitſch takes a limited univerſe (call it U, as I have done) an extent greater than the utmoſt extent of all the names, otherwiſe indefinite. And here he falls into ſome confuſion: X and Y being the names, he ſays U muſt be of greater extent than X+Y: now if we had X)Y, U need only be of greater extent than Y. If from the genus Y be taken all the ſpecies X, the remainder is denoted by Y—X. Accordingly, the contrary of X is U—X.

Mr. Drobitſch then lays down eight forms of predication, of which, however, he only uſes the ordinary ones. And I cannot find out that the limited univerſe, or the contrary, has any uſe except to furniſh means of notation. The eight forms are;—firſt, X=y, or my X)Y; ſecondly, X=Y, or X)Y+Y)X; thirdly, x=y, or XY; fourthly, u=Y, or Y)X; fifthly, X<U—Y, which tells us that X is all contained in what is left of the univerſe after Y is removed, or is X.Y; ſixthly, X=z<Z<U—Y, a very roundabout way of ſaying that X is *ſub*contrary of Y, or X.Y+xy; ſeventhly, x=U—Y or X:Y; eighthly,

$x = X - Y$, which tells us that Y is a ſubidentical of X, or $Y)X + X:Y$.

This is in fact a mixture of two ſyſtems, both in principle and notation. The forms are A_1, A', O_1 (and O'), E_1, I_1, D, D_1 (and D'), and C_1. Alſo C is virtually given: but E', I', C', do not appear. The ordinary rules under which the mathematicians uſe $=$ and $<$, remain true in this logical uſe of them: and thus there is an elegant mode of exhibiting the inference in ſyllogiſms. For inſtance, in *Cameſtres* we have $P = m$, $S < U - M \therefore < U - m \therefore < U - P$; or $S < U - P$.

It would have been more conſiſtent to have made $=$, $<$, and $>$, (introducing this laſt) ſerve all purpoſes. But it has happened very often that a ſyſtem of notation, already exhibited, has been extended by a better one, and mended only, inſtead of being reconſtructed. Ploucquet had uſed the large and ſmall letters, and $>$ for denial: the latter ſymbol a ſtrange one, if mathematical analogy were intended. Mr. Drobitſch has ingeniouſly contrived that $<$ ſhould repreſent denial, and has been led to what might have uſefully amended all he had to begin with. Taking little x to repreſent a part of the extent of X, &c. and U for the extent of the univerſe, the following notation might have been adopted:

Firſt when $<$ and $>$ both include their limit, $=$. We ſhould have

A_1	$X < Y$ or $Y > X$	A'	$Y < X$ or $X > Y$
O_1	$x < U - Y$ or $U - Y > x$	O'	$y < U - X$ or $U - X > y$
E_1	$X < U - Y$ or $U - Y > X$	E'	$X > U - Y$ or $U - Y < X$
I_1	$x < Y$ or $Y > x$	I'	is inexpreſſible.

To expreſs I', we muſt invent a ſymbol for a part of $U - X$.

Next, when $<$ and $>$ do not include their limits, we have

D_1	$X < Y$ or $Y > X$	C_1	$X < U - Y$ or $U - Y > X$
D	$X = Y$ or $Y = X$	C	$X = U - Y$ or $U - Y = X$
D'	$X > Y$ or $Y < X$	C'	$X > U - Y$ or $U - Y < X$

P is inexpreſſible.

I am inclined to think that the repreſentation of quantity and location both under one ſymbol is objectionable, if that ſymbol be one already appropriated in mathematics to quantity only. I would on no account accuſtom myſelf to read $A < B$ as A is leſs than (becauſe a part of) B. Mr. Drobitſch is much more complete than his predeceſſors in his enumeration of the various kinds of ſorites.

October 29, 1847. While this ſheet was paſſing through the preſs, I became acquainted with "A ſyllabus of logic, in which the views of Kant are generally adopted, and the laws of ſyllogiſm ſymbolically expreſſed. By Thomas Solly, Eſq." Cambridge, 1839, 8vo. The ſymbolical expreſſion here given is of a peculiar character: the algebraic ſigns are adopted in a ſenſe which preſerves the rules of ſign, while the ſymbols repreſent the terms of the ſyllogiſm, or elſe the notions of particular and univerſal. Thus, if p ſtand for particular, u for univerſal, and m for one of the terms of a ſyllogiſm, $m = u$ or $m - u = 0$ implies that m is a univerſal term, and $(m-u)(n-p)=0$ implies the alternative that either m is univerſal, or n is particular. By means of ſuch alternative relations, the conditions of validity of the various figures are expreſſed. Mr. Solly contends for ſix forms in each figure, by introducing all forms which have weakened concluſions, and proves *à priori*, from his equations, that ſix and no more are poſſible in each figure. If I had admitted weakened forms, there would have been ſixteen more ſyllogiſms, which might be deduced, either from the eight univerſals, or from the ſixteen particulars.

THE END.

C. WHITTINGHAM, CHISWICK.

Subscription Edition

OF

SMITH'S ANCIENT HISTORY.

(FOR TERMS SEE PAGE I.)

"*The instruction derived from the particular history of any one nation or state increases in geometrical ratio to the student's knowledge of universal history.*"—SIR F. PALGRAVE.

OPINIONS OF THE PRESS.

Saturday Review.

"His points of transition are well chosen, and his wide and various panorama of principalities, powers, and dominions, clearly arranged. He has availed himself liberally of the new lights, thrown by recent discovery and philology, upon the annals of the East, and in all that relates to the Oriental empires and African kingdoms or republics (if, indeed, that term can be applied to the timocratic, or perhaps *ploutocratic* Carthage) his work is far in advance of any Ancient History in our language."

Athenæum.

"In relating not only all the leading events of the epochs here referred to, but also the remarkable incidents of the periods between the respective epochs, one great and rare power was required—that of condensation. There are cases in which an historian not only has superabundance of materials, but he is bound to use them all. Such a case has existed here; and Mr. Smith has been equal to its exigencies. Mr. Smith's lucidity of description is enhanced by excellent maps and charts, whereby the student will all the more readily comprehend the relations of nations and the movements on a field of battle."

Evangelical Christendom.

"The style is admirable, clear, expressive, animated, and often singularly forcible and happy. In his description of 'decisive battles' Mr. Smith shows the same talent for military engineering which is so conspicuous in the pages of Dr. Arnold, and the human interest as well as the ethical value of the work is sustained throughout by its just and noble sympathies. In a day when literature is so full of insinuated doubt and outspoken unbelief, we cannot too cordially hail a contribution to historic authorship so thoroughly imbued with Christian faith and feeling, so devoutly observant of the hand of God, and so fitted to impress with the continual presence and overruling in human affairs of Him who 'is wise in counsel and excellent in working.'"

Educational Times.

"It is written in a picturesque and graphic style, and is undoubtedly not open to the charge so frequently brought against works of this class, of being dull reading.

"It contains the results of the extensive and important researches which have been made since Rollin's time, and which have almost revolutionised our knowledge of the history of the Ancient World."

LONDON: JAMES WALTON, 137, GOWER STREET.

Handbook of Astronomy. By Dr. Lardner.

Third Edition. Revised and completed to 1867. By EDWIN DUNKIN, F.R.A.S., Superintendent of the Altazimuth Department, Royal Observatory, Greenwich. 138 Illustrations. Small 8vo, 7*s.* 6*d.*

"It is not very long since a lecturer was explaining some astronomical facts to his pupils; and in order to set the matter clearly before them, he referred to more than one large and important volume on the subject, but without a decidedly satisfactory result. One of the pupils, however, produced from his pocket a small unpretending work (Dr. Lardner's *Handbook*), and that which a lengthy paragraph in the large work had failed to make clear, was completely elucidated in a short pithy sentence in the small book. It has often been remarked that Dr. Lardner, beyond most others, was enabled to present the gist of a matter before the reader in the fewest words. A Third Edition of the well-known *Handbook of Astronomy* is now before us, edited by Mr. Dunkin of the Greenwich Observatory, who has added to the text all that has lately been discovered, so as to bring the work down to the present time. We can cordially recommend it as most useful to all those who desire to possess a complete manual of the science and practice of astronomy in a portable and inexpensive form."—*Astronomical Register.*

Erichsen's Science and Art of Surgery.

A Treatise on Surgical Injuries, Diseases, and Operations. Fifth Edition, thoroughly revised and much improved. Many additional Illustrations. 8vo.

[*Early in December.*

On Syphilis and Local Contagious Dis-

orders. By BERKELEY HILL, M.B., Surgeon to Out-Patients at the Lock Hospital, Assistant-Surgeon to University College Hospital. 8vo, 16*s.*

[*Just Published.*

Quain's Anatomy. Seventh Edition. By

DR. SHARPEY, DR. ALLEN THOMSON, and DR. CLELAND. Illustrated by 800 Figures on Wood, for the most part new and on a larger scale. 2 Vols. 8vo, 1*l.* 11*s.* 6*d.*

"On this book may be safely rested the reputation of our British school of anatomists. It is not equalled in completeness, accuracy, and perfect adjustments of parts by any other similar work with which we are acquainted."—*British Medical Journal.*

"The General Anatomy is again from the able pen of Dr. Sharpey, and maintains its position as the best treatise on the subject in the English language. In its present shape this work is one of the most complete and accurate anatomical text-books in existence."—*Edin. Medical Journal.*

Kirkes's Handbook of Physiology. Sixth

Edition. Edited by W. MORRANT BAKER, F.R.C.S., Demonstrator of Anatomy and Operative Surgery, and Warden of the College at St. Bartholomew's Hospital. Many Illustrations. Small 8vo, 12*s.* 6*d.*

"No book in the English language excels, if indeed any equal, this one as a text-book of physiology. It gives the student that which he ought to know in this science concisely, plainly, and sufficiently; and any one who thoroughly masters its contents may be regarded as possessing a solid foundation of physiological knowledge."—*British Medical Journal.*

The Essentials of Bandaging, including the

Management of Fractures and Dislocations, with Directions for Using other Surgical Apparatus. By BERKELEY HILL, M.B. Lond., F.R.C.S., Instructor in Bandaging in University College Hospital, and Surgeon to Out-patients at the Lock Hospital. Illustrated by 110 Engravings. Fcap. 8vo, 6*d.*

Chemistry.

CHEMISTRY FOR SCHOOLS. An Introduction to the Practical Study of Chemistry. By C. HAUGHTON GILL, Assistant-Examiner in Chemistry at the University of London, late Lecturer on Chemistry in University College School. With Illustrations. Small 8vo. [*Early in* 1869.

DR. HOFMANN'S MODERN CHEMISTRY, Experimental and Theoretic. Small 8vo, 4*s.* 6*d.*

"It is in the truest sense an introduction to chemistry; and as such it possesses the highest value—a value which is equally great to the student, new to the science, and to the lecturer who has spent years in teaching it."—*Reader.*

BARON LIEBIG'S WORKS.

"Side by side, as long as husbandry shall last, will these three names shine in co-equal glory:—Antoine Lavoisier, Humphry Davy, Justus Liebig. To Lavoisier belongs the noble initiation of the work; to Davy, its splendid prosecution; to Liebig, its glorious consummation. Embracing in his masterly induction the results of all foregone and contemporary investigation, and supplying its large defects by his own incomparable researches, Liebig has built up on imperishable foundations, as a connected whole, the code of simple general laws on which regenerated agriculture must henceforth for all time repose."—*International Exhibition Report.*

THE NATURAL LAWS OF HUSBANDRY. 8vo, 10*s.* 6*d.*

FAMILIAR LETTERS ON CHEMISTRY, in its Relations to Physiology, Dietetics, Agriculture, Commerce, and Political Economy. Fourth Edition. Small 8vo, 7*s.* 6*d.*

LETTERS ON MODERN AGRICULTURE. Small 8vo, 6*s.*

BUNSEN'S GASOMETRY: comprising the leading Physical and Chemical Properties of Gases, together with the Methods of Gas Analysis. By DR. ROSCOE. 8vo, 3*s.* 6*d.*

Natural Science.

HANDBOOK OF ASTRONOMY. By DR. LARDNER. Third Edition. Revised and completed to 1867, by EDWIN DUNKIN, F.R.A.S., Superintendent of the Altazimuth Department, Royal Observatory, Greenwich. 38 Plates and 100 Wood Engravings. Small 8vo, 7*s.* 6*d.*

Natural Science, *continued.*

NATURAL PHILOSOPHY FOR SCHOOLS. By DR. LARDNER. 328 Illustrations. Fourth Edition. 1 vol. 3*s.* 6*d.* cloth.

"This will be a very convenient class-book for junior students in private schools. It is intended to convey, in clear and precise terms, general notions of all the principal divisions of Physical Science, illustrated largely by diagrams. These diagrams exhibit the forms and arrangement of apparatus, and the manner of performing the most important experiments."—*British Quarterly Review.*

HANDBOOK OF NATURAL PHILOSOPHY. By DR. LARDNER. 1334 Illustrations. Complete in 4 vols. 20*s.* Sold separately as under:

MECHANICS. With 357 Illustrations. 1 vol. 5*s.*

HYDROSTATICS, PNEUMATICS, and HEAT. 292 Illustrations. 1 vol. 5*s.*

OPTICS. With 290 Illustrations. 1 vol. 5*s.*

ELECTRICITY, MAGNETISM, and ACOUSTICS. Edited by PROFESSOR FOSTER. 400 Illustrations. 1 vol. 5*s.*

A GUIDE TO THE STARS. In Eight Planispheres, showing the Aspect of the Heavens for every Night in the Year. With an Explanatory Introduction. 8vo, 5*s.*

AN ELEMENTARY TREATISE ON MECHANICS, for the Use of the Junior University Students. By RICHARD POTTER, A.M., late Professor of Natural Philosophy in University College, London. Fourth Edition. 8vo, 8*s.* 6*d.* cloth.

POTTER'S TREATISE ON OPTICS. Part I. All the requisite Propositions carried to First Approximations, with the construction of Optical Instruments: for the Use of Junior University Students. Third Edition. 8vo, 9*s.* 6*d.* cloth.

Part II. The Higher Propositions, with their application to the more perfect forms of Instruments. 8vo, 12*s.* 6*d.*

PHYSICAL OPTICS; or the NATURE AND PROPERTIES OF LIGHT. A Descriptive and Experimental Treatise. By RICHARD POTTER. 100 Illustrations. 8vo, 6*s.* 6*d.*

Natural Science, *continued.*

A FIRST BOOK OF NATURAL PHILOSOPHY; an Introduction to the Study of Statics, Dynamics, Hydrostatics, Optics, and Acoustics, with numerous Examples. By SAMUEL NEWTH, M.A., Fellow of University College, London. Tenth Thousand, with large additions. 12mo, 3*s.* 6*d.*

*** This work embraces all the subjects in Natural Philosophy required at the Matriculation Examination of the University of London. The present edition has been to a large extent re-written, the points found to present most difficulty have been simplified, fresh matter has been introduced and new examples have been added.

ELEMENTS OF MECHANICS, including Hydrostatics, with numerous Examples. By SAMUEL NEWTH, M.A. Fourth Edition, revised and enlarged. Small 8vo, 8*s.* 6*d.* cloth.

*** The First Part contains all the subjects in Mechanics and Hydrostatics required for the B.A. and B.Sc. Examinations of the University of London.

HENRY'S GLOSSARY OF SCIENTIFIC TERMS for GENERAL USE. 12mo, 3*s.* 6*d.*

"To students of works on the various sciences, it can scarcely fail to be of much service. The definitions are brief, but are, nevertheless, sufficiently precise and sufficiently plain; and in all cases the etymologies of the terms are traced with care." —*National Society's Monthly Paper.*

COMMON THINGS EXPLAINED. By DIONYSIUS LARDNER, D.C.L. Containing, Air—Earth—Fire—Water —Time—The Almanack—Clocks and Watches — Spectacles — Colour — Kaleidoscope— Pumps — Man—The Eye —The Printing Press—The Potter's Art—Locomotion and Transport—The Surface of the Earth, or First Notions of Geography. (*From "The Museum of Science and Art."*) With 233 Illustrations. 5*s.*, cloth lettered.

STEAM AND ITS USES; inluding the Steam Engine, the Locomotive and Steam Navigation. By DIONYSIUS LARDNER, D.C.L. (*From "The Museum of Science and Art."*) 1 vol., 89 Illustrations. 2*s.*

POPULAR GEOLOGY. By DIONYSIUS LARDNER, D. C. L. (*From "The Museum of Science and Art."*) 201 Illustrations. 2*s.* 6*d.*

Natural Science, *continued.*

POPULAR ASTRONOMY. By DIONYSIUS LARDNER, D.C.L. Containing, How to Observe the Heavens —Latitudes and Longitudes — The Earth—The Sun — The Moon — The Planets: are they Inhabited?—The New Planets—Leverrier and Adams's Planet—The Tides—Lunar Influences —The Stellar Universe—Light—Comets—Cometary Influences—Eclipses —Terrestrial Rotation—Lunar Rotation —Astronomical Instruments. (*From "The Museum of Science and Art."*) 182 Illustrations. 4*s.* 6*d.*

THE MICROSCOPE. By DIONYSIUS LARDNER, D.C.L. (*From "The Museum of Science and Art."*) 1 vol. 147 Engravings. 2*s.*

POPULAR PHYSICS. By DR. LARDNER. Containing, Magnitude and Minuteness — Atmosphere — Thunder and Lightning — Terrestrial Heat— Meteoric Stones—Popular Fallacies— Weather Prognostics—Thermometer— Barometer — Safety Lamp — Whitworth's Micrometric Apparatus—Electro-Motive Power — Sound — Magic Lantern — Camera Obscura — Camera Lucida—Looking Glass—Stereoscope —Science and Poetry. (*From "The Museum of Science and Art."*) With 85 Illustrations. 2*s.* 6*d.*, cloth lettered.

THE BEE AND WHITE ANTS. Their Manners and Habits. By DR. LARDNER. With Illustrations of Animal Instinct and Intelligence. (*From "The Museum of Science and Art."*) 1 vol., with 135 Illustrations. 2*s.*, cloth lettered.

Economy in Food.

DR. EDWARD SMITH'S PRACTICAL DIETARY FOR FAMILIES, SCHOOLS, AND THE LABOURING CLASSES. Fourth Thousand. Small 8vo, 3*s.* 6*d.*

CHAP. I. Elements of Food which the Body requires.—II. What Elements can be supplied by Food.—III. Qualities of Foods:—Dry Farinaceous Foods— Fresh Vegetables —Sugars— Fats, Oil, &c.—Meats—Fish—Gelatin —Eggs— Cows' Milk — Cheese — Tea and Coffee—Alcohols—Condiments.— IV. Dietary for Families. Infancy— Childhood—Youth—Adult and Middle Life—Old Age.—V. Dietary in Schools.

Latin and Greek Authors, with English Notes.

TACITUS, GERMANIA, AGRICOLA, and FIRST BOOK OF THE ANNALS. With English Notes. Edited by Dr. WILLIAM SMITH. Fifth Edition. 12mo, 5*s*.

By the same Editor,

PLATO. The Apology of Socrates, the Crito and Part of the Phædo; with Notes in English from Stallbaum. Schliermacher's Introduction. Fourth Edition. 12mo, 5*s*.

English.

DR. R. G. LATHAM ON THE ENGLISH LANGUAGE. Fifth Edition. 8vo, 18*s*.

"Let him, though he knows no characters but those of his mother tongue, read through the book as well as he can, and we will answer that he will at the close, find himself in the possession of much larger and clearer notions of general grammar, and especially of comparative etymology, than he would have supposed possible at the outset."—*English Journal of Education.*

LATHAM'S HAND-BOOK OF THE ENGLISH LANGUAGE. Fifth Edition. Small 8vo, 7*s*. 6*d*.

LATHAM'S ENGLISH GRAMMAR FOR SCHOOLS. Nineteenth Thousand. Small 8vo, 4*s*. 6*d*.

"A work in which grammar, no longer an assemblage of conventional rules of speech, becomes a philosophical analysis of our language, and an elementary intellectual exercise adapted to the highest purposes of instruction."—*Minutes of Council of Education (St. Mark's College), Vol. I.* (1845).

LATHAM AND MABERLY'S SMALLER ENGLISH GRAMMAR. Third Edition. Fcap. 8vo, 2*s*. 6*d*.

LATHAM'S GRAMMAR FOR CLASSICAL SCHOOLS. Third Edition. Fcap. 8vo, 2*s*. 6*d*.

LATHAM'S LOGIC IN ITS APPLICATION TO LANGUAGE. 12mo, 6*s*.

Cambridge Local Examinations.

MASON'S ENGLISH GRAMMAR, including the Principles of Grammatical Analysis. By C. P. MASON, B.A., Fellow of University College, London. Tenth Edition. 12mo, 2*s*. 6*d*.

*** This work is recommended by the University of Cambridge as a text-book for Candidates preparing for the Local Examinations conducted by the University.

MASON'S FIRST STEPS IN ENGLISH GRAMMAR for JUNIOR CLASSES. Price 9*d*. cloth.

MASON'S GRAMMATICAL ANALYSIS of SENTENCES. 12mo, 1*s*.

THE FIRST ENGLISH READER. By the Rev. G. D. ABBOTT. Third Edition. 12mo, with Illustrations. 1*s*. cloth limp.

THE SECOND ENGLISH READER. By the Rev. G. D. ABBOTT. Third Edition. 12mo, 1*s*. 6*d*. cloth.

French.

Professor Merlet's Works.

MERLET'S FRENCH GRAMMAR. New Edition. 12mo, 5*s*. 6*d*. bound.

MERLET'S LE TRADUCTEUR; Selections, Historical, Dramatic, and Miscellaneous, from the best French Writers, on a plan calculated to render reading and translation peculiarly serviceable in acquiring the French language. Eighteenth Edition. 12mo, 5*s*. 6*d*. bound.

MERLET'S FRENCH COMPOSITION. Consisting of Extracts from English Authors to be turned into French, with Notes indicating the Differences in Style between the two Languages. A List of Idioms, with Explanations, Mercantile Terms, Correspondence, and Essays. 12mo, 3*s*. 6*d*.

MERLET'S FRENCH SYNONYMES, explained in Alphabetical Order, with Copious Examples. 12mo, cloth, 2*s*. 6*d*.

De Morgan's Works, *continued.*

THE BOOK OF ALMANACS. With an Index of Reference by which the Almanac may be found for every year, whether in Old Style or New, from any Epoch, Ancient or Modern, up to A.D. 2000, with means of finding the day of New or Full Moon from B.C. 2000 to A.D. 2000. Compiled by PROFESSOR DE MORGAN. 5*s.* cloth, lettered.

"A more useful chronological handbook could scarcely be conceived. It will save an immensity of calculation, and is in many other respects invaluable as a chronological guide and instructor."—*Oxford Herald.*

Logic.

ART OF REASONING: a Popular Exposition of the Principles of Logic, Inductive and Deductive; with an Introductory Outline of the History of Logic, and an Appendix on recent Logical Developments, with Notes. By SAMUEL NEIL. Crown 8vo. 4*s.* 6*d.* cloth.

AN INVESTIGATION OF THE LAWS OF THOUGHT, on which are founded the Mathematical Theories of Logic and Probabilities. By GEORGE BOOLE, LL.D., Professor of Mathematics in Queen's College, Cork. 8vo, 14*s.* cloth.

Mathematical Works and Tables.

MATHEMATICAL EXAMPLES; a Graduated Series of Elementary Examples in Arithmetic, Algebra, Logarithms, Trigonometry, and Mechanics. By SAMUEL NEWTH, M.A., Fellow of University College, London. Second Edition, revised. Crown 8vo, 8*s.* 6*d.* cloth.

SOLD SEPARATELY WITHOUT THE ANSWERS.

Arithmetical Examples. Crown 8vo, 2*s.* 6*d.* cloth.

Algebraical Examples. Crown 8vo, 2*s.* 6*d.* cloth.

Trigonometrical Examples, including Examples in the Use and Theory of Logarithms. Crown 8vo, 2*s.* 6*d.* cloth.

A COURSE OF ARITHMETIC, as Taught in the Pestalozzian School, Worksop. By J. ELLENBERGER. Post 8vo, 5*s.* cloth.

Mathematical Works, *continued.*

LESSONS ON FORM; or, an Introduction to Geometry, as given in a Pestalozzian School, Cheam, Surrey. By CHARLES REINER. 12mo, with numerous Diagrams. 3*s.* 6*d.* cloth.

"It has been found in the actual use of these lessons, for a considerable period, that a larger average number of pupils are brought to study the Mathematics with decided success, and that all pursue them in a superior manner."—*Rev. Dr. Mayo.*

LESSONS ON NUMBER, as given in a Pestalozzian School at Cheam, Surrey. By CHARLES REINER. The Master's Manual. New Edition. 12mo, cloth, 5*s.*

TABLES OF LOGARITHMS COMMON and TRIGONOMETRICAL TO FIVE PLACES. Under the Superintendence of the Society for the Diffusion of Useful Knowledge. Fcap. 8vo, 1*s.* 6*d.*

FOUR-FIGURE LOGARITHMS AND ANTI-LOGARITHMS. On a Card, 1*s.*

BARLOW'S TABLES OF SQUARES, CUBES, SQUARE ROOTS, CUBE ROOTS, and RECIPROCALS of all Integer Numbers up to 10,000. Stereotype Edition, examined and corrected. Royal 12mo, 8*s.*

London Latin and Greek Grammars.

THE LONDON LATIN GRAMMAR; including the Eton Syntax and Prosody in English, accompanied with Notes. Twenty-fifth Edition. Fcap. 8vo, 1*s.* 6*d.* cloth.

THE LONDON GREEK GRAMMAR. Designed to exhibit, in small compass, the Elements of the Greek Language. Ninth Edition. 12mo, 1*s.* 6*d.* cloth.

Greek and Latin.

NEW LATIN DELECTUS; being Sentences for Translation from Latin into English, and English into Latin; arranged in a systematic Progression. By Dr. ALEXANDER ALLEN. Fifth Edition, revised. 12mo, 4*s.* cl.

Greek and Latin, *continued.*

NEW LATIN READING BOOK; consisting of Short Sentences, Easy Narrations, and Descriptions, selected from Cæsar's Gallic War; in Systematic Progression. With a Dictionary. Third Edition, revised. 12mo, 2*s.* 6*d.*

"The plan of this work differs in one important point from other works of a similar kind. The sentences have been selected exclusively from Cæsar's Commentary on the Gallic War, instead of being taken from different authors, as has usually been the case. There is an obvious advantage in this plan; the same words are continually repeated by the same author in a simple narrative; and the pupil thus becomes accustomed to his style, and finds the work of translation grow easier every day, which cannot be the case when the extracts are taken from many different authors, whose style must of course vary exceedingly."—*Preface.*

A NEW GREEK DELECTUS; being Sentences for Translation from Greek into English, and English into Greek; arranged in a systematic Progression. By Dr. RAPHAEL KÜHNER. Translated and Edited from the German, by the late Dr. ALEXANDER ALLEN. Seventh Edition, revised. 12mo, 4*s.* cloth.

"It is an analytical and synthetical praxis on the forms of the Greek language, communicating, by the way, considerable knowledge of the syntax, and information on points connected with the Greek writers."—*Preface.*

THE GREEK TESTAMENT ROOTS, in a Selection of Texts, giving the power of Reading the whole Greek Testament without difficulty. With Grammatical Notes and a Parsing Lexicon, associating the Greek Primitives with English Derivatives. By G. K. GILLESPIE, A.M. Post 8vo, 7*s.* 6*d.* cloth.

ANALYTICAL LATIN EXERCISES. By C. P. MASON, B.A., Fellow of University College, London. 12mo, 3*s.* 6*d.* cloth.

These Exercises are based, as regards the Accidence, on the Crude Form system, which is so applied that the work may be used in connection with any of the best school grammars in current use. The study of the accidence is succeeded by a systematic development of the formation and meaning of compound and derived words. The general object of the book is to give the learner a clear understanding of the formation and sense of the words of the Latin language.

Interlinear Translations.

LOCKE'S SYSTEM OF CLASSICAL INSTRUCTION. Interlinear Translations. 1*s.* 6*d.* each.

Latin.

Phædrus's Fables of Æsop.
Virgil's Æneid. Book I.
Parsing Lessons to do.
Cæsar's Invasion of Britain.

Greek.

Homer's Iliad. Book I.
Xenophon's Memorabilia. Book I.
Herodotus's Histories. Selections.

French.

Sismondi; the Battles of Cressy and Poictiers.

Also, to accompany the Latin and Greek Series.

The London Latin Grammar. 12mo, 1*s.* 6*d.*
The London Greek Grammar. 12mo, 1*s.* 6*d.*

HARDY AND ADAMS'S ANABASIS OF XENOPHON. Expressly for Schools. With Notes, Index of Names, and a Map. 12mo, 4*s.* 6*d.* cl.

TERENCE, ANDRIA. With English Notes, Summaries, and Life of Terence. By NEWENHAM TRAVERS, B.A., late Assistant-Master in University College School. Fcap. 8vo, 3*s.* 6*d.*

Italian.

FIRST ITALIAN COURSE; being a Practical and Easy Method of Learning the Elements of the Italian Language. Edited from the German work by FILIPPI, after the method of Dr. AHN, and adapted by numerous additions for the use of English learners. By W. BROWNRIGG SMITH, M.A., Second Classical Master of the City of London School. 12mo, 3*s.* 6*d.* cloth.

For Coin Collectors.

A NUMISMATIC MANUAL, or, Guide to the Collection and Study of Greek, Roman, and English Coins. Illustrated by Engravings of many hundred types, by means of which even imperfect and obliterated pieces may be easily deciphered. By J. Y. AKERMAN, F.S.A. 8vo, 21*s.* cloth.

Sir George Ramsay's Works.

I.

THE MORALIST AND POLITICIAN; or, Many Things in Few Words. Fcap. 8vo, cloth, *5s.*

"Many of the political aphorisms are sagacious and true, and deserve the attention of our statesmen on account of their logical soundness."—*Public Opinion.*

"A book which reminds us in its style of some parts of Coleridge's 'Aids to Reflection,' without affecting to emulate its power. Without being profound, it is thoughtful and sensible." — *Notes and Queries.*

"It will not please idle readers, but thinkers will appreciate it highly." — *Churchman.*

II.

INSTINCT AND REASON; or, The First Principles of Human Knowledge. Crown 8vo, *5s.*

III.

PRINCIPLES OF PSYCHOLOGY. 8vo, *10s. 6d.*

Drawing.

LINEAL DRAWING COPIES for the Earliest Instruction. Comprising upwards of 200 subjects on 24 sheets, mounted on 12 pieces of thick pasteboard, in a Portfolio. By the Author of "Drawing for Young Children." *5s. 6d.*

These copies consist of Geometrical Figures and Forms of simple objects, with the Roman and Writing Alphabets in Capital and Small Letters. They are printed white on black ground. They may also be used in teaching the Letters, in teaching Writing, and in giving Lessons on the Elements of Form and Geometry.

EASY DRAWING COPIES for ELEMENTARY INSTRUCTION. Simple Outlines without Perspective. 67 subjects, in a Portfolio. By the Author of "Drawing for Young Children." *6s. 6d.*

Sold also in Two Sets.

Set I. Twenty-six subjects mounted on thick pasteboard, in a Portfolio. Price *3s. 6d.*

Set II. Forty-one subjects mounted on thick pasteboard, in a Portfolio. Price *3s. 6d.*

The copies are sufficiently large and bold to be drawn from by forty or fifty children at the same time.

PERSPECTIVE: ITS PRINCIPLES AND PRACTICE. By G. B. Moore, Teacher of Drawing in University College. In 2 Parts. Text and Plates. 8vo, *8s. 6d.* cloth.

Monetary History of England.

THE MYSTERY OF MONEY Explained and Illustrated by the Monetary History of England from the Norman Conquest to the present time. 8vo, *7s. 6d.* cloth.

The volume consists of a historical survey, concise but complete, of the Monetary System of England, from the Norman Conquest to the present time. . . . . The book is the work of a gentleman who has been engaged in trade and banking since the year 1801.

Singing.

THE SINGING MASTER COMPLETE. 1 vol. 8vo, *6s.* cloth.

CONSISTING OF

I. First Lessons in Singing, and the Notation of Music. 8vo, *1s.*

II. Rudiments of the Science of Harmony. 8vo, *1s.* [1s.

III. The First Class Tune Book. 8vo,

IV. The Second Class Tune Book. 8vo, *1s. 6d.*

V. The Hymn Tune Book. 70 Popular Psalm and Hymn Tunes. 8vo, *1s. 6d.*

*** *Any Part may be purchased separately.*

Scientific Charts.

PICTORIAL ILLUSTRATIONS OF SCIENCE AND ART. With Explanatory Notes. Large Printed Sheets each appropriated to one Subject, and containing from 50 to 100 Engraved Figures.

The purpose of the Editor is to furnish teachers with a powerful means of Instruction, by putting at one view before the Pupil representations of the objects, phenomena, and processes comprised in the subjects to which these sheets are appropriated. The Explanatory Notes, though brief, are sufficient to remind the Teacher of the points necessary to be impressed on the Pupil, and references are given to Text books where he may find such further developments as he may think necessary.

Complete, 9 sheets folded in a cover, *4s. 6d.*; also in Parts at *1s. 6d.* The size of the sheet is 22 by 28 inches. Any sheet may be purchased separately, price *6d.*

Part I. *1s. 6d.*

1. Mechanic Powers.
2. Machinery.
3. Watch and Clock Work.

Part II. *1s. 6d.*

4. Elements of Machinery.
5. Motion and Force.
6. Steam Engine.

Part III. *1s. 6d.*

7. Hydrostatics.
8. Hydraulics.
9. Pneumatics.

Materia Medica.

DR. GARROD'S ESSENTIALS OF MATERIA MEDICA and THERAPEUTICS. Third Edition, revised and enlarged; *adapted to the Second Edition of the British Pharmacopœia.* Small 8vo, 12*s.* 6*d.* [*Just published.*

"Dr. Garrod's Work is not only an explanation of, and in some respects a Commentary on the new Pharmacopœia, but it is a complete Treatise on Materia Medica. We must not omit to notice a very useful feature in the shape of a posological table, in which may be seen at a glance the appropriate dose of all the articles of the Materia Medica."—*Edin. Medical Journal.*

Medicine.

HANDBOOK OF THE PRINCIPLES AND PRACTICE OF MEDICINE. By JAMES ANDREW, M.D., Physician to St. Bartholomew's Hospital. Small 8vo. [*Preparing.*

ON SYPHILIS AND LOCAL CONTAGIOUS DISORDERS. By BERKELEY HILL, M.B., Surgeon to Out-Patients at the Lock Hospital. 8vo. 16*s.* [*Just published.*

ON THE WASTING DISEASES OF CHILDREN. By EUSTACE SMITH, M.D., Physician Extraordinary to his Majesty the King of the Belgians. Small 8vo, 7*s.* 6*d.* [*Just published.*

ON URINARY AND RENAL DISEASES. By WILLIAM ROBERTS, M.D., Physician to the Royal Infirmary, Manchester. A Practical Treatise. Illustrated by numerous Cases and Engravings. Small 8vo, 12*s.* 6*d.*

"We commend the work most cordially as admirably adapted to supply in medical literature a treatise which every intelligent physician must have felt to be needed." — *American Journal of Medical Science.*

DR. GARROD ON GOUT AND RHEUMATIC GOUT. Second Edition, with extensive alterations. Small 8vo, 15*s.*

"Dr. Garrod has in this edition incorporated the results of his increased experience of the nature and treatment of gout; and has added a chapter on the diseases to which gouty persons are peculiarly liable."—*British Medical Journal.*

DR. MURPHY'S PRINCIPLES AND PRACTICE OF MIDWIFERY. Second and greatly Improved Edition. Many Illustrations. Small 8vo, 12*s.* 6*d.*

"The object of Dr. Murphy has been to make his Lectures as practically useful as possible."—*British Medical Journal.*

Medicine, *continued.*

DR. HILLIER'S HANDBOOK OF SKIN DISEASES, FOR PRACTITIONERS and STUDENTS. Small 8vo, 7*s.* 6*d.* cloth.

"The treatment laid down by the author is simple, rational, and in accordance with the results of an extended experience."—*American Journal of Medical Science.*

DISEASES OF CHILDREN, TREATED CLINICALLY: Founded upon Lectures delivered at the Hospital for Sick Children. By THOMAS HILLIER, M.D., Physician to the Hospital for Sick Children. Small 8vo, 8*s.* 6*d.* [*Just published.*

"It is a thoroughly sound piece of observation and practical application of experience. It is so thoroughly clinical that it is impossible to review it. But from the therapeutical point of view, which chiefly interests us, we may recommend it with great confidence. — *The Practitioner.*

DR. WALSHE ON DISEASES OF THE HEART AND GREAT VESSELS. Third Edition. Revised and greatly Enlarged. Small 8vo, 12*s.* 6*d.*

DR. WALSHE ON DISEASES OF THE LUNGS. Third Edition. Revised and greatly enlarged. Small 8vo, 12*s.* 6*d.*

DR. EDWARD SMITH ON HEALTH AND DISEASE, as Illustrated by the Cyclical Changes in the Human System. Small 8vo, 10*s.* 6*d.*

"One of the most remarkable, valuable, and useful books we have ever met with." —*Dublin Medical Journal.*

DR. EDWARD SMITH ON CONSUMPTION, in its Early and Remediable Stages. Small 8vo, 10*s.* 6*d.* cloth.

DR. GEORGE HARLEY ON JAUNDICE; its Pathology and Treatment. With the application of Physiological Chemistry to the Detection and Treatment of Diseases of the Liver and Pancreas. 8vo, 7*s.* 6*d.*

"This is a suggestive and valuable book."—*Ranking's Abstract.*

DR. GEORGE HARLEY ON DIABETES. Small 8vo, 2*s.* 6*d.*

DR. GEORGE HARLEY ON ALBUMINURIA, WITH or WITHOUT DROPSY. Small 8vo, 2*s.* 6*d.*

Popular Physiology.

ANIMAL PHYSIOLOGY FOR SCHOOLS. By Dr. LARDNER. With 190 Illustrations. Second Edition. 1 vol. large 12mo, 3*s.* 6*d.* cloth.

BRADBURY, EVANS, AND CO., PRINTERS, WHITEFRIARS.